Clinical Knowledge Management:
Opportunities and Challenges

Rajeev K. Bali
Coventry University, UK

IDEA GROUP PUBLISHING
Hershey • London • Melbourne • Singapore

Acquisitions Editor:	Renée Davies
Development Editor:	Kristin Roth
Senior Managing Editor:	Amanda Appicello
Managing Editor:	Jennifer Neidig
Copy Editor:	Jennifer Young
Typesetter:	Cindy Consonery
Cover Design:	Lisa Tosheff
Printed at:	Yurchak Printing Inc.

Published in the United States of America by
    Idea Group Publishing (an imprint of Idea Group Inc.)
    701 E. Chocolate Avenue
    Hershey PA 17033
    Tel: 717-533-8845
    Fax: 717-533-8661
    E-mail: cust@idea-group.com
    Web site: http://www.idea-group.com

and in the United Kingdom by
    Idea Group Publishing (an imprint of Idea Group Inc.)
    3 Henrietta Street
    Covent Garden
    London WC2E 8LU
    Tel: 44 20 7240 0856
    Fax: 44 20 7379 3313
    Web site: http://www.eurospan.co.uk

Copyright © 2005 by Idea Group Inc. All rights reserved. No part of this book may be reproduced, stored or distributed in any form or by any means, electronic or mechanical, including photocopying, without written permission from the publisher.

Product or company names used in this book are for identification purposes only. Inclusion of the names of the products or companies does not indicate a claim of ownership by IGI of the trademark or registered trademark.

        Library of Congress Cataloging-in-Publication Data

Clinical knowledge management : opportunities and challenges / Rajeev K. Bali, editor.
  p. cm.
 Summary: "This book establishes a convergence in thinking between knowledge management and knowledge engineering healthcare applications"--Provided by publisher.
 Includes bibliographical references and index.
 ISBN 1-59140-300-6 (hardcover) -- ISBN 1-59140-301-4 (softcover) -- ISBN 1-59140-302-2 (ebook)
 1. Medical informatics. 2. Medical telematics. 3. Knowledge management. 4. Medicine--Decision making--Data processing. I. Bali, Rajeev K., 1972-
 R858.C56 2005
 610'.68--dc22
                                  2005004517

British Cataloguing in Publication Data
A Cataloguing in Publication record for this book is available from the British Library.

All work contributed to this book is new, previously-unpublished material. The views expressed in this book are those of the authors, but not necessarily of the publisher.

For my family

Ipsa Scientia Potestas Est
(Knowledge itself is power)

SIR FRANCIS BACON
MEDITATIONES SACRÆ. DE HÆRESIBUS (1597)

Clinical Knowledge Management: Opportunities and Challenges

Table of Contents

Foreword viii
    Swamy Laxminarayan, Idaho State University, USA

Preface xi
    Rajeev K. Bali, Coventry University, UK

Section I: Key Opportunities and Challenges in Clinical Knowledge Management

Chapter I
Issues in Clinical Knowledge Management: Revisiting Healthcare Management 1
    Rajeev K. Bali, Coventry University, UK
    Ashish Dwivedi, University of Hull, UK
    Raouf Naguib, Coventry University, UK

Chapter II
It's High Time for Application Frameworks for Healthcare 11
    Efstathios Marinos, National Technical University of Athens, Greece
    George Marinos, National Technical University of Athens, Greece
    Antonios Kordatzakis, National Technical University of Athens, Greece
    Maria Pragmatefteli, National Technical University of Athens, Greece
    Aggelos Georgoulas, National Technical University of Athens, Greece
    Dimitris Koutsouris, National Technical University of Athens, Greece

Chapter III
Management and Analysis of Time-Related Data in Internet-Based Healthcare Delivery 33
    Chris D. Nugent, University of Ulster at Jordanstown, Northern Ireland
    Juan C. Augusto, University of Ulster at Jordanstown, Northern Ireland

Chapter IV
Interactive Information Retrieval as a Step Towards Effective Knowledge
Management in Healthcare 52
    Jörg Ontrup, Bielefeld University, Germany
    Helge Ritter, Bielefeld University, Germany

Chapter V
The Challenge of Privacy and Security and the Implementation of Health
Knowledge Management Systems 72
    Martin Orr, Waitemata District Health Board, New Zealand

Section II: Organisational, Cultural and Regulatory Aspects of Clinical Knowledge Management

Chapter VI
Knowledge Cycles and Sharing: Considerations for Healthcare Management 96
    Maurice Yolles, Liverpool John Moores University, UK

Chapter VII
Key Performance Indicators and Information Flow: The Cornerstones of Effective
Knowledge Management for Managed Care 116
    Alexander Berler, National Technical University of Athens, Greece
    Sotiris Pavlopoulos, National Technical University of Athens, Greece
    Dimitris Koutsouris, National Technical University of Athens, Greece

Chapter VIII
Multimedia Capture, Collaboration and Knowledge Management 139
    Subramanyam Vdaygiri, Siemens Corporate Research Inc., USA
    Stuart Goose, Siemens Corporate Research Inc., USA

Chapter IX
Biomedical Image Registration for Diagnostic Decision Making and
Treatment Monitoring 159
    Xiu Ying Wang, The University of Sydney, Australia and Heilongjiang
       University, China
    David Dagan Feng, The University of Sydney, Australia and Hong Kong
       Polytechnic University, Hong Kong, China

Chapter X
Clinical Knowledge Management: The Role of an Integrated Drug Delivery
System 182
    Sheila Price, Loughborough University, UK
    Ron Summers, Loughborough University, UK

Chapter XI
Medical Decision Support Systems and Knowledge Sharing Standards 196
    Srinivasa Raghavan, Krea Corporation, USA

Section III: Knowledge Management in Action: Clinical Cases and Application

Chapter XII
Feasibility of Joint Working in the Exchange and Sharing of Caller Information Between Ambulance, Fire and Police Services of Barfordshire 219
    Steve Clarke, The University of Hull, UK
    Brian Lehaney, Coventry University, UK
    Huw Evans, University of Hull, UK

Chapter XIII
Organizing for Knowledge Management: The Cancer Information Service as an Exemplar 234
    J. David Johnson, University of Kentucky, USA

Chapter XIV
Clinical Decision Support Systems: Basic Principles and Applications in Diagnosis and Therapy 251
    Spyretta Golemati, National Technical University of Athens, Greece
    Stavroula G. Mougiakakou, National Technical University of Athens, Greece
    John Stoitsis, National Technical University of Athens, Greece
    Ioannis Valavanis, National Technical University of Athens, Greece
    Konstantina S. Nikita, National Technical University of Athens, Greece

Chapter XV
Towards Knowledge Intensive Inter-Organizational Systems in Healthcare 271
    Teemu Paavola, LifeIT Plc, Finland and Helsinki University of Technology, Finland
    Pekka Turunen, University of Kuopio, Finland
    Jari Vuori, University of Kuopio, Finland

Chapter XVI
An Overview of Efforts to Bring Clinical Knowledge to the Point of Care 285
    Dean F. Sittig, Medical Informatics Department, Kaiser Permanente Northwest, USA, Care Management Institute, Kaiser Permanente, USA and Oregon Health & Sciences University, USA

Chapter XVII
Social Capital, An Important Ingredient to Effective Knowledge Sharing: Meditute, A Case Study 297
    Jay Whittaker, University of Ballarat, Australia
    John Van Beveren, University of Ballarat, Australia

Glossary	315
About the Authors	322
Index	332

Foreword

I am honored to be invited by the editor, Dr. Raj Bali, to write the foreword for this book. In today's information technology world, we are facing daunting challenges in realizing an all aspiring and an all-encompassing paradigm of "data-information-knowledge-intelligence-wisdom". In the early nineties, under the aegis of the United States National Information Infrastructure, the Internet facilitated the creation of an "information-for-all" environment. Despite the unstructured nature of its existence, the Internet has seen an unprecedented global growth in its role as a promoter of information solutions to the citizens of the world. In contrast to the developments we witnessed in the past decade, the features of the next generation Internet have shifted emphasis from the "information-for-all" environment to a "knowledge-for all" paradigm. Some have even called it the Internet 3. Healthcare is undoubtedly one of the major areas in which we are beginning to see revolutionary changes that are attributable to the emergence of the knowledge engineering concepts. Bali and his eminent authors have done great justice to the book's contents, by pooling together many different dimensions of knowledge management into this book.

"Knowledge" is the key phraseology that has become the guiding mantra of future systems. As aptly stated by the National Library of Medicine's report on the next generation of their program on the Integrated Advanced Information Management Systems (IAIMS), "if the challenges of the 20th century IAIMS was tying together all of the heterogeneous systems that an organization owned, the principle challenge of the next generation of IAIMS efforts is effective integration of information, data, and knowledge residing in systems owned and operated by other organizations." There is no doubt that, in recent times, we are beginning to see that knowledge revolution. Advances in the field of medical informatics are a clear testimony of newer technology developments facilitating the storage, retrieval, sharing, and optimal use of biomedical information, data, and knowledge for problem solving. These are reflected in the design and implementation of comprehensive knowledge-based networks of interoperable health record systems. They provide information and knowledge for making sound decisions about health, when and where needed.

This book delves into the technologies of knowledge management beginning from the concepts of knowledge creation and extending to the abstraction and discovery tools, as well as integration, knowledge sharing and structural influences that need to be considered for successful decision making and global coordination.

There are three major and somewhat overlapping areas of knowledge engineering applications, which have dominated the healthcare sector: education, patient care, and research. Knowledge stimulates creation of new knowledge and the management and dissemination of such new knowledge is the key to the building of modern educational infrastructure in medicine and healthcare. Whether it is the utility of the electronic cadaver in anatomy education, or the capturing of evidence-based medical content, or the design of a rule-based expert system in disease diagnosis, technology developments have stayed focused on creating the knowledge discovery tools, with insights mainly borrowed from the Artificial Intelligence methodologies. These include machine learning, case-based reasoning, genetic algorithms, neural nets, intelligent agents, and stochastic models of natural language understanding, as well as the emerging computation and artificial life. The central dogma in healthcare research is to ensure the patient to be the principle focus, from diagnosis and early intervention to treatment and care. Especially with the advent of the Internet, clinical knowledge management is a topic of paramount importance. As Bali et al. have pointed out in the opening chapter of this book, "future healthcare institutions will face the challenge of transforming large amounts of medical data into clinically-relevant information for diagnosis, to make recognition of it by deriving knowledge and to effectively transfer the knowledge acquired to the caregiver as and when required."

Creation of new knowledge from existing knowledge is what makes the field grow. Bali and his authors present in the book a number of discussions of the available technologies to stimulate the future expansion. Knowledge repositories are increasingly getting larger in size and complex in structure, as seen for example, in the hospital information systems. Such massive data explosions require efficient knowledge management strategies, including the critical need to develop knowledge retrieval and data mining tools. The latter mostly consist of appropriate software-based techniques to find difficult-to-see patterns in large groups of data. The effective analysis and interpretation of such large amounts of data collected are being enhanced by applying machine vision techniques while at the same time we are looking at machine learning mechanisms to provide self-learning instructions between processes. These are all some of the modern day innovations that are providing the capabilities to extract new knowledge from the existing knowledge. Healthcare is benefitting immensely from these applications, making it possible for healthcare professionals to access medical expert knowledge where and when needed.

Medical knowledge stems from scores of multiple sources. The design principles for the management of knowledge sharing and its global impact are a complex mix of issues characterized by varying cultural, legal, regulatory, and sociological determinants. What is especially important is to improve the overall health of the population by improving the quality of healthcare services, as well as by controlling the cost-effectiveness of medical examinations and treatment (Golemati et al.). Technology's answer to this lies in the vast emergence of clinical decision support systems in which knowledge management strategies are vital to the overall design. I am very pleased that the authors have done an excellent job by taking a succinct view of what the issues are and the priorities of what needs to be addressed in this 'fast lane' knowledge world at large and the literature resource in particular. My congratulations to Editor Bali and his entire team.

Professor Swamy Laxminarayan, Fellow AIMBE
Chief of Biomedical Information Engineering
Idaho State University, USA
s.n.laxminarayan@ieee.org

Preface

When in doubt, tell the truth....

MARK TWAIN

The key to the success of the clinical healthcare sector in the 21st century is to achieve an effective integration of technology with human-based clinical decision-making processes. By doing so, healthcare institutions are free to disseminate acquired knowledge in a manner that ensures its availability to other healthcare stakeholders for such areas as preventative and operative medical diagnosis and treatment. This is of paramount importance as healthcare and clinical management continues its growth as a global priority area.

A few basic statistics: the average physician spends about 25 percent of his or her time managing information and is required to learn approximately two million clinical specifics (The Knowledge Management Centre, 2000); in the UK, each doctor receives about 15 kg of clinical guidelines per annum (Wyatt, 2000); up to 98000 patients die every year as a result of preventable medical errors—in the USA, it is estimated that the financial cost of these errors is between $37.6 billion to $50 billion (Duff, 2002); adverse drug reactions result in more than 770,000 injuries and deaths each year (Taylor, Manzo, & Sinnett, 2002); in 1995, more than 5 percent of patients had adverse reactions to drugs while under medical care—43 percent of which were serious, life threatening, or fatal (Davenport & Glaser, 2002). To further compound these pressures, biomedical literature is doubling every 19 years. These statistics illustrate how difficult it is for healthcare institutions and stakeholders to successfully meet information needs that are growing at an exponential rate.

Knowledge Management (KM) as a discipline is said not to have a commonly accepted or *de facto* definition. However, some common ground has been established which covers the following points. KM is a multi-disciplinary paradigm (Gupta, Iyer & Aronson, 2000) that often uses technology to support the acquisition, generation, codification, and transfer of knowledge in the context of specific organisational processes. Knowledge can either be tacit or explicit (explicit knowledge typically takes the form of company documents and is easily available, whilst tacit knowledge is subjective and cognitive). As tacit knowledge is often stored in the minds of healthcare professionals, the ultimate objective of KM is to transform tacit knowledge into explicit knowledge to allow effective dissemination. The definition of KM by Gupta, Iyer, & Aronson (2000) is one such description amongst many—whichever KM definition one accepts, one unmovable truth remains: healthcare KM has made a profound impact on the international medical scene.

Of course, the notion and concept of management *per se* is nothing new in the clinical environment. Innovations and improvements in such disciplines as organizational behavior, information technology, teamwork, informatics, artificial intelligence, leadership, training, human resource management, motivation, and strategy have been equally applicable and relevant in the clinical and healthcare sectors as they have been in others. Clinicians and managers have used many of these disciplines (in combination) many times before; they may have, inadvertently and partially, carried out knowledge management *avant la lettre*.

Proactive and considered use of these previously unintentionally used KM components and techniques would reap enormous clinical and organizational benefits. Respected healthcare institutions the world over, including the UK's National Health Service (NHS), have adopted (or are in the process of adopting) the KM ideal as the driving force of current or planned modernization and organizational change programs.

Modernization and change on such a massive scale is similar to the choices faced following the purchase of a nearly derelict house. Do you carry out cosmetic changes? Do you choose to extend? Or is the work so large that it would be more efficient to pull down the existing structure and start from scratch? With this last option, one also has the opportunity to apply a more solid foundation, one that is underpinned by modern materials, which meet contemporary guidelines and regulations. This is the challenge faced by the healthcare sector. "Blowing up" existing structures affords stakeholders the opportunity to sift through the *disjecta membra*, discarding the obsolete, in order to pick out the most relevant aspects for a future-proof healthcare system.

And what of technology? Until recently, the focus of Information and Communication Technology (ICT) solutions for healthcare was on the storage of data in an electronic medium, the prime objective of which was to allow exploitation of this data at a later point in time. As such, most healthcare ICT applications were purpose built to provide support for retrospective information retrieval needs and, in some cases, to analyze the decisions undertaken. Technology, incorporating such tools and techniques as artificial intelligence and data mining, now has the capability and capacity (Deveau, 2000) to assist the healthcare knowledge explosion, real-life examples of which will be found in this book.

The upsurge of clinical-related research can be traced to new scientific domains (such as bioinformatics and cybernetics) which evolved from trans-disciplinary research; from this, clinical systems found an increased interest in recycling knowledge acquired from previous best practices. KM professionals, and indeed experts from knowledge engineering (KE) domains, had attempted to bring together different methodologies for knowledge recycling.

KE practitioners have tended to focus on codified knowledge (i.e., to discover new ways of effectively representing healthcare related information). Practitioners from KM have mainly concentrated on the macro and policy aspects and how healthcare-related information can best be disseminated to support knowledge recycling and the creation of new knowledge. This contrasting approach by practitioners of these two domains is leading to the emergence of a new knowledge age in clinical healthcare. The plethora of new technologies offering more efficient methods of managing clinical services mean that practitioners, academics and managers find it increasingly difficult to catch up with these new innovations and challenges.

It should be noted that any progress in the arena of clinical KM requires the support and cooperation of clinicians, healthcare professionals and managers, academics and, of course, patients. This has to be carefully balanced against the legislation and regulations laid down by national and international Governments as "even minor organizational changes may have unexpected harmful effects" (Tallis, 2004). Further, all stakeholders have to work together to banish the view that "once academics get hold of something, nothing would happen" (Tallis, 2004). Clinical KM programs and initiatives are therefore a skilful blend of necessary regulation, opinion, viewpoint, partnership, recognition of issues and challenges (and how best to overcome them) and the willingness to learn from the experiences, and mistakes, of other implementations.

Organization of the Book

With the obvious global need for understanding in this evolving area, this book provides a valuable insight into the various trends, innovations, and organizational challenges of contemporary clinical management. Such is the interest in the clinical KM arena that I was overwhelmed with expressions of interest to submit chapters. I have managed to carefully select the most appropriate of these but, in doing so, left out many almost as deserving. All contributions underwent a double blind review process in order to ensure academic rigor. Readers can therefore be assured that only the very highest quality of contributions were accepted for the final publication.

The book contains 17 chapters, split into three sections; the first contains chapters which describe opportunities and challenges for the clinical KM sector. The second section is comprised of chapters that discuss some of the organisational, cultural and regulatory aspects of clinical KM. The final section describes clinical and healthcare cases that demonstrate some of the key aspects of KM theory in action.

Section I: Key Opportunities and Challenges in Clinical Knowledge Management

A chapter by Bali, Dwivedi and Naguib begins this section of five chapters and examines the factors necessary for the successful incorporation of KM techniques into the clinical setting. Alternative healthcare management concepts are reviewed and discussed.

The next chapter by Marinos, Marinos, Kordatzakis, Pragmatefteli, Georgoulas and Koutsouris discusses the efficacy of framework adoption for the healthcare sector. The authors argue that the software industry requires healthcare standardization to promulgate frameworks that will make software development more efficient.

Nugent and Augusto's chapter on time-related data for Internet-based healthcare provides the rationale for Internet usage for a holistic approach to distributed healthcare management. The work is supported by case study examples.

The chapter by Ontrup and Ritter demonstrates how contemporary information retrieval methodologies can open up new possibilities to support KM in the clinical sector. They go on to discuss novel text mining techniques and algorithms that can increase accessibility to medical experts by providing contextual information and links to medical literature knowledge.

Orr's chapter talks about privacy and security challenges during healthcare KM system implementation. The chapter uses a number of evolving simple visual and mnemonic models that are based on observations, reflections, and understanding of the literature.

Section II: Organisational, Cultural and Regulatory Aspects of Clinical Knowledge Management

Having introduced some of the key issues and challenges for clinical KM, this section presents chapters that highlight organisational, cultural and regulatory aspects. In the first of six such chapters, Yolles considers the notion of knowledge cycles and sharing. He argues that while knowledge is necessary for people to execute their jobs competently, there is also a need to have the potential for easy access to the knowledge of others.

The next chapter by Berler, Pavlopoulos and Koutsouris discusses the importance of key performance indicators for healthcare informatics. The authors explore trends and best practices regarding knowledge management from the viewpoint of performance management

Vdaygiri and Goose present novel methods and technologies from the corporate world. They explain how such technologies can contribute to streamlining the processes within healthcare enterprises, telemedicine environments, and home healthcare practices.

The next chapter by Wang and Feng continues the theme of novel technology and focuses on the fundamental theories of biomedical image registration. The authors explain the fundamental connection between biomedical image registration and clinical KM that could improve the quality and safety of healthcare.

The role of an integrated drug delivery system is tackled by Price and Summers. The move towards electronic data capture and information retrieval is documented together with cross-organisational working and sharing of clinical records. The authors identify key drivers for change and explain the crucial role that all stakeholders play to bring about effective and efficient patient care.

Finally in this section, Raghavan discusses the concept of medical decision support system and the knowledge sharing standards among such systems. The evolution of decision support in the healthcare arena is explained together with the need for knowledge sharing among medical decision support systems.

Section III: Knowledge Management in Action: Clinical Cases and Application

This section of six chapters builds upon the preceding sections and presents clinical KM cases in action. Clarke, Lehaney and Evans start the section with their chapter on the exchange and sharing of knowledge between the emergency services of a UK county. The highly participative study takes into account technological potential and constraints, organisational issues, and geographic factors. Lessons learned include the need to adopt a more closely integrate operational and strategic planning in the area and to make more explicit use of known and tested methodologies.

Johnson's chapter on the Cancer Information Service discusses how societal trends in consumer and client information behavior impact on clinical knowledge management. The service serves as a critical knowledge broker, synthesizing, and translating information for clients before, during, and after their interactions with clinical practices; thus enabling health professionals to focus on their unique functions.

Golemati, Mougiakakou, Stoitsis, Valavanis and Nikita describe basic principles and applications for clinical decision support systems. The authors discuss how such systems make use of advanced modeling techniques and available patient data to optimize and individualize patient treatment. The chapter concludes by stating that knowledge-oriented decision support systems aim to improve the overall health of the population by improving the quality of healthcare services, as well as by controlling the cost-effectiveness of medical examinations and treatment.

Knowledge intensive inter-organizational systems for healthcare are the basis for the chapter by Paavola, Turunen and Vuori. The chapter promulgates recent findings and understanding on how information and knowledge systems can be better adopted to support new ways of work and improve productivity in public funded healthcare. The authors advise that issues related to clinical KM such as the varying information and knowledge processing needs of clinicians from various medical expertise domains should be examined carefully when developing new clinical information systems.

Sittig's chapter provides an overview of the efforts to develop systems capable of delivering information at the point of care. The author examines systems in four distinct areas: library-type applications, real-time clinical decision support systems, hybrid systems and finally computable guidelines, all of which combine to provide an effective point of care.

The section ends with Whittaker and Van Beveren's case study chapter that introduces social capital as a concept useful in identifying the ingredients necessary for knowledge sharing in healthcare. The authors highlight the importance of social capital where information and knowledge systems are used in the sharing process. They conclude that the use of social capital to analyse knowledge sharing initiatives will lead to more holistic approaches.

I have managed to solicit chapters from countries as diverse as Finland, Japan, Germany, Australia, Northern Ireland, Greece, New Zealand, the USA, and the UK, demonstrating the importance being afforded to this increasingly key area on the international stage. I hope that academics, clinical practitioners, managers, and students will value this text on their bookshelves as, in the ensuing pages, there is much food for thought—*bon appétit*.

Rajeev K. Bali, Ph.D.
Warwickshire, UK
August 2004

References

Davenport, T.H. & Glaser, J. (2002). Just-in-time-delivery comes to knowledge management. *Harvard Business Review, 80*(7), 107-111.

Deveau, D., (2000). Minds plus matter: Knowledge is business power. *Computing Canada, 26*(8), 14-15.

Duff, S. (2002). It's easier to tell the truth. *Modern Healthcare, 32*(23), 12-13.

Gupta, B., Iyer, L.S., & Aronson, J.E. (2000). Knowledge management: Practices and challenges. *Industrial Management & Data Systems, 100*(1), 17-21.

The Knowledge Management Centre (2002). About the Knowledge Management Centre [Online document], October 2000. Retrieved November 12, 2002, from *http://www.ucl.ac.uk/kmc/kmc2/AboutKMC/index.html*

Tallis, R. (2004). Time: Yes, it can be a healer. *The Times (T2)*, 6 August, 2004, 9.

Taylor, R., Manzo, J., & Sinnett, M. (2002). Quantifying value for physician order-entry systems: A balance of cost and quality. *Healthcare Financial Management, 56*(7), 44-48.

Wyatt, J.C. (2000). 7. Intranets. *Journal of the Royal Society of Medicine, 93*(10), 530-534.

Acknowledgments

While I would dearly love to take sole credit for this book, it would be both impossible and a complete falsehood. The book has truly been a coming together of academic and practitioner minds, without whom this book would merely have remained a scintilla of an idea quietly percolating away in the deepest recesses of my mind. I am therefore indebted to many people for various reasons.

I thank all contributors of this book for their excellent chapters; many contributors also served as reviewers, and additional thanks are due to these hard-working soles for giving up so much of their valuable time and collective energies. Thanks also to everybody who submitted proposals for giving me that most rare and coveted of headaches: a plethora of high quality and relevant submissions from which to choose.

Thanks to the publishing team at Idea Group Inc. (IGI): Michele Rossi, my development editor for her gentle, persuasive and, above all, much valued guidance and support; Jan Travers for initiating the publishing process and Mehdi Khosrow-Pour for green-lighting the project.

Sincere thanks to Professor Swamy Laxminarayan, chief of biomedical information engineering at Idaho State University in the USA for writing such a fine foreword and for his kind words and unstinting support in recent years.

Professor Raouf Naguib, head of the Biomedical Computing Research Group (BIOCORE) at Coventry University in the UK was a great source of encouragement and provided me with extensive insights into the crazy world of academia. Virtually his first words to me came in the form of advice: to focus on that which I did best, words which obviously stuck with me. I additionally thank Raouf for encouraging me to form my Knowledge Management for Healthcare (KMH) research subgroup, which generated immediate interest and recognition from international academic and healthcare institutions and which continues to go from strength to strength.

Thanks to my former student Dr. Ashish Dwivedi for his seminal work in the area of clinical and healthcare knowledge management and for forming the granite-like foundation of the KMH subgroup. I appreciate also the expressions of interest and words of support from my numerous interactions with conference delegates in the USA, Singapore, Mexico and the UK.

Last, but by no means least, I thank my family for their support during the management of this, my latest project.

Rajeev K. Bali, Ph.D.
Warwickshire, UK
August 2004

Section I

Key Opportunities and Challenges in Clinical Knowledge Management

Chapter I

Issues in Clinical Knowledge Management:
Revisiting Healthcare Management

Rajeev K. Bali, Coventry University, UK

Ashish Dwivedi, The University of Hull, UK

Raouf Naguib, Coventry University, UK

Abstract

The objective of this chapter is to examine some of the key issues surrounding the incorporation of the Knowledge Management (KM) paradigm in healthcare. We discuss whether it would it be beneficial for healthcare organizations to adopt the KM paradigm so as to facilitate effective decision-making in the context of healthcare delivery. Alternative healthcare management concepts with respect to their ability in providing a solution to the above-mentioned issue are reviewed. This chapter concludes that the KM paradigm can transform the healthcare sector.

Introduction

In today's information age, data has become a major asset for healthcare institutions. Recent innovations in Information Technology (IT) have transformed the way that healthcare organizations function. Applications of concepts such as Data Warehousing and Data Mining have exponentially increased the amount of information to which a healthcare organization has access, thus creating the problem of "information explosion". This problem has been further accentuated by the advent of new disciplines such as Bioinformatics and Genetic Engineering, both of which hold very promising solutions which may significantly change the face of the entire healthcare process from diagnosis to delivery (Dwivedi, Bali, James, Naguib, & Johnston, 2002b).

Until the early 1980s, IT solutions for healthcare used to focus on such concepts as data warehousing. The emphasis was on storage of data in an electronic medium, the prime objective of which was to allow exploitation of this data at a later point in time. As such, most of the IT applications in healthcare were built to provide support for retrospective information retrieval needs and, in some cases, to analyze the decisions undertaken. This has changed healthcare institutions' perspectives towards the concept of utility of clinical data. Clinical data that was traditionally used in a supportive capacity for historical purposes has today become an opportunity that allows healthcare stakeholders to tackle problems before they arise.

Healthcare Management Concepts

Healthcare managers are being forced to examine costs associated with healthcare and are under increasing pressure to discover approaches that would help carry out activities better, faster and cheaper (Davis & Klein, 2000; Latamore, 1999). Workflow and associated Internet technologies are being seen as an instrument to cut administrative expenses. Specifically designed IT implementations such as workflow tools are being used to automate the electronic paper flow in a managed care operation, thereby cutting administrative expenses (Latamore, 1999).

One of the most challenging issues in healthcare relates to the transformation of raw clinical data into contextually relevant information. Advances in IT and telecommunications have made it possible for healthcare institutions to face the challenge of transforming large amounts of medical data into relevant clinical information (Dwivedi, Bali, James, & Naguib, 2001b). This can be achieved by integrating information using workflow, context management and collaboration tools, giving healthcare a mechanism for effectively transferring the acquired knowledge, as and when required (Dwivedi, Bali, James, & Naguib, 2002a).

Kennedy (1995, p. 85) quotes Keever (a healthcare management executive) who notes that "Healthcare is the most disjointed industry…in terms of information exchange… Every hospital, doctor, insurer and independent lab has its own set of information, and …no one does a very good job of sharing it." From a management perspective, these new challenges have forced healthcare stakeholders to look at different healthcare management concepts that could alleviate the problem of information explosion. The following are some of the new paradigms and concepts that have caught the attention of healthcare stakeholders.

Evidence Based Medicine (EBM)

EBM is defined as the "conscientious, explicit and judicious use of current best evidence in making decisions about the care of individual patients" (Cowling, Newman, & Leigh, 1999, p. 149). A typical EBM process starts with an identification of knowledge-gaps in current healthcare treatment processes, followed by a search for the best evidence. This is then succeeded by a process to aid in the selection of appropriate electronic data/information sources and IT applications that focus on clinical competencies in the context of the best evidence generated.

The next step is to carry out a critical appraisal of the best evidence identified by carrying out checks for accuracy and diagnostic validity of the procedure/treatment identified by the best evidence generated. The costs and benefits of alternative procedures (i.e., the current best evidence procedure/treatment being recommended) are then considered. The last step is its application to patients' healthcare which calls for integration of the best evidence with the General Practitioners' (GP) clinical expertise so as to provide best treatment and care (Cowling et al., 1999).

Model of Integrated Patient Pathways (MIPP/IPP)

Schmid and Conen (2000) have argued that the model of integrated patient pathways (MIPP/IPP) is a more comprehensive concept for healthcare institutions. As the acronym suggests, IPPs aim to enable better support for healthcare institutions by focusing on the creation of clinical guidelines for commonly accepted diagnostic and therapeutic procedures at a defined level of quality. This would lead to cost-efficient treatment. It could be argued that IPP calls for in-house development of standardized clinical treatment procedures for some pre-defined diagnoses and treatments.

Schmid & Conen (2000) elaborate that IPP aims to ensure that patients receive the right treatment which is based upon best practice guidelines that have sufficient evidence to warrant the label of "best practice" and which have been proven to be clinically adequate. They argue that when a hospital tries to implement IPP, it will automatically go through a circular chain process that calls for identifying sources of best practice, converting them to worldwide implementation practices and then, based upon their performance, converting them to benchmarks. Deliberation on current health reform is centered on two competing objectives: expanding access and containing costs. The challenge is to find an acceptable balance between providing increased access to healthcare services while at the same time conserving healthcare resources.

Pryga and Dyer (1992) have noted that, in the USA, hospitals receive a fixed amount per patient for each Medicare patient admission. As such, they have an objective of providing essential medical services whilst physicians are remunerated on the basis of the clinical service provided. The situation emerges where the physician and healthcare managers can have conflicting goals; such a dilemma is bound to affect formulation of best care practices particularly for preventive care.

Clinical Governance (CG)

Clinical governance (CG) was first introduced in the UK by way of a National Health Service (NHS) white paper (Firth-Cozens, 1999) and calls for an integrated approach to quality, team development, clinical audit skills, risk management skills, and information systems. A typical CG process can be delineated into a sequential process that calls for (a) the means to disseminate knowledge about relevant evidence from research, (b) best treatments rather than focusing just on recognition of poor treatments, (c) better appreciation of what IT led solutions can do for clinical governance, and (d) knowing what data/information is available so as to provide baselines for best care and treatments.

Melvin, Wright, Harrison, Robinson, Connelly, and Williams (1999) have remarked that the NHS has witnessed the incorporation and development of many approaches that support and promote effective healthcare, but in practice, none of them have been successful. Research by Zairi and Whymark (1999) submits that the problem lies in the lack of proper systems to support the measurement of organizational effectiveness (i.e., clinical) in a healthcare delivery context.

According to Sewell (1997), one of the biggest challenges in having concise summaries of the most effective clinical practices is establishing what is meant by "quality in healthcare" (i.e., a measurement standard for clinical effectiveness). Sewell (1997) elaborates that measurement standards in clinical practice will change from each context and that this is attributed to the linkage between measurement standards and values and the expectations of the individual healthcare stakeholders (which, in turn, originate from the shared values and expectations to which all the healthcare stakeholders subscribe).

Melvin, et al. (1999) have noted that, in the UK, the NHS has started to support the concept of clinical governance by identifying individual best effective clinical practices. This process provides concise summaries of the most effective clinical practices in all key clinical areas. Summaries that are successfully substantiated are then disseminated throughout the NHS. Sewell (1997) has noted that the USA, Canada, Australia and New Zealand have adopted a formal accreditation system for the healthcare sector based upon the ISO 9000 approach.

Community Health Information Networks (CHIN)

Modern day healthcare organizations have realized that in the future their survival would depend upon their ability to give the caregiver access to such information that would enable the caregiver to deliver personalized clinical diagnosis and treatment in real-time in very specific clinical contexts, a process termed Information Therapy (Dwivedi et al., 2002a). This vision has been translated into concepts such as Integrated Delivery System (IDS) and Community Health Information Networks (CHIN) (Lang, 1997; Mercer, 2001; Morrissey, 2000).

IDS refers to a Healthcare Information System (HIS), a business model based on computing technologies such as Object Orientation (OO) "to share key data, with partners and providers, that will allow faster and more accurate decision making ... to deliver care to a broader population with fewer requirements for expensive and scarce resources" (Lang, 1997, p.18).

CHINs are integrated healthcare institutions based upon a combination of different technology platforms connected to enable support for data sharing amongst different healthcare providers (Mercer, 2001). Both IDS and CHIN are very similar in nature and both refer to an integrated network for allowing the delivery of personalized healthcare. CHINs were founded on the premise that patient information should be shared by competitors (Morrissey, 2000). The main aim of CHIN was to enable hospitals and other healthcare stakeholders to electronically exchange patient encounter summaries and medical records between emergency departments and related departments.

Another factor responsible for emphasis on CHIN was the perception in the healthcare industry that, for small-scale players to survive as individual entities, it was essential for them to form some sort of technological alliances (Huston & Huston, 2000). The original technological objective of CHIN was to enhance data-sharing capabilities amongst different healthcare stakeholders. The original technological infrastructure supported the creation of "point to point" connections. This did not succeed primarily due to limitations in technology coupled with the high amount of financial resources required to establish the "point to point" technological infrastructure (Morrissey, 2000).

The objective behind the incorporation of the CHIN concept is that it allows users to collect data which could be used to formulate "best practice protocols for effective treatment at a low-cost", that is, clinical best evidence practices for both healthcare diagnosis and delivery (Kennedy, 1995). It was anticipated that the advent of CHINs in conjunction with Internet technologies would empower healthcare stakeholders to provide healthcare to patients in real time whilst being in geographically distinct locations (Kennedy, 1995).

KM Taxonomies

KM has become an important focus area for organizations (Earl & Scott, 1999). It has been argued that KM evolved from the applications of expert systems and artificial intelligence (Liebowitz & Beckman, 1998; Sieloff, 1999). Almost all of the definitions of KM state that it is a multi-disciplinary paradigm (Gupta, Iyer & Aronson, 2000) that has further accentuated the controversy regarding the origins of KM. One of the main factors behind widespread interest in KM is its role as a possible source of competitive advantage (Havens & Knapp, 1999; Nonaka, 1991). A number of leading management researchers have affirmed that the Hungarian chemist, economist and philosopher Michael Polanyi was among the earliest theorists who popularized the concept of characterizing knowledge as "tacit or explicit" which is now recognized as the accepted knowledge categorization approach (Gupta et al., 2000; Hansen, Nohria & Tierney, 1999; Zack, 1999).

The cornerstone of any KM project is to transform tacit knowledge to explicit knowledge so as to allow its effective dissemination (Gupta et al., 2000). This can be best met by developing a KM framework. Authors such as Blackler (1995) have reiterated that the concept of knowledge is complex and, in an organizational context, its relevance to organization theory has not yet been sufficiently understood and documented. This is one of the fundamental reasons why KM does not have a widely accepted framework that can enable healthcare institutions in creating KM systems and a culture conducive to KM practices.

KM is underpinned by information technology paradigms such as Workflow, Intelligent Agents and Data Mining. According to Manchester (1999), a common point about software technologies such as (1) information retrieval, (2) document management and (3) workflow processing is that they blend well with the Internet and related technologies (i.e., technologies which focus on dissemination of information). Deveau (2000, p. 14) submits that: "KM is about mapping processes and exploiting the knowledge database. It's taking people's minds and applying technology." Deveau (2000) also noted that information technology puts the organization in a position to state the currently available information in the organizational knowledge base. At this point, the role of IT ends and the role of KM commences.

As KM deals with the tacit and contextual aspects of information, it allows an organization to know what is important for it in particular circumstances, in the process maximizing the value of that information and creating competitive advantages and wealth.

Applicability of the KM Paradigm in Healthcare

A KM solution would allow healthcare institutions to give clinical data context, so as to allow knowledge derivation for more effective clinical diagnoses. In the future, healthcare systems would see increased interest in knowledge recycling of the collaborative learning process acquired from previous healthcare industry practices. This chapter puts forward the notion that this sector has been exclusively focused on IT to meet the challenges described above and reiterates that this challenge cannot be met by an IT led solution.

KM initiatives should be incorporated within the technological revolution that is speeding across healthcare industry. There has to be balance between organizational and technological aspects of the healthcare process, that is, one cannot exist without the other (Dwivedi et al., 2001a). This chapter emphasizes the importance of clinicians taking a holistic view of their organization. Clinicians therefore need to have an understanding of IT in a healthcare context and a shared vision of the organization. Clinicians and healthcare administrators thus need to acquire both organizational and technological insights if they are to have a holistic view of their organization.

The KM paradigm can enable the healthcare sector to successfully overcome the information and knowledge explosion, made possible by adopting a KM framework that is specially customized for healthcare institutions in light of their ICT implementation level. Adoption of KM is essential for healthcare institutions as it would enable them to identify, preserve and disseminate "best context" healthcare practices to different healthcare stakeholders.

Prefatory Analysis of Alternative Healthcare Concepts

The failure of some healthcare management concepts propelled a new stream of thought that advocated the incorporation of the KM paradigm in healthcare (Health Canada, 1999; Mercer, 2001). KM could allow healthcare organizations to truly take advantage of the driving forces behind the creation of the CHIN concept. However, very few organizations have adopted a comprehensive healthcare KM system. The main reason attributed is the failure of healthcare stakeholders in properly creating a conducive organizational

Table 1. Prefatory analysis of alternative healthcare concepts

Concept	Support for People	Support for Process	Support for Technology	Limitations
CG	Present	Insufficient	Present	Policy initiative
EBM	Insufficient	Insufficient	Present	Tacit Processes?
CHIN	Insufficient	Absent	Present	Limited Trials
IHCDS	Insufficient	Insufficient	Present	Technology focus
IPP	Insufficient	Present	Present	Tacit Knowledge?
KM	Present	Present	Present	Not validated

culture. Based on a literature review above, a preliminary conceptual analysis of alternative healthcare management concepts is presented in Table 1. As can be seen from the table, healthcare stakeholders are searching for alternative paradigms that support collaboration in order to synergistically learn from others' experiences, training and knowledge within specific organizational cultures. Healthcare institutions have realized that existing concepts such as EBM and CG do not enable healthcare stakeholders to achieve this challenge as they do not holistically support effective integration of IT within specific organizational cultures and processes. Contemporary concepts such as EBM, CHIN, ICHDS and IPP focus on IT at the expense of having too little emphasis on people. This is further aggravated by the presence of dysfunctional organizational processes in the majority of healthcare institutions.

Conclusion

For any healthcare organization to succeed, it needs to excel in a number of key processes (i.e., patient diagnosis, care treatment, etc.) that are necessary for it to achieve its mission. If the processes are repetitive, automation is possible via the use of IT. Modern IT applications in healthcare are not sufficient in meeting the information needs of current healthcare institutions as they lack the ability to deliver precise, accurate and contextual information to the desired caregiver at the desired time.

This chapter has presented an analysis of alternative healthcare management concepts with respect to their ability in providing a solution to the issue of information management. Furthermore, this chapter has examined the feasibility of the KM paradigm in solving the problem of information explosion in healthcare and has found validation for the proposition that the current focus on technological solutions will aggravate the problem of explosion in clinical information systems for healthcare institutions.

The chapter has also presented the key requirements for creating a KM framework, which can act as a template in enabling healthcare institutions in their attempts to initiate KM projects. This chapter concludes that any potential solution has to come from a domain that synergistically combines people, organizational processes and technology, thereby enabling healthcare stakeholders to have a holistic view of the entire healthcare continuum. This chapter further concludes that KM is the only paradigm that combines the above-mentioned perspectives (i.e., people, organizational processes, and technology) into healthcare and as such, KM is the next indispensable step for integrated healthcare management.

References

Blackler, F. (1995). Knowledge, knowledge work and organizations: An overview and interpretation. *Organization Studies, 16*(6), 1021-1046.

Cowling, A., Newman, K. & Leigh, S. (1999). Developing a competency framework to support training in evidence-based healthcare. *International Journal of Health Care Quality Assurance, 12*(4), 149-159.

Davis, M. & Klein, J. (2000). Net holds breakthrough solutions. *Modern Healthcare*, Supplement, Eye on Info. Feb 7, 2000, 14-15.

Deveau, D. (2000). Minds plus matter: Knowledge is business power. *Computing Canada, 26*(8), 14-15.

Dwivedi, A., Bali, R.K., James, A.E. & Naguib, R.N.G. (2001a). Telehealth systems, Considering knowledge management and ICT issues. *Proceedgins of the IEEE-EMBC 23rd Annual International Conference of the IEEE Engineering in Medicine and Biology Society (EMBS)*, [CD-ROM], Istanbul, Turkey.

Dwivedi, A., Bali, R.K., James, A.E. & Naguib, R.N.G. (2001b). Workflow management systems, The healthcare technology of the future? *Proceedings of the IEEE EMBC-2001 23rd Annual International Conference of the IEEE Engineering in Medicine and Biology Society (EMBS)*, [CD-ROM], Istanbul, Turkey.

Dwivedi, A., Bali, R.K., James, A.E. & Naguib, R.N.G. (2002a). The efficacy of using object oriented technologies to build collaborative applications in healthcare and medical information systems. *Proceedings of the IEEE Canadian Conference on Electrical and Computer Engineering (CCECE) 2002*, Winnipeg, Canada (Vol. 2, pp. 1188-1193).

Dwivedi, A., Bali, R.K., James, A.E., Naguib, R.N.G. & Johnston, D. (2002b). Merger of knowledge management and information technology in healthcare: Opportunities and challenges. *Proceedings of the IEEE Canadian Conference on Electrical and Computer Engineering (CCECE) 2002*, Winnipeg, Canada (Vol. 2, pp.1194-1199).

Earl, M. & Scott, I. (1999). What is a chief knowledge officer? *Sloan Management Review, 40*(2), 29-38.

Firth-Cozens, J. (1999). Clinical governance development needs in health service staff. *Clinical Performance and Quality Health Care, 7*(4), 155-160.

Gupta, B., Iyer, L.S. & Aronson, J.E. (2000). Knowledge management: Practices and challenges. *Industrial Management & Data Systems, 100*(1), 17-21.

Hansen, M.T., Nohria, N. & Tierney, T. (1999). What's your strategy for managing knowledge? *Harvard Business Review, 77*(2), 106-116.

Havens, C. & Knapp, E. (1999). Easing into knowledge management. *Strategy & Leadership, 27*(2), 4-9.

Health Canada (1998). Vision and strategy for knowledge management and IM/IT for Health Canada [online]. Retrieved March 20, 2002, from *http://www.hc-sc.gc.ca/iacb-dgiac/km-gs/english/vsmenu_e.htm*

Huston, T.L. & Huston, J.L. (2000). Is telemedicine a practical reality? *Association for Computing Machinery, Communications of the ACM, 43*(6), 91-95.

Kennedy, M. (1995). Integration fever. *Computerworld, 29*(14), 81-83.

Lang, R.D. (1997). The promise of Object Technology. *Journal of Healthcare Resource Management, 15*(10), 18-21.

Latamore, G.B. (1999). Workflow tools cut costs for high quality care. *Health Management Technology, 20*(4), 32-33.

Liebowitz, J. & Beckman, T. (1998). *Knowledge organizations: What every manager should know.* USA: St. Lucie Press.

Manchester, P. (1999). Technologies form vital component, INFRASTRUCTURE, The IT infrastructure will capture, store and distribute the information that might be turned into knowledge. *Financial Times,* London, Nov 10, 1999, 8.

Melvin, K., Wright, J., Harrison, S.R., Robinson, M., Connelly, J. & Williams, D.R. (1999). Effective practice in the NHS and the contribution from public health, a qualitative study. *British Journal of Clinical Governance, 4*(3), 88-97.

Mercer, K. (2001). Examining the impact of health information networks on health system integration in Canada. *Leadership in Health Services, 14*(3), 1-30.

Morrissey, J. (2000). The evolution of a CHIN. *Modern Healthcare*, Supplement, Eye on Info, May 29, 2000, 42-43.

Nonaka, I. (1991). The knowledge-creating company. *Harvard Business Review, 69*(6), 96-104.

Pryga, E. & Dyer, C. (1992). The AHA's national health reform strategy: An overview. *Trustee, 45*(5), 10.

Schmid, K. & Conen, D. (2000). Integrated patient pathways, "mipp"—A tool for quality improvement and cost management in health care. Pilot study on the pathway "Acute Myocardial Infarction". *International Journal of Health Care Quality Assurance, 13*(2), 87-92.

Sewell, N. (1997). Continuous quality improvement in acute health care, creating a holistic and integrated approach. *International Journal of Health Care Quality Assurance, 10*(1), 20-26.

Sieloff, C. (1999). If only HP knew what HP knows: The roots of knowledge management at Hewlett-Packard. *Journal of Knowledge management, 3*(1), 47-53.

Zack, M.H. (1999). Managing codified knowledge. *Sloan Management Review, 40*(4), 45-58.

Zairi, M. & Whymark, J. (1999). Best practice organisational effectiveness in NHS Trusts Allington NHS Trust case study. *Journal of Management in Medicine, 13*(5), 298-307.

Chapter II

It's High Time for Application Frameworks for Healthcare

Efstathios Marinos, National Technical University of Athens, Greece

George Marinos, National Technical University of Athens, Greece

Antonios Kordatzakis, National Technical University of Athens, Greece

Maria Pragmatefteli, National Technical University of Athens, Greece

Aggelos Georgoulas, National Technical University of Athens, Greece

Dimitris Koutsouris, National Technical University of Athens, Greece

Abstract

This chapter discusses the impact of adopting application frameworks in the healthcare (HC) domain. It argues the shortcomings of existing HC applications and systems, examines the benefits of application frameworks use during and after the software development, and presents such an application framework. The authors hope that this chapter will put on the table the discussion about the necessity of application frameworks in HC because they strongly believe that the software industry can tremendously benefit from the work done so far in the area of the HC standardization, in order to provide HC-specific application frameworks that will make software development easier and more efficient.

Copyright © 2005, Idea Group Inc. Copying or distributing in print or electronic forms without written permission of Idea Group Inc. is prohibited.

Introduction

Healthcare Information Systems (HIS) development is by definition very challenging since it involves many domains apart from the health domain (e.g., Enterprise Resource Planning, Financial and Billing Services, etc.). The usual scenario for a healthcare (HC) organization is to have a lot of different applications/systems in order to cover all required functionality. As a result, heterogeneity has been a key problem for HIS as Kuhn and Giuse report (2002, p. 66). Understanding the needs of HC in the HIS area, CEN-TC 251 introduced the Healthcare Information Systems Architecture (HISA). HISA describes a "middleware of common services" that should exist within every HIS. In the HISA scenario, applications "acting" within the HIS are using these common services. CEN and other standardization organizations such as the International Organization for Standardization (ISO), Health Level Seven (HL7) and Object Management Group (OMG) have put a lot of effort towards the direction of providing standards, mainly not in contradiction to each other (even if they are not always perfectly aligned) as Jagannathan et al. report (Jagannathan, Wreder, Glicksman & alSafadi, 1998, p. 24), covering other issues of the HC domain as well as implementations of the middleware of common services.

Although today, a set of standards to use in various circumstances exist, the lack of HC standards use is a reality. That is not the only reason for the weaknesses of current HIS. As Beale underlines (2002, p.1), another important reason is the usual practice of having the domain concepts hard-coded into the software and database models. The shortcomings of such practices are especially evident in the HC sector, where the total number of concepts and the observed rate of change are very high. Limited interoperability and application lifespan result in overwhelming costs for the maintenance and extension of healthcare information systems, minimizing the benefits of IT infrastructure for the healthcare providers.

Our intent is to "put on the table" the discussion about the necessity for Application Frameworks for HIS. The term Application Framework uses for the purpose of this chapter, Johnson and Foote's (1988) definition for object-oriented frameworks: It is a set of classes that embodies an abstract design for solutions to a family of related problems. We are especially concerned about the applications layer of the HISA approach, because there are already implementations and approaches concerning the middleware of common services such as the Distributed Healthcare Environment (DHE) and OpenEMed. The adoption of an Application Framework allows the rapid designing and development of robust, portable and secure HC applications, tailored to the needs of each specific HC provider. The main features of an application framework for HIS should be inherent support of global HC standards, such as HL7 and DICOM, inherent support of open standards and interoperable protocols, such as XML and SOAP and the ability to design and build highly customizable and adaptive applications and user interfaces. In order to provide future-proof applications, domain concepts, business rules, and workflow, as well as user interface design should not be hard-coded, but maintained externally to the applications.

In the following pages one will find reference to HC related standards and outstanding work, concise reference to IT related standards and architectures but no reference to the significant work done in the field of security and trust due to space limitation except for a short reference to Web Services security issues. The characteristics of existing HIS are examined, thus concluding in the need for application framework for HIS. Additionally, we briefly present an application framework for HIS and subsequently assess the experience gained by its development.

HC Related Standards, Outstanding Work

ISO TC-215. ISO Technical Committee 215 (ISO TC-215) has so far published nine standards in the area of health informatics: ISO/TS 17090 part 1, part 2, part 3 are about Public Key Infrastructure. ISO/TR 18307 is about interoperability and compatibility in messaging and communication standards. ISO/TS 18308 is about the "Requirements for an electronic health record architecture". ISO 18104 concerns the "Integration of a reference terminology model for nursing". ISO 18810 deals with "Clinical analyser interfaces to laboratory information systems". Finally, ISO 22857 provides "Guidelines on data protection to facilitate trans-border flows of personal health information".

CEN TC-251. CEN/TC-251 is a European Union (EU) standardization organization dedicated to the field of Health Information and Communications Technology. There are four working groups dedicated to different areas: WG1 is assigned the development of European standards to facilitate communication between information systems for health-related purposes, WGII handles semantic organization of information and knowledge, WGIII quality safety and security, and WGIV development and promotion of standards that enable the interoperability of devices and information systems in health informatics.

WGIV has released the European Prestandard (ENV) 12967.01 (CEN-TC 251, "ENV-12967.01: Healthcare Information System Architecture - Part 1: Healthcare Middleware Layer"), that became a standard at 1998. The ENV 12967.01 (p. 5) aims to establish both **"general principles for the architecture of HC information systems** as well as the scope of a set of **Healthcare Common Services** (HCS), provided by the middleware layer of the healthcare information system [bolds added]". ENV 12967.01 concerns only the middleware of common services.

Please note that the *middleware of common services* is not directly related to the term middleware usually meaning the *application server* located on the middle tier of 3-tier architecture. Figure 1 demonstrates the relation between the HISA and software-architecture perspectives.

ENV 12967.01 is independent of any specific technological environment and does not imply the adoption of any specific organizational, design or implementation solution; furthermore, it's applicable to the information systems of any type of healthcare organization. The specification of the characteristics of the identified HCS is limited to the sole formalization of their external behavior, in terms of their function and of the information to be made available to the rest of the system. Such behavior is defined at the conceptual level only, through formalisms and notations suitable to identify the scope of the services. ENV 12967.01 specifies only a fundamental set of classes of

Figure 1. ENV 12967.01 architecture over classic three-tier architecture

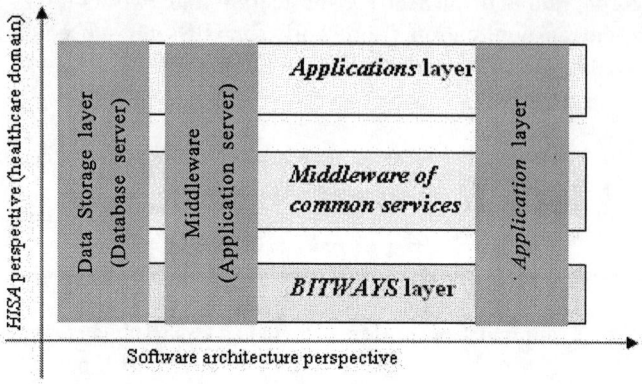

information and services which are necessary or useful in HIS which is considered to be of major relevance. Thus, the set of services and information classes identified in ENV 12967.01 is not exhaustive and may be extended in individual installations, according to specific requirements, depending on the particular nature of the information system, as well as on national and local needs.

A HISA full-compliant middleware, called DHE has been developed within RICHE and EDITH EU projects. DHE under the umbrella of *Health Advanced Networked System Architecture* (HANSA*)* has demonstration implementations in 23 hospitals in the EU as mentioned in HANSA's Web site.

HL7. HL7 is one of the most active ANSI-accredited Standards Developing Organizations operating in the HC. HL7's mission is (HL7 Web site, 2004, "*about*" section): *"To provide standards for the exchange, management and integration of data that support clinical patient care and the management, delivery and evaluation of healthcare services. Specifically, to create flexible, cost effective approaches, standards, guidelines, methodologies, and related services for interoperability between healthcare information systems."*

The *HL7 Version 2 Messaging Standard—Application Protocol for Electronic Data Exchange in Healthcare Environments*—is considered to be the workhorse of data exchange in healthcare and is the most widely implemented standard in HC information around the world. There are different versions of HL7 messaging standard with most recent the version 2.5 (HL7 version 2.5, 2003) which was approved as ANSI standard June 26, 2003, while HL7 version 3 is in progress. The HL7 version 3 uses an object-oriented development methodology and a *Reference Information Model* (RIM) to create messages that are to be exchanged among the subsystems of a health institution. In HL7 version 3, the major change from previous versions is the adoption of the RIM. The process for building version 2.x messages was entirely *ad hoc* and there was no explicit methodology as Beeler et al. note (1999). Within the RIM, there is no notion of *optionality*. The presence or absence of a data element value is defined in the *Hierarchical Message Description (HMD)* associated with a trigger event. This approach leads to both more messages and more precise definitions. Specific users of the RIM are expected to utilize relevant portions of the RIM as needed, adopting its content to their own information modelling needs and notations ("RIM, The Health Level Seven", 2004). Other ANSI or ISO accredited standards organizations such as X12 and CEN-TC251 have expressed interest in using or referencing the *RIM* in their own standards development work, while openEHR's Reference Model (e.g. the openEHR Demographic Reference Model), already provides correspondences both to HL7 RIM and CEN's ENV 13606.

Apart of the RIM specification, another foundation component of version 3 is the *vocabulary*. For HL7 (V3 Guide, section 2.3), "the vocabulary domain is the set of all concepts that can be taken as valid values in an instance of a coded field or attribute".

Especially for HL7 version 3, attention has been paid to XML and Web Services and relative work is currently being done (e.g., Web Services Profile).

openEHR. Another reference work in the health domain is the openEHR work. It started as the GEHR EU project in 1992, and as mentioned in the "Origins" section of openEHR Web site ("The Interface between the GEHR", 2004), had actually influenced the CEN TC-251 work on Electronic Health Record (EHR) (the first CEN pre-standard for EHR, ENV 12265). Later at 1999, after a number of EU and Australian projects, openEHR was formed as an open source foundation (actually an online community) whose aim is to promote and facilitate progress towards electronic healthcare records of high quality.

Although openEHR concerns the EHR, its constraint-based domain models approach, establishes the separation of knowledge and information levels in information systems as author indicates (Beale, 2002, p. 2). In order to define valid information structures at the knowledge level model, the concept of *Archetypes* is used. Archetypes enable users in a domain to formally express their concepts, enable information systems to guide and validate user input, guarantee interoperability at the knowledge level, and provide a well-defined basis for efficient querying of complex data.

OMG Healthcare Task Force. "The Object Management Group (OMG) is an open membership, not-for-profit consortium that produces and maintains computer industry specifications for interoperable enterprise applications" as stated in its Web site ("About the Object Management Group", 2004). OMG is widely known due to its relation with CORBA. Dedicated OMG Task Forces are working on standardizing domain facilities in industries such as HC.

The OMG Healthcare Domain Task Force has defined standard interface requirements between HC-related services and functions. These middleware specifications include:

- The *Person Identification Service* (PIDS) specification defines the interfaces that organize person ID management functionality to meet HC needs. The specification is designed to support assignment of IDs within a particular domain and correlation of IDs among multiple domains, searching and matching of people independent of matching algorithms, and federation of PIDS services.
- The *Lexicon Query Service* specifies a set of common, read-only methods for accessing the content of medical terminology systems. It can be used to implement a common interface to any of the major medical coding schemes and achieve unambiguous concept representation in a distributed HC environment.

- The *Clinical Observations Access Service* (COAS) is a set of interfaces and data structures with which a server can supply clinical observations. It provides a variety of assessment mechanisms, so that complex clinical information can be efficiently searched and retrieved.

- The *Resource Access Decision* (RAD) Facility is a mechanism for obtaining authorization decisions and administrating access-decision policies. It establishes a common way for an application to request and receive an authorization decision.

- The *Clinical Image Access Service* (CIAS) is a set of interfaces and data structures with which a server can provide unified access to clinical images from DICOM or non-DICOM image sources. This specification utilizes and extends the COAS.

OpenEMed is a set of open-source components based on the aforementioned OMG specifications, HL7 and other data standards. The OpenEMed components are written in Java and target interoperable service functionality that reduces the time it takes to build HC related systems. It includes sample implementations of the PIDS, COAS, RAD, and the *Terminology Query Service* and requires CORBA infrastructure to run. OpenEMed also provides tools for assisting the healthcare application development, such as libraries providing persistence to a number of different databases, general GUI tools for managing XML driven interfaces, tools for handling SSL, and so on.

DICOM. The increased use of computer-based systems in medicine acknowledged the need for a standardized method to transfer images and the associated medical data among systems from different vendors. DICOM standard (DICOM P.S. 3.1, 2003) was developed based on this need. The goals of DICOM standard are the creation of an open environment among vendors, the interchange of medical images and related data, and the facilitation of interoperability among systems. DICOM standard concentrates on the following needs: digital image generation, digital transfer/archiving (PACS), post-processing of medical data, cross-vendor compatibility and communication via networks/media as outlined in Gorissen's presentation (1997).

In order to facilitate interoperability of medical-imaging equipment, DICOM specifies a set of network communication protocols, the syntax and semantics of commands and associated information which can be exchanged, a set of media storage services, a file format to facilitate access to the images and related information stored on media and information, that must be supplied with an implementation for which conformance to the standard is affirmed.

IT Related Standards, Architectures

It's not in the intentions of this chapter to provide an extensive reference to existing information technology architectures and standards. In the following paragraphs we will discuss both client/server and distributed architectures. We will also make a short reference to the most common ways of realizing distributed architecture: CORBA, DCOM, and XML Web Services.

In the *client/server* scenario we have two distinct parts: The server, where the database software is hosted and the clients where the application runs. The defining characteristic of this design as MacDonald identifies (2003, p. 6), is that "although the application is shared with multiple clients, it's really the clients rather than the server that perform all the work". The fallbacks of the client-server architecture (with respect to performance) are as follows: it can't accommodate easy client interaction, it doesn't allow for thin clients, there is a possibility for bottleneck on the server-side database, and in most cases the deployment is an important problem.

The basic concept behind *distributed applications* is that at least part of the functionality runs remotely, as process in another computer. The client-side application communicates with the remote components, sending instructions or retrieving information. Distributed architecture has some important advantages over the *client/server* architecture with more important the issue of scalability: new computers can be added providing additional computing power and stability because one computer can fail without derailing the entire application. Other advantages are the support to thin clients, the cross-platform code integration and the distributed transactions (MacDonald, 2003, pp. 9-11).

During the last decade, evolution of both networking technologies and the structure of companies and organizations have led to the emergence of complex distributed systems consisting of diverse, discrete software modules that interoperate to perform some tasks. The interoperation of such remote modules has been achieved by *distributed object environments*, such as CORBA and DCOM. The vision behind such environments is that there is no distinction between local and remote objects, as seen from the programmer's point of view. This is achieved in the case of CORBA by using the Object Request Broker (ORB), while in the case of DCOM using the windows infrastructure (that is the reason why CORBA is cross-platform in contradiction with DCOM that targets windows systems only).

In the last years, a new way of realizing the distributed architecture has been presented: "XML Web Services makes it easier for systems in different environments to exchange information through an interface (e.g., SOAP 1.2)" as noted in W3C's "Web Services Activity Statement" ("Web Services Activity statement", "Introduction" section). The power of Web services, in addition to their great interoperability and extensibility thanks to the use of XML, is that they can then be combined in a loosely-coupled way in order to achieve complex operations. Programs providing simple services can interact with each other in order to deliver sophisticated added-value services. The only assumption between the XML Web service client and the XML Web service is that recipients will understand the messages they receive. As a result, programs written in any language, using any component model, and running on any operating system can access XML Web services.

Security in Web Services. SOAP uses XML to specify remote method invocation calls, usually over HTTP. In such architecture an obvious security choice is to use *transport level* security mechanisms provided by SSL/TLS or IPSec. However, these mechanisms do not provide complete protection especially for the next generation of Web services, which will be able to run on new protocols and which will include federated applications. In addition to the basic security requirements (confidentiality, integrity and authenticity of data exchanged), Web services require data validation, accountability, and distributed authentication and authorization, none of these being provided by SSL. These security requirements can be met with the use of the emerging XML security technologies (especially XML Signature, "XML-Signature Syntax and Processing", 2002, and XML Encryption, "Encryption Syntax and Processing", 2002), which apply security at the message layer of the Open Systems Interconnection (OSI) stack, providing end-to-end security in Web services environments. This security model is described in the *WS-Security Specification* version 1.0, launched by the *Organization for the Advancement of Structured Information Standards (OASIS)* Web Services Security TC in March 2004. This specification proposes a standard set of SOAP extensions that can be used when building secure Web services to implement message content integrity and confidentiality. The specification is flexible and is designed to be used as the basis for securing Web services within a wide variety of security models including Public Key Infrastructure (PKI), Kerberos, and Secure Sockets Layer (SSL). For example, a PKI can be used to provide authentication, digital signatures, and key distribution.

WS-Security Specification version 1.0 has been the basis for the development of a complete security solution for future-proof Web services. IBM and Microsoft have collaborated to propose a **comprehensive** Web Services security plan and roadmap for developing a set of Web Service Security specifications that address security issues for messages exchanged in a Web service environment, in a compatible, extensible and interoperable manner. The proposed plan as presented (Security in a Web Services World, 2002) covers a wide range of security issues, namely: WS-Security, WS-Policy, WS-Trust, WS-Privacy, WS-Secure Conversation, and WS-Authorization.

Characteristics of Existing HC Information Systems

Current HC information systems are built using platform and device specific features, while the health domain concepts are hard-coded directly into the software and database models (Beale, 2002, p. 1). Moreover, they are characterized by limited support of global healthcare standards, the use of obsolete technologies and inflexible architecture design (de Velde, 2000, p. 1; Kuhn & Giuse, 2001, p. 66).

The shortcomings of such practices are especially evident in HC, where the total number of concepts and the observed rate of change are very high. Taking into account that, for example, a medical-classic term set, the SNOMED-CT, codes some 357,000 atomic concepts, one can only imagine the amount of work to be done in order to deliver robust healthcare applications, let alone the boom of cost in case of changes. Lenz and Kuhn (2004, p.1) report that this change can be brought by either internal causes, such as new diagnostic or therapeutic procedures or external, such as economic pressure. The inherent inflexibility results in limited application lifespan and overwhelming costs for the maintenance and extension of HC information systems.

Current systems are inadequately integrated into the established operational workflows and hospital procedures, minimizing user acceptance and delivered productivity. Moreover, inadequate use of global standards results in applications that exhibit limited interoperability, increasing the cost of intercommunication between heterogeneous HIS. Such a need is presented in the "Introduction" section of the ENV 12967.01 standard, where is identified that HC organizational structures usually consist of networks of HC centers distributed over the territory, characterized by a high degree of heterogeneity and diversity, from organizational, logistic, clinical and even cultural perspectives. The situation becomes even more complicated by the fact that a large number of isolated and incompatible applications are already installed and operational in many healthcare organizations, satisfying specific user needs as Kuhn and Giuse report (2002, p.66).

Kuhn and Giuse (2002, pp. 65-67) affirm that HIS at the majority of institutions today are far from the stated goals of supporting HC. Several problems contribute to this situation: Questions concerning integration and data input are still unsolved, the market remains volatile and few successful systems have been deployed, it is difficult to demonstrate return on investment while health IT departments lack adequate financial support and there seem to be more failures and concrete difficulties than success stories suggest.

The Need for HIS Application Frameworks

A crucial challenge for building applications for the health domain is that the number of the concepts that should be covered is very large and that each individual concept can be quite complex, as Beale, Goodchild and Heard identify (2002). Furthermore, health data should be available to process at the semantic level in order to empower decision support and evidence-based medicine. Finally, all these requirements should be satisfied in a cost-effective manner that does not overburden the budget of HC providers.

Before proposing ways to meet the challenge of developing applications for HIS, it's useful to see the bigger picture of HIS. HISA distinguishes three layers within a HIS: the *Bitways* layer (infrastructure), the Middleware of common services layer and the *Applications* layer. Such an approach fits nicely with the current situation where many HIS are comprised of many specialized applications or even sub-systems dedicated to specific areas such as outpatient scheduling, billing, ward management, Radiology Information Systems (RIS), Laboratory Information Systems (LIS), and so on. Even if some common services are treated inside a sub system, it is not inhibitory to realize the above HIS architecture, as long as the *middleware of common services* provides them.

Within the *Applications layer* reside concept-specific applications, providing the necessary interface for the users in order to insert or view data, generate reports, documents, and so on. A plethora of such applications is either available on the market or custom-developed. Usually, these applications are stand-alone and the usual practice in order to intercommunicate with other applications within the HIS is to provide a "bridge" to the "outside world". In the best-case scenario, such applications use a standard protocol (e.g., version 2.x HL7 messages), or a custom-made protocol mutually accepted within the scope of HIS.

In order to develop HIS we need: the development of the *Bitways* layer (infrastructure), the development of the *middleware of common services* and the development of the applications located in the *Applications* layer. As far as the *middleware of common services* is concerned, a lot of work has been done both for standardization (e.g., CEN's work on the architecture specification, OMG's work on the *common services* interfaces specifications, HL7's work on defining messages to be exchanged within the HIS), and software development (e.g., *DHE*, and *OpenEMed*).

What seems to be missing is the means to create applications for a HIS. That's partially true since there are strong and very popular development tools for all kinds of applications: windows, Web or even smart-phone applications. Development environments/ platforms like .net, Java, win32 as well as development suites targeting those environments like Microsoft's Visual Studio, Borland's Delphi, JBuilder, Eclipse, and so on, are only some of the means currently available for delivering useful, well-designed and functional applications. However, these are mainly *Rapid Application Development* (RAD) tools for generic purposes. There are no domain-specific development tools available today, which is to say that the idea of Application Frameworks for HC is not widespread. *In order not to reinvent the wheel* (in terms of the HIS applications development and HC in general), **the software industry can tremendously benefit from the work done so far in the area of the HC informatics standardization, in order to provide HC-specific Application Frameworks** (to be used by RAD tools) **that will make software development easier and more efficient**. For the purposes of this chapter, we will concentrate only on *Application Frameworks for HIS*; adopting *HISA's* perspective namely that *HISA* targets any type of HC organization.

The main features of such an Application Framework for HIS should be:

- The ability to design and build highly customizable and adaptive applications;
- The inherent support of global healthcare standards, such as HL7, DICOM, and so on; and
- The inherent support of open standards and interoperable protocols, such as XML and SOAP.

In order to achieve the first goal, an approach like the one *openEHR* (Beale, 2002), is evangelizing might be very useful: "separation of knowledge and information levels in information systems". Actually, we go a little further: data concepts, business rules and workflow, as well as user interfaces should not be hard- coded in the applications produced. Key consequence of such a practice is that software and data depend only on small, non-volatile models. Applications can thus be developed and deployed quickly, even without waiting for the knowledge concepts to be thoroughly defined. Such an approach is closer to today's design of information systems practice that as Kuhn and Giuse (2002, p. 71) say "...is moving from completely modelling a system before implementing it, to interactive rapid prototyping". This methodology also, allows new or modified domain concepts, changes in operational workflows (e.g., hospital procedures), or changes in security/privacy policy rules, to be quickly and seamlessly integrated at a low cost without major software recompilation and user annoyance or ideally without any recompilation of the code.

The inherent support of global HC standards should also be an *a priori* characteristic of an Application Framework for HIS. Roughly analogous to the need of supporting HC standards within a HC application framework is the need of supporting for example ODBC or the SOAP within a RAD: it's inconceivable not to; try to imagine for example someone that uses Microsoft's Visual studio, having to read and implement the SOAP! In reality, that means that the developer will be exhausted by SOAP technicalities missing the business process details. Kuhn and Giuse (2002, p.65) refer to Regenstrief medical record system (one of the success stories in health informatics) and to the developers' statement that: "we believe that our success represents persistent efforts to build interfaces directly to multiple independent instruments and other data collection systems, using medical standards such as HL7, LOINC, and DICOM". So, from the developers' point of view (who will be the actual users of such a framework) it is clear that it is more than enviable to have handy HC standards implementations.

The same stands for IT standards and interoperable protocols. Actually, once more as Kuhn and Giuse (2002, p. 72) summarize, "…the broad acceptance of web standards such as XML has led to their integration into the ongoing standardization process".

Using frameworks in order to provide applications offers significant benefits. Applications can extend and specialize extension points—-called "hot spots" as mentioned by Markiewicz and Lucena (2001)—provided by the framework, according to their special needs (Johnson, 1997). The modularity, reusability, extensibility and inversion of control that a framework offers can encapsulate an amount of complexity that framework clients (applications) can reuse (Fayad & Schmidt, 1997). In such a scenario as Fayad and Hamu explains (2000), applications are built on top of frameworks, which provide an extensive library of business objects supporting the intended application domain and use its services. The use of frameworks can reduce the overall cost of creation and maintenance of a health-related application. This cost is enormous because of the special requirements of HC, namely increased security, interoperability, extensibility, portability, and the ability to evolve. Application frameworks can effectively handle these aspects. The development of a framework is generally more expensive than a single application but can payoff through the repeated generation of applications within the proposed domain as Markiewicz and Lucena say (2001).

Besides the above benefits from application frameworks, it's significant to note another side-effect of adopting frameworks in application development: Most times, the use of a framework denotes the use of an Information Model or something analogous, since the framework is not an *ad-hoc* development result. That means that at the end, we will have a repository of business objects definitions or at least, a formal description of concepts. That is very critical in order to translate between different ontologies. In case this "formal description of concepts" originates from the *middleware of common services* (possible as a common service itself), it can even result in consistently merging the ontologies used. Although using reference models for realizing the *middleware of common services* layer is a common place that is not the case when it comes to the applications located in the *applications* layer.

An application framework providing a clear description of the underlying data structures as well as built-in validity and consistency rules for data verification can also help reduce data entry errors and minimize the excessive cost of cleaning medical databases for knowledge discovery purposes. Krzystof (2001) claims that data preparation for any knowledge discovery process usually consumes at least half the entire project effort and is a major contributor to its cost. This situation is even more challenging for medical databases that are characterized by incomplete, imprecise and inconsistent data, as data collection is usually a by-product of the primary patient care activities. Decision support systems are powerful tools for improving medical decision making and preventing medical errors. It is reported that medical errors are a leading cause of death, while the majority of adverse events that contribute to patient morbidity or mortality are preventable. Decision support systems can help reduce medical errors by questioning the actions of healthcare professionals, offering advice and examining a range of possibilities that humans cannot possibly remember (Kohn, Corrigan, & Donaldson, 1999). The accuracy of developed decision support systems depends heavily on the precision and correctness of the available data and the sufficient modelling of the underlying domain concepts.

While the technical advantages of application frameworks are sufficiently covered by literature, their business perspective is largely neglected. It is reported that the adaptive nature of application frameworks enables flexible response to new and rapidly changing market opportunities, making a framework a strategic tool for attacking vertical markets (Codenie, De Hondt, Steyaert, & Vercammen, 1997). Return on investment from a developed application framework may come from selling it to other companies or by future savings in development effort within a company itself, such as higher software quality and shorter lead times (Fayad, Schmidt, & Johnson, 1999).

Although business opportunities from developing a framework may be intuitive, the lack of a reliable business model for framework development has made management departments reluctant to adopt framework-based solutions. The Value Based Reuse Investment (VBRI) approach (Fayad et al., 1999) tries to fill the gap and views reuse investments as a continuous process of formulating strategic options and then evaluating them with respect to their potential for value creation. It also suggests that the "potential of

Figure 2. Framework.Health services placement over the 3-tier architecture

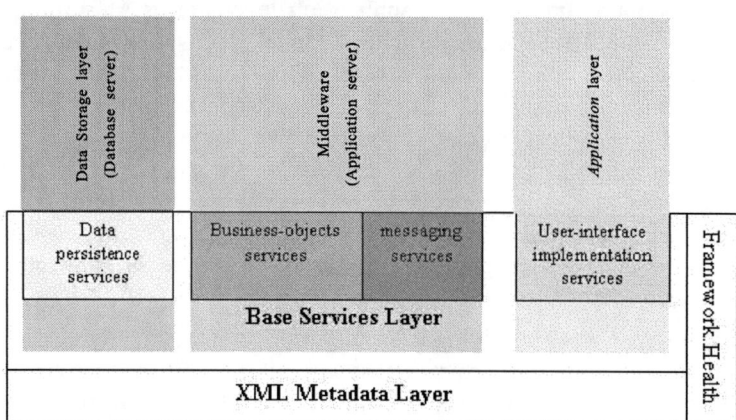

application framework technology for value creation can only be realized if it is exploited at all levels of the Reuse Capability Model" and not only at the lower levels of the model, such as at the level of individual projects, which is already practiced by many organizations worldwide. This means that reuse should be employed at an organization-wide level and over multiple projects. Moreover, reuse strategies should take into account families of related products and create new business opportunities in a rapidly changing market environment.

Developing an Application Framework for HIS

For the last two years, the authors of this document have toiled developing an application framework for HC domain (Framework.Health) able to be integrated within Microsoft's Visual Studio .Net.

The framework provides all the software modules needed for the handling of dynamically defined and reusable HC business objects (e.g., patient demographics, diagnoses, exams). The built-in infrastructure of the framework includes an *XML metadata layer* and the *base services layer* (see Figure 2).

The *XML metadata layer* provides a common domain concept repository that allows the semantic interoperability among the application tiers and supports all operations performed by the *base services layer*. The HC business objects populating the domain concept repository are derived from HL7 v 2.4 standard entities. These concepts, along with business rules, operational workflow and user interface design, are not hard-coded into the applications, but maintained externally into the *XML metadata layer*. This layer also defines a set of *XML languages* for formally describing data types, data structures and messages, as well as user interface behavior and application workflows that are needed to build the data and software model.

The *base services layer* exposes all the functionality needed for building HC applications, namely:

1. The *data persistence* services provide a uniform and flexible way of accessing the underlying data storage. The data model is explicitly derived from the business objects described in XML resource files (e.g., a relational database schema is generated dynamically from these files). Any changes to the business objects are easily incorporated into the data model implementation. Moreover, the framework provides a set of software libraries for accessing the data storage, which accelerate the application development and provide a uniform and reliable way of communicating with the data storage tier.

2. The framework provides a set of libraries that aid the construction and manipulation of business entities using an object-oriented programming language. These libraries make use of the metadata layer to create the appropriate structure for each

entity that the application uses and provide a common interface for the manipulation of the data stored in these structures, (e.g., HL7 *segments* and *data types* are instantiated as objects). These libraries also provide coupling with the *data persistence services*.

3. The *messaging services* include XML-based message definition, message creation and validation. The framework provides a solid infrastructure for secure message transfer, as well as message processing and persistence tools. A specific messaging implementation is provided for HL7 v2.4 messages.

4. Adaptive *user interface construction* involves all the mechanisms for user interface definition through the use of XML and dynamic UI creation based on the XML description files. One can find a complete discussion of the used approach in (Marinos, Marinos & Koutsouris, 2003).

The overall framework architecture consists of three distinct parts: the *Data Model*, the *Controller* and the *Presentation Layer*. The architecture is depicted on Figure 3.

Data Model. The framework provides a reference implementation of the HL7 standard data model. This model includes data types, segments (a logical grouping of data fields) and messages (the atomic units of data transferred between systems); please refer to (HL7 version 2.5, 2003) for a consist definition of the terms "data types", "segments" and "messages". These entities are not hard-coded into the framework, but externally defined in XML format. Figure 4 shows the XML definition of the CE (Coded Element) HL7 data type.

Similar definitions apply to *segments* and *messages* and are customizable according to each specific implementation (e.g., according to HL7 conformance profiles). Figure 5 shows an example of a segment definition.

XML definitions also describe the mapping of such entities to database schema elements, as shown in Figure 6. The elements and attributes of the XML fragment shown on Figure 6, instruct some specialized framework components on how to create the database

Figure 3. Framework health architecture

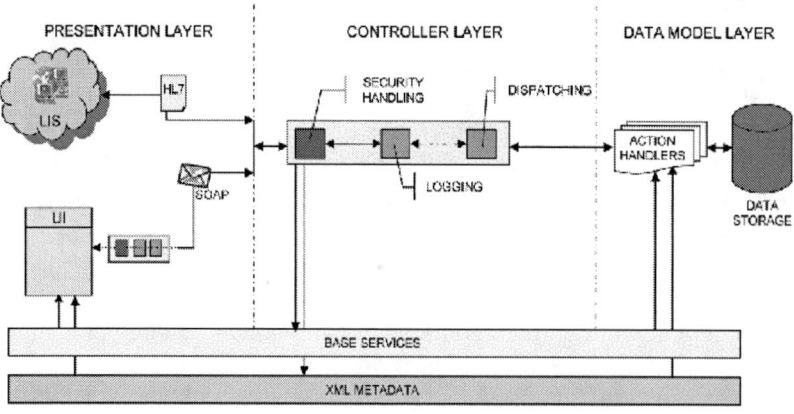

Figure 4. Data type definition example

```
<datatype name ="CE"
          type ="complex"
          descr="2.9.3 - Coded Element">
   <field seq="1" name="Identifier" dt="ST" />
   <field seq="2" name="Text" dt="ST" />
   <field seq="3" name="NameOfCodingSystem"
          dt="IS"  tbl="0396" />
   ...
</datatype>
```

Figure 5. Segment definition example

```
<segment name="PID">
   <field seq="1" len="4" name="SetIDPID"
          dt="SI" min="0" max="1" />
   <field seq="2" len="20" name="PatientID"
          dt="CX" min="0" max="1" />
   <field seq="5" len="250" name="PatientName"
          dt="XPN" min="1" max="unbounded"/>
   ...
</segment>
```

Figure 6. Mapping to database tables

```
<table    ref="PID" id="1000"
          name="PID_TABLE" >
   <column ref="SetIDPID"
           name="SetIDPID_COLUMN" />
   <column ref="PatientID"
           name="PatientID_COLUMN"
           import="true" />
   <column ref="PatientName"
           link="true" />
   ...
</table>
```

schema and how to map business entities to database tables during query and transaction activities. E.g. the *link="true"* directive advises the framework that the *PatientName* field of the *PID* segment is not saved in the *PID*_TABLE table, but is stored in the table reserved for the *XPN* data type (the field's data type).

The *Data Model* tier contains also the actual processing units that handle message content, called *Action handlers*. These are autonomous software modules that use the base services functionality and may produce new messages that respond to initial requests or notify interested parties about a system event.

Controller. The *Controller* module receives and processes all ingoing/outgoing messages. The *Controller* implements a messaging pipeline using the SOAP protocol, where messages are processed at each node of the *Pipeline* and forwarded to the next node. This involves authentication and authorization of the message sender, logging and tracing of all message flow within the system, applying business rules and dispatching the messages to the appropriate *Action handlers*.

The *Pipeline* is fully configurable through an XML configuration file. For example, additional processing nodes can be added (e.g., a message compression/decompression module), as well as additional *Action handlers*. For example, changes in business processes or operational workflows result in the need for exchanging new messages. New *Action handlers* can be implemented for processing these messages and the *Controller* can be easily configured to dispatch the appropriate message to the dedicated *Action handler*.

Presentation Layer services. The *Presentation layer services* of the framework contain a set of tools and methods targeting the development of adaptive XML-based user interface modules. Using these tools and methods one can assemble user interface modules from XML documents. We call these documents user interface documents. An example of such a document is shown on Figure 7.

The key elements of the user interface document are:

1. *Serializable User Interface Components*: The framework provides a set of common visual controls like menus, text boxes, buttons, list boxes, and so on that support XML serialization and a set of specialized user interface controls for handling the framework's data model that also support XML serialization. These controls support direct binding to framework's message structures. Serializable forms and dialogs are also available as part of the framework. An example definition of such components is shown on part A of Figure 7.

2. *Serializable non-visual Components*: A *user interface document* contains also XML descriptions targeting non-visual components of the framework that also support XML serialization. These components undertake a set of operations such as sending and receiving documents to/from the *Controller*, sending and receiving documents to/from different devices, such as printers or file systems and transforming documents. An example definition of such components is shown on part B of Figure 7. A complete user interface module is a collection of objects capable of accepting and producing XML descriptions. We call these objects *Xml objects*. New *Xml objects* can be created by anyone planning to extend the user interface infrastructure of the framework. The basic interface of an *Xml object* contains only three operations: the *Describe* operation supporting the initial definition and construction of the object based on an XML description, the *LoadXml* operation supporting the loading of an XML description from the object, and the *Xml* operation supporting the generation of an XML description by the object.

3. *Data Exchange between Xml Objects*: The exchange of data between different *Xml Objects* makes use of XSLT transformations to restructure XML descriptions generated from one type of object to descriptions suitable for loading from another type of object. An example of a transformation is shown on part C of Figure 7.

4. *XML Description of Data Flow*: An XML vocabulary has been defined in order to describe the data flow among *Xml Objects*. We call these descriptions *Xml Jobs*. The *Xml Jobs* behave as event handlers for the *Xml Objects*. An example of an *Xml Job* is shown on part D of Figure 7.

This approach results in an open and extensible set of tools capable of describing and implementing client side business logic modules portable to different platforms and environments.

Figure 7. A user interface document

```
<guimodule name="PIDInfo">
  <form name="PIDInfo" caption="Patient Information" activecontrol="RefreshBtn">
    <label name="PatientIDLabel" left="8" top="20" width="176" height="17" caption="Patient
    Internal ID: " visible="-1"/>
    <label name="PatientNameLabel" left="8" top="48" width="156" height="17" caption="Patient
    Name:" visible="-1"/>
    <textbox name="PatientID" left="188" top="16" width="340" height="24" tabstop="-1"
    tabindex="0" visible="-1" align="left" initialtext="" enabled="0" editlength="0"
    edittype="none" passwordchar="" validation="none" field-type="none" displaymask=""
    optional="0"/>
    <textbox name="PatientFirstName" left="8" top="72" width="320" height="24" tabstop="-1"                A
    tabindex="1" visible="-1" align="left" initialtext="" enabled="0" editlength="0" edit-
    type="none" passwordchar="" validation="none" field-type="none" displaymask="" optional="0"/
    >
    <textbox name="PatientSurName" left="8" top="104" width="320" height="24" tabstop="-1"
    tabindex="2" visible="-1" align="left" initialtext="" enabled="0" editlength="0"
    edittype="none" passwordchar="" validation="none" field-type="none" displaymask=""
    optional="0"/>
    <button name="RefreshBtn" left="8" top="248" width="75" height="25" caption="Refresh"
    enabled="-1" visible="-1" tabindex="3" tabstop="-1" onclick="getPatientInfo"/>
    <button name="CancelBtn" left="152" top="248" width="75" height="25" caption="Cancel"
    enabled="-1" visible="-1" tabindex="4" tabstop="-1"/>
  </form>

  <messagebuilder name="SelectPatient" id="QRY_A01"></messagebuilder>                                     B
  <weblink name="HL7Link" address="http://147.102.33.19/HL7Ora/MessageHandler.aspx"></weblink>

  <transformation name="prepareAdmitMessage">
    <xsl:stylesheet version="1.0" xmlns:xsl="http://www.w3.org/1999/XSL/Transform">
    <xsl:output method="xml" version="1.0" encoding="UTF-8" indent="yes"/>
    <xsl:template match="/">
      <QRY_A01>
        <QRD>
          <WhoSubjectFilter>
            <ID>
              <xsl:value-of select="//textbox[@name='PatientID']/@text"/>                                 C
            </ID>
          </WhoSubjectFilter>
        </QRD>
      </QRY_A01>
    </xsl:template>
    </xsl:stylesheet>
  </transformation>

  <job name="getPatientInfo">
    <input name="PIDInfo"></input>
    <function name="prepareAdmitMessage"></function>
    <function name="NewAppointment"></function>                                                           D
    <output name="HL7Link"></output>
    <function name="getPIDInfo"></function>
    <output name="PIDInfo"></output>
  </job>
</guimodule>
```

From our experience in developing and using our framework, we would like to emphasize the following:

1. Using XML not only as the format for exchanged data but also for the native environment for business logic description, UI description, and business objects definition results in a rich metadata repository, providing true customization capabilities, enabling independent UI development and allowing the transition to future technologies.

2. Inherent support of messaging standards assists in achieving the interoperability requirement.
3. The adoption of messaging communication protocol (realized through the adoption of Web Services) leads to a loosely-coupled architecture that also contributes to the extensibility and better maintenance of the delivered software.

Our approach in the application framework development up to now concentrates more on providing basic functionality to the developer in order to deliver quickly robust applications. As a result, our framework lacks the wide support to other standards like DICOM. However, now we have acknowledged the value of such support in order to cover the entire HC domain.

What the Future Holds

Nowadays it is commonly accepted that standardization at all levels is more than critical. Efforts for harmonization among several standardization bodies in the area of HC, give hope for achieving one global set of standards in the future.

Rouggery (2003) at the purpose section of *Web Services Profile* presenting its scope, more or less describes the future as follows: "The ideal situation that we will be looking at is a HC environment where "plug-n-play" interoperability via Web Services is a reality. In this environment Independent Software Vendors (ISV) and corporate developers implementing HL7 interfaces can write generic and reusable classes, subroutines, and modules consistent with the guidelines set forth by the HL7 standard for Web Services standards in order to handle HL7 message traffic from a potentially unlimited number of connecting applications and services. Applications that "expose" HL7 messages (e.g., Web Services servers) will do so according to the HL7 Web Services Profile (WSP) guidelines; "consumers" of HL7 messages (e.g., Web Services clients) can be written without prior knowledge of the application that they will ultimately end up talking to. In addition, this "plug-n-play" environment will take advantage of supporting discovery protocols such as UDDI to break the rigidity of the current hard-coded message routing infrastructures of most Healthcare enterprises". We can not be certain, that the widely adopted standard for HC messaging in the future will be the HL7; but it's very likely to witness such a scenario. Web Services for sure will play a very important role in the future of HC informatics.

Conclusion

The healthcare sector is characterized by a continuously growing competition for survival under difficult economic conditions of scarce resources and limited growth with the constant threat of tightened regulations, public scrutiny, and negative media exposure. Consequently, HC providers are increasingly seeking opportunities in the information technology (IT) field to operate more effectively and efficiently, and to reduce the overall costs of healthcare delivery while improving the quality (Ashry & Taylor, 2000).

The HC software community is moving towards the notions of openness (based on officially approved and globally acceptable standards) and modularity to tackle the integration challenge (Blobel & Holena, 1997, p. 24). Open architectures based on officially approved and globally acceptable standards can contribute to interoperability. The high rate of change creates the need for systems that have the ability to evolve and adapt to the new conditions without excessive cost. Modular software design enables best of breed modules—handling emerging and even unpredictable user or unit requirements—to be gradually added into the systems and successfully glued within the overall, distributed architecture.

In order to achieve the above goal, there is a need for a systematic way of developing applications targeting the *Applications layer*. An effective means to realize it is the use of Application Frameworks. Such Application Frameworks should make use of the existing standards in order to reap the benefits of the significant work done in the domain, presenting the knowledge that lies within the standards and thus introducing them in daily use.

The use of Application Frameworks will result in achieving interoperability within HIS both on a functional and on conceptual level, reinforcing the role Hospital Information Systems play in healthcare.

Acknowledgments

The authors would like to thank Aggelos Androulidakis and George Karkalis for their contribution to the framework development, as part of the software development and system design team, as well as Miss Chryssa Marinou for her valuable help with English language issues.

References

Ashry, N. & Taylor, W. (2000). Requirements analysis as innovation diffusion: A proposed requirements analysis strategy for the development of an integrated hospital information support system [Electronic version]. *Proceedings of the 33rd Hawaii International Conference on System Science, Hawaii.* Retrieved April 14, 2003, from IEEE Computer Society Digital Library.

Beale, T. (2002) Archetypes: Constraint-based domain models for future-proof information systems. Retrieved March 10, 2004, from the openEHR Web site *http://www.openehr.org/downloads/archetypes/archetypes_new.pdf*

Beale, T., Goodchild, A. & Heard, S. (2002). Design principles for the EHR (version 2.4) April 1, 2002. Retrieved January 13, 2004, from the openEHR Web site *http://www.openehr.org/downloads/design_principles_2_4.pdf*

Beeler, G., et al. (1999). Message development framework (version 3.3). Retrieved March 15, 2004, from the HL7 Web site *http://www.hl7.org/Library/mdf99/mdf99.pdf*

Blobel, B. & Holena, M. (1997). Comparing middleware concepts for advanced healthcare system architectures. *International Journal of Medical Informatics, 46*, 69-85

Case, J., Mckenzie, L., Wellnet, A. & Schadow, G. (2003). HL7 Reference Information Model (version 02-02). Retrieved March 15, 2004, from the HL7 Web site *http://www.hl7.org/Library/data-model/RIM/C30202/rim.htm*

CEN TC 251 (1998). *prENV 12967-1 - HISA Part 1: Healthcare Middleware Layer.* [Standard]

CEN TC-251, (n.d.). *About TC-251*. Retrieved March 15, 2004, from the CEN TC-251 Web site *http://www.CEN/TC251.org*

Codenie, W., De Hondt, K., Steyaert, P. & Vercammen, A. (1997). From custom applications to domain-specific frameworks [Electronic Version]. *Communications of the ACM, 40*(10), 71-77.

De Velde, V. (2000). Framework for a clinical information system. *International Journal of Medical Informatics, 57*, 57-72.

Digital Imaging and Communication in Medicine (DICOM), PS 3.1, (2003). National Electrical Manufacturers Association. Retrieved February 20, 2004, from NEMA Web site *http://medical.nema.org/dicom/2003.html*

Fayad, M. & Hamu, L. (2000). Enterprise frameworks: Guidelines for selection [Electronic Version]. *ACM Computing Surveys, 32*(1). Retrieved September 15, 2003, from the ACM portal Web site *http://portal.acm.org/citation.cfm?doid=351936.351940*

Fayad, M. & Schmidt, D. (1997). Object-oriented application frameworks [Electronic Version]. *Communications of the ACM, 40*(10), 32-38.

Fayad, M., Schmidt, D. & Johnson, R. (1999). *Object-oriented frameworks: Problems & experiences in building application frameworks.* John Wiley.

Gorissen, B. (1997). *DICOM in a Nutshell*, Retrieved February 1, 2004, from the Philips Medical Systems Web site *ftp://ftp-wjq.philips.com/medical/interoperability/out/DICOM_Information/d2course.pdf*

HANSA. (n.d.). Home page. Available at *http://www.sintec.ro/hansa/index.html*

HL7 (n.d.). *What is HL7?* Retrieved March 15, 2004, from the HL7 Web site *http://www.hl7.org/about*

HL7 (2003). *HL7 messaging standard version 2.5* [ANSI Standard].

HL7 (2004). *HL7 V3 Guide*, version 3 ballot 6 foundation documents. Retrieved March 1, 2004, from HL7 Web site members' area *http://www.hl7.org/v3ballot6/html/foundationdocuments/welcome/index.htm*

IBM Corporation, Microsoft Corporation (April 7, 2002). *Security in a Web Services world: A proposed architecture and roadmap.* Whitepaper Version 1.0. Retrieved

October 20, 2003, from IBM's Web site *http://www-106.ibm.com/developerworks/ webservices/library/ws-secmap/#2.1*

Jagannathan, V., Wreder, K., Glicksman, R. & alSafadi, Y. (1998). Objects in healthcare-focus on standards. *StandardView, 6*(1), 22-26.

Johnson, R. (1997), Frameworks = (components + patterns) [Electronic Version]. *Communications of the ACM, 40*(10), 39-42.

Johnson, R. & Foote, B. (1988). Designing reusable classes. *Journal of Object-Oriented Programming, 1*(2).

Kohn, L., Corrigan, J. & Donaldson, M. (1999). *To err is human: Building a safer health system.* Washington, DC: National Academy Press.

Krzystof, C. (2001). *Medical data mining and knowledge discovery.* Heidelberg: Physica-Verlag.

Kuhn, K. & Giuse, D. (2001). From hospital information systems to health information systems: Problems, challenges, perspectives [Electronic Version]. *Yearbook of Medical Informatics 2001*, pp. 63-76. Stuttgart: Schattauer.

Lenz, R. & Kuhn, K. (2004). Towards a continuous evolution and adaptation of information systems in healthcare. *International Journal of Medical Informatics, 73*(1), 75-89.

MacDonald, M. (2003). *Microsoft .Net distributed applications: Integrating XML Web services and .Net remoting.* Washington, DC: Microsoft Press.

Marinos, E., Marinos, G. & Koutsouris, D. (2003). Towards an XML-based user interface for electronic health record. *Proceedings of the 25th Annual International Conference of the IEEE Engineering in Medicine and Biology Society*, Cancun, Mexico, September 17-21.

Markiewicz, M. & Lucena, C. (2001). Object-oriented framework development [Electronic Version]. *Crossroads, 7*(4), 3-9.

Object Management Group (n.d.). About the object management group. Retrieved March 12, 2004, from the OMG Web site *http://www.omg.org/gettingstarted/ gettingstartedindex.htm*

OMG HealthCare Task Force (n.d.). Home page. Retrieved January 15, 2004, from *http://healthcare.omg.org/*

OpenEMed (n.d.). Home page. Retrieved March 1, 2004, from *http://openemed.org/*

OpenEHR (n.d.). The Interface between the GEHR Project and Technical Committee TC/251-Medical Informatics of CEN. Retrieved March 2, 2004, from OpenEHR Web site *http://www.openehr.org/intro_origins_gehr_gehrcen.htm*

Organization for the Advancement of Structured Information Standards (2004). Web Services security: SOAP message security. Retrieved March 2004 from *http:// www.oasis-open.org/committees/download.php/5941/oasis-200401-wss-soap-message-security-1.0.pdf*

W3C Web Services Activity (n.d.). *Web Services activity statement.* Retrieved March 2, 2004, from W3C Web site *http://www.w3.org/2002/ws/Activity*

World Wide Web Consortium (2002). XML encryption syntax and processing. Retrieved December 10, 2002 from the W3C Recommendation *http://www.w3.org/TR/xmlenc-core/*

World Wide Web Consortium (2002). XML-signature syntax and processing. Retrieved February 12, 2002 from the W3C Recommendation *http://www.w3.org/TR/xmldsig-core/*

Chapter III

Management and Analysis of Time-Related Data in Internet-Based Healthcare Delivery

Chris D. Nugent, University of Ulster at Jordanstown, Northern Ireland

Juan C. Augusto, University of Ulster at Jordanstown, Northern Ireland

Abstract

Technological advancements have and will continue to revolutionize the way healthcare is being delivered and the interaction between patients and healthcare professionals. A key component to these changes will be an extension to the utilization of the Internet, further exploiting the extensible and interoperable features it offers. In addition to providing a distributed communications infrastructure, the ability to incorporate within Internet-based systems a means of intelligent data analysis in supporting healthcare management has now become a reality. In this chapter, we provide the rationale for usage of the Internet as a core infrastructure for a holistic approach to distributed healthcare management and supplement this through the identification of the potential role of temporal reasoning in addressing the time relevance of patient centered clinical information. Our work is exemplified through case studies where Internet-based systems and temporal reasoning may be employed.

Copyright © 2005, Idea Group Inc. Copying or distributing in print or electronic forms without written permission of Idea Group Inc. is prohibited.

Introduction

With the recent advancements in computing power, digital signal processing techniques and data communications a new era in the delivery of medicine and patient management has evolved. As the underpinning communication systems and infrastructures continue to develop and move from being fixed or terrestrial in nature to being Internet and mobile based, we can expect the flexibility and application of Information and Communication Technologies (ICT) in healthcare to expand.

As the percentage of the elderly within the population increases, we now face the challenge of healthcare service initiatives driven by the goals of individual autonomy and quality of life. The results have subsequently produced a manifest shift from institutional to community care. To improve upon current levels of care provided requires the cost-effective application of ICT. As shown in Figure 1, (Malaysian Telemedicine Blueprint, 2001) this requires the redevelopment of healthcare infrastructures and their service provision by shifting the allocation of resources from secondary and tertiary-care institutions towards the preventative management at the primary care level, as well as providing services to individuals at home where they are likely to be most cost-effective.

Central to the uptake of ICT solutions within healthcare will be the usage of the Internet. The Internet has the ability to harness the communications between patients and healthcare professionals and also between healthcare professionals. Such an infrastructure is driven by vast amounts of clinical information, however, care must be taken that efficient processing takes place to optimize usage of the information collated. One of the many attributes related to healthcare information is "the notion of time". The application of computational approaches that can effectively manage time related healthcare information has recently been found as not having their full potential exploited (Augusto, 2003a). It has been the aim of this work to focus on these time related healthcare information issues and show how techniques, capable of effectively managing time related information, may be employed in the healthcare arena.

Figure 1. The shift of allocation of resources during the transition from Industrial Age medicine to Information Age healthcare

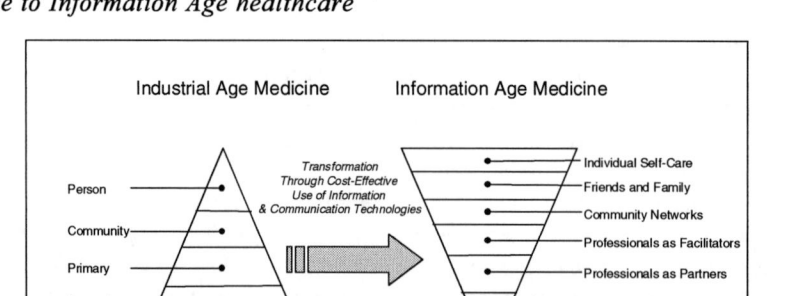

Given the breadth of the subject area addressed in this work, introductory and pertinent issues have only be presented to provide a more readable article for a broader target audience. References to further technical information and guidance are provided throughout. The remainder of the work is divided as follows; Section 2 provides an overview of Internet-based healthcare and decision support systems. Section 3 provides a classical overview of how temporal reasoning based systems can accommodate for time centered data and specifically Internet-based healthcare. Section 3 provides three case studies offering general examples as to how temporal reasoning systems may be employed. Section 4 provides details on future challenges and visions in this area and overall conclusions are drawn in Section 5.

The Internet and Decision Support Systems

A key component to the facilitation of ICT within healthcare is the usage of Internet driven technologies (Bushko, 2002). The Internet, coupled with tools such as videoconferencing, mobile communications, and digital technologies offer the potential to significantly improve the processes supporting patient care. Although new and still with a long way to go before being massively used, Internet-based healthcare is not a futuristic dream but a reality that is already benefiting a range of communities and healthcare professionals around the globe. This will inevitability result in patients (and healthcare professionals) being more frequently confronted with ICT healthcare applications. This, coupled with the fact that patients are now demanding a larger role in the management of their own health and healthcare (Beun, 2003), will result in the relationship between patients and healthcare professionals changing.

A major factor related to the role the Internet may offer has been its wide scale uptake. Figures would now suggest that nearly one in every two people in Western Europe and the United States have Internet access. This uptake has been propagated into the health domain with General Practitioners and Hospitals now also having increased access to the Internet. This subsequently increases the long term potential for Internet driven healthcare related activities. Figure 2 shows a summary, at a European level, of the percentages of Internet access in the aforementioned healthcare domains (Beolchi, 2003).

Examples of Internet-Based Healthcare

Although beyond the scope of this article to present an in depth coverage of how the Internet is currently utilized in the healthcare domain, it is nonetheless possible to identify a number of succinct areas in which it is used by both patients and healthcare professionals:

Information Provision: The Internet can be effectively used to deliver healthcare information. Already, the Internet provides an immense resource for patient-healthcare

Figure 2. Internet access for general practitioners and hospitals at a European level

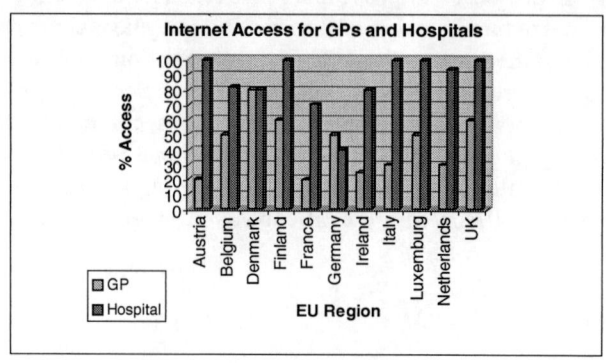

information, with thousands of disease support groups, clinical advice forums and disease-specific information resources. According to Provost, Perri, Baujard, and Noyer (2003), patients use the Internet to search for health and drug information, disease description, medical literature, online medical consultation, and to seek second opinions on a medical diagnosis. Health professionals on the other hand use the Internet to search for information on drugs, medical literature and recommend Web sites and support groups to patients (Provost et al., 2003). The Internet, can in addition, be seen as a valuable aid to computer assisted learning. A multitude of resources exist, (for example, Physicians Online) which can provide access to a wide range of valuable multimedia learning resources.

Electronic Care Communications: Electronic care communications, or healthcare administration systems, provide the potential for the use of secure local and wide area networks to speed up: scheduling appointments; referrals between healthcare professionals; communication of outcomes and discharge letters. The Internet can also be useful to spread policies (Policies, Medical Practice Management) and can be used as an assistance tool in the entry point in hospitals. For example, in Chonnam National University Hospital of Korea (Korea Chonnam Hospital), an institution with about 4,000 patient visits per day, the Internet is used to manage all patient information.

e-Prescribing: e-Prescribing has been defined as the use of an automated data entry system to generate a prescription (Krohn, 2003), however it is appreciated that in the modern ICT based healthcare arena e-Prescribing refers to the use of both computing and telecommunications technology to capture, store, print and often transmit prescription data. Contemporary e-Prescription systems provide connectivity between general practitioners, pharmacists and healthcare agencies enabling secure and accurate transmission of new and repeat prescriptions. At present, e-prescribing is being addressed at national levels with many systems already operational (Middleton, 2000) and others being piloted prior to uptake (Mundy & Chadwick, 2003).

Delivery and Support of Direct Care: The Internet has been comprehensively reported to support many home-based care models (Beolchi, 2003) supporting aspects such as medication management (Nugent, Finlay, & Black, 2001), vital signs assessment and cardiac monitoring to name but a few. Internet-based home care can be an efficient cost-effective means to reach distant groups of people. In Japan, where there are more than 200 islands with 80 of them having less than 300 people and 60 having no healthcare clinics, Internet-based healthcare plays an important role in preventive medicine and early diagnosis (Isechi, Oda, Shinkura, Akiba, Fujikawa, & Yamazaki, 2004).

Decision Support and the Internet

Many "intelligent" tools have made their way into medical applications. Indeed, medicine has always been one of the most important and sought-after application areas of decision support systems and artificial intelligence. Such systems have been helping healthcare professionals for more than four decades with the first instance of usage of computers in medicine considered to be the analysis of electrocardiogram data in the late 1950s (Stallman & Pipberger, 1961). Computerized approaches to decision support can provide intermediate patient assessments. In the most general case they are useful for the analysis of complex and large amounts of data quickly. This results in repeatable processing of clinical and patient data, and offers benefits of consistency and lower levels of observer variability (Willems, Abreu-Lima, Arnaud, Van Bemmel, et al., 1988). Successful approaches to medical decision support have varied from the application of decision trees and expert systems (van Bemmel & Musen, 1997) to Neural Networks and Genetic Programming.

The establishment of Internet-based healthcare systems has produced a realm of opportunities for the uptake and application of medical decision support systems. With Internet-based healthcare intermediate automated decision support can now be facilitated. A problem associated with such an approach is the identification of the most relevant information upon which to perform the analysis and the format of the resulting output to be presented to the clinician/patient. The amount of information generated is vast, in addition to its complexity and multi-faceted structure.

The element of time, associated with clinical information can be identified as an important aspect in the delivery and automation of decision support that can be facilitated by Internet-based healthcare delivery. It has recently been reported that the element of time has not had its full potential exploited in terms of the benefits it can offer in medical decision support systems (Augusto, 2003a). Although Internet-based systems and medical decision support systems are slowly merging, consideration must also be given to the tools required to analyze time related information and give consideration as to how they can be incorporated into Internet-based healthcare delivery.

The Benefit of Temporal Reasoning in Remote Decision Support for Medicine

We address here some general issues related to Temporal Reasoning (TR) which will be associated with the application domain in the following sections. Firstly, we wish to highlight the importance of TR in daily life. Although usually unconsciously, we need to make considerations on a daily basis regarding the preferred order that some of our activities have to be performed in, or for how long some of these activities will last. Frequently deadlines have to be met, such as at work we are constantly undertaking activities which have prescribed time limits, in our personal lives we engage in activities that are constrained by time; to collect children from school or to reach the supermarket before it closes. These activities sustain our perception of time ubiquity and provide a clear practical motivation behind the research in TR.

The importance of addressing the way in which humans deal with time-related notions was identified in the 1970s as an important goal to be addressed by the Artificial Intelligence (AI) community as a step towards providing intelligent behavior from computing systems. Now, TR is a well-established area of research within AI. Research in the area has been steadily increasing over the past two decades and as a result much understanding about the related problems has been gained and many techniques and associated tools are now available.

There are a plethora of concepts to be considered when designing dynamic systems. An elementary issue relates to the temporal references to be used (hours, minutes, calendar

Figure 3. Qualitative temporal relationships between durative and non-durative temporal references (a) interval-interval case (b) point-point case (c) point-interval case

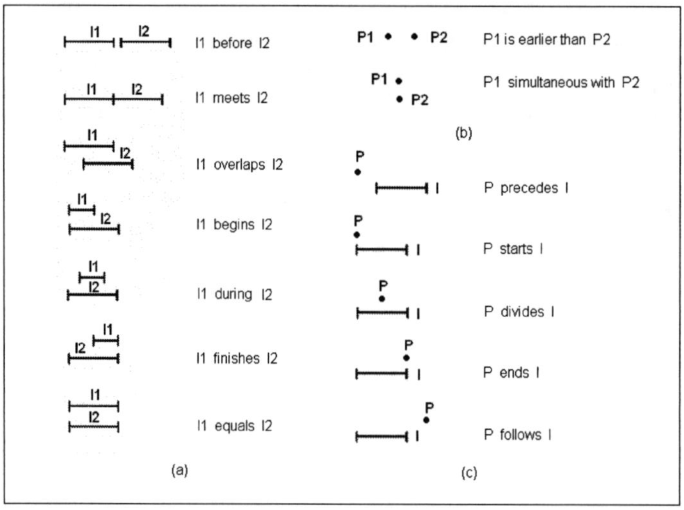

dates, etc.). Also, should the primitive occurrence of them be instantaneous (e.g., turning a light on or off) or durative (e.g., keeping a light on for a period of time), or should both be used? Some of the more basic temporal frameworks are based either on time points, on intervals or on a combination of both. Figure 3 (a) depicts at an intuitive level, the possible meaningful situations two intervals can be in. Part (b) shows the two meaningful relative positions for two points, while part (c) shows the possible situations between a point and an interval.

Explicit details of how the aforementioned can be integrated in a TR system can be found in Allen (1984) and Meiri (1991). Temporal references provide a way to directly represent situations involving different types of sequential or concurrent time-related concepts like states holding for a given period or events that can provoke a transition in the system from one state to the next one. Specific research has been focused on repetitive patterns that may be used to reason about processes (Cukierman & Delgrande, 2000).

The durative time references included in Figure 3 are called intervals and rely on precise knowledge about their duration. However, it is usually the case that the knowledge available is not as precise as desired, for example, it may be known when an event started but not when it finished or vice versa; these are called semi-intervals (see Freksa, 1992). If for example the end of an interval is not known then it may be important to infer when, possibly within some boundaries, the point at which the semi-interval can be supposed to persist to (see Sadri, 1987; Augusto 2003b).

It is impossible to give here a full coverage of all the subtleties involved in TR. We have introduced some basic concepts and invite the interested reader to review the following to obtain a more complete picture of the subtleties, challenges and benefits involved in using TR: Newton-Smith (1980); Shoham (1987); Sandewall (1994); Galton (1995) and Augusto (2001).

Why Temporal Reasoning is Important in Medicine?

Reflecting the many possible instances of time-related information in medicine there has been quite an intense research activity in the area during the last decade. See for example some of the special issues focused on the topic: Keravnou (1991); Keravnou (1996); Combi & Shahar (1997); Shahar & Combi (1999) and a more recent survey by Augusto (2003a).

A long list of general activities has been identified as key initial steps in the process to provide explicit temporal awareness for systems related to medicine (Shahar, 1999). Some of these activities are:

a. *Determining bounds for absolute occurrences*, some information may be missing, for example we know when some symptoms finished but not when they started. Hence it is possible by using other information to infer, at least within some time boundaries, the point in time at which the symptoms started.

b. *Persistence derivation*, if we know that some symptoms started to develop, how long can we assume that they will last?

c. *Inconsistency detection and clipping of uncertainty*, in real life we know that there are conditions which are mutually exclusive, for example, it is not possible to have "high blood pressure" and "normal" blood pressure at the same time. However, we may have information indicating a patient having high blood pressure from days one to three and normal blood pressure from days five to eight; but we do not know what the blood pressure was during day four. It may be equally plausible that the persistence of the period of high blood pressure extends into day four as well as the persistence into the past that the period of normal blood pressure may have started on day four, leading to an inconsistency about what the system can actually infer about day four. Gathering more information or applying extra inference mechanisms can lead to the elimination of the ambiguity.

d. *Deriving new occurrences from other occurrences*, sometimes knowledge can be inferred instead of gathered directly from the patient's interview or medical records. For example, if we are informed a patient has been taking specific medication and it is known that this particular medication has known side effects.

e. *Deriving temporal relations between occurrences*, sometimes information about which events have occurred is available but not in which order they occurred. It is then important to have a means to disambiguate these qualitative relationships, for example, if they were consecutive or concurrent, as they may play an important role in the ensuing diagnosis. A headache followed by a stomachache may report unrelated events but the headache may be a later consequence of the digestive problem.

Shahar and Combi (1999) presented a range of sophisticated systems dealing with various subtle topics addressing these integration issues. However, they also recognized that these were only the first steps to tackle the formidable task of providing flexible and rich integration between time representation and use in medicine. We share that view and the purpose of this article is then to unravel further challenges which should be addressed in order to obtain the next generation of intelligent assistants for medicine, but in this case, focusing on distance-based decision support.

Why Temporal Reasoning is Important for Internet-Based Medicine?

There are many areas of Internet-based technology that can benefit from TR related technologies. As a way of illustration consider the scenarios previously discussed:

- *Information Provision* (training of personnel and students). It is important for medical staff undergoing training to be capable of identifying and distinguishing different clinical contexts based on their evolution in time.

- *Electronic Care Communications* (hospital management). The ideal scenario for this application would be that the system is used to allocate resources in an efficient manner so that in transit or arriving patients can be directed to the correct places for treatment and to ensure that all the resources required are available at the right time.

- *E-prescribing* (medication management). Internet-based prescribing can bring efficiency in the timely location and preparation of medication in response to the demands of a patient's medication prescription as issued by their Doctor.
- *Delivery and support of direct care* (online medical diagnosis system). One of the challenges in this scenario is the replacement of *in situ* examinations as a way to collect evidence about symptoms. It is, however, also important to provide a means to replace oral questions about the development of the symptoms through time.

Intelligent support in Internet-based medical assistance is a challenging area that demands a synergistic multidisciplinary effort involving areas of research like human-computer interaction, security, databases, and medical informatics. It also demands further multidisciplinary interaction between those whose knowledge of time-oriented systems will find in the area of medicine a challenging but highly rewarding field of application. Researchers and practitioners from medical informatics, TR in AI, temporal databases, active databases, real-time databases, visualization of dynamic systems, and real-time systems should have some knowledge and experience to share in this fertile area.

Although we believe all the areas listed above are important we will exemplify how this context will affect key aspects of the traditional interaction between the professional and patient during effective diagnosis. More specifically, we highlight the importance of handling time-related concepts in Internet-based medicine and how its use may affect the accuracy of diagnosis. We have found that existing systems do not handle the richness of time-related concepts during the diagnosis stage and the variety of temporal references required. For example, some symptoms are described as occurring on a particular day, like "The symptoms started last Monday". Some of them are identified precisely once the duration is known, for example, "He had fever for three days". Sometimes the duration is not precisely identifiable and there is some degree of uncertainty the system should be able to handle, for example, "Started yesterday evening and stopped at some point during the night". It is important to recognize repetitive processes, for example, "He has headaches each time that he goes to music class" and frequencies of occurrence, for example, "He has been taking this medicine three times each day". Rich calendric references should be handled, like seasons, in order to discover potential causes of disease, for example allergies. Here we consider why, where and when these issues are important in the context of Internet-based medical assistance.

Interaction Shifts with Internet-Based Diagnosis

The use of the Internet as an intermediate level between health centers and patients brings a shift in the interaction and the usual tasks involved. The interaction is no longer a physical meeting but instead there is a media that may restrict, sometimes significantly, what each person involved in the communication perceives from each other.

According to the degree of technology involved many traditional sources of information may not be available. For example, the patient cannot be touched, visualization of the patient's body may be restricted, use of some technological advances like scanners, radiographies, and ECGs may not be possible. Hence, making the most of what is available will be vital.

Here the interaction between (i) a rich interface that allows extraction of information from the patient in terms of the symptoms and (ii) suitable algorithms that can relate a, most possibly, incomplete description of symptoms to a meaningful subset of possible scenarios, will be crucial for the effectiveness of such systems. The interface becomes a critical part of the system. In a routine visit to a clinician, natural language, body language and other usual means of communications between humans are available. With Internet-based consultations we can consider some substitutes like video and sound but some of them may or may not be available. There may be occasions when these media would not be usable, for example, due to privacy issues. Until image, video and sound are widely available at the level of quality required to replace a face to face clinical examination we focus on the more basic and less sophisticated ways of collecting information via dynamically generated web based forms that can be used as the base for interaction either in synchronous or asynchronous communication between patient and health professional.

Other areas of computer science become relevant like Natural Language processing and appropriate interfaces that are friendly enough for the patient while gathering as much information for the clinicians as possible. At the diagnosis level different subtleties will help to identify between a possible dangerous situation and a non-dangerous one or between two diseases that may require very different treatment even when they share similar symptoms, for example, flu and hepatitis. Being able to successfully detect the described symptoms with pre-known patterns of disease will require mechanisms like: a) disambiguating relative orders between events and descriptions, b) inferring possible durations for them when they are not given explicitly, c) dealing with degrees of uncertainty in terms of the temporal scope of a given set of events and conditions, d) using the partial list obtained at any time during the interaction to assess which is the most likely scenario which in turn will help to select which questions to ask next or which information to gather in order to maximize efficiency during the diagnosis process.

There are quite a few hypotheses that must be taken into account to supply the system with extra information that is available or gathered by other means. One basic point is that patients should have a history, the normal approach for storing time-related information being temporal databases (Tansel, Clifford, Gadia, Jajodia, Segev, & Snodgrass, 1993; Etzioni, Jajodia, & Sripada, 1998). Also, the system should be time sensitive in the sense that each subsequent visit should provide a different context. For example, if the purpose of the later visit is to incorporate further information about a previous description of symptoms, the system should react accordingly and should present information differently and/or different information. Once all the symptoms have been entered and those that are relevant to the hypothetical syndrome are identified, the system may advise on how to monitor for their evolution in time.

The rules for diagnosis should be "time-aware" and the interface with the patient should allow some way to clearly indicate key time-based references, for example, the frequency, duration and proximity of symptoms. The inference engine should instantiate internally these temporal references with patterns and use the time of occurrence as a reference during the reasoning. A specific device can be used for daily monitoring so as to reduce errors, but more sophisticated users or applications will demand a more sophisticated interaction and description of events and conditions. For example, if a patient is trying to describe symptoms to ascertain if they have a medical condition, the temporal distance between their occurrence and the relative order of their occurrence can make a difference in the final diagnosis. In an asynchronous context, patients may be monitored and the system, based on the trends of the input data in time, may decide to take a cautionary step of contacting a health professional.

Exemplar Scenarios of TR in Healthcare

To detail further of how TR in Internet-based healthcare may be employed, the following sections provide an overview of three potential candidates.

Example 1: Use of TR in Smart Homes

Here we consider a living environment for a person who supports and promotes independent living by including a number of technological solutions embedded into their home environment. The information gathered from such technology can be monitored remotely via an Internet-based control center and used to provide a means of interaction in the home environment between the person and their surrounding environment. (A more technical description of this scenario can be found in Augusto and Nugent, 2004.)

In the scenario shown in Figure 4, a person's movements can be monitored via motion sensors in each room. The key issue, in terms of analysis of the information generated

Figure 4. Layout of apartment indicating embedded technology to support independent living

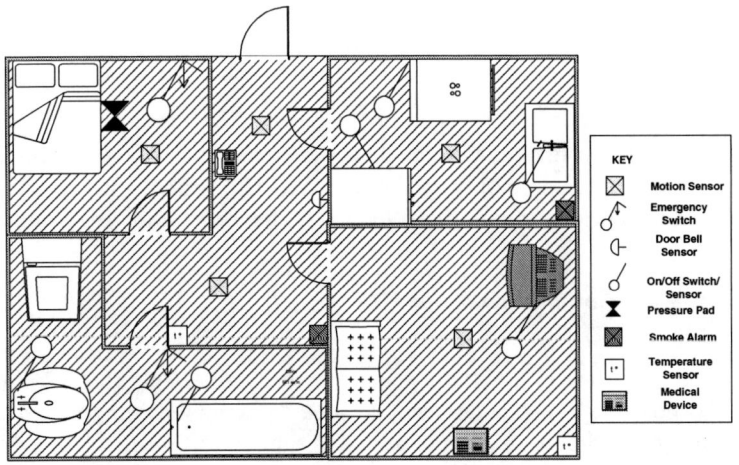

by the smart house and in determining if an abnormal situation has arisen, is to primarily ascertain the current location of the person. Following this, the sequence of their events can be monitored along with their interaction with domestic appliances. The house, as shown, is equipped with emergency switches, smoke and temperature sensor/alarms, medical vital signs devices, sensors to detect movement out of bed and sensors to detect if taps have been turned on or off. A middleware software structure provides the necessary interoperability solution for the devices to ensure different data transmission formats and standards can communicate between the house and the control center.

The diversity of the types of information generated by the sensors provides a number of dimensions related to the person's activities; (i) their whereabouts (ii) their interaction with appliances and (iii) the duration of these events. The key element to take into consideration is that as the person moves around the house and interacts with the domestic appliances, analysis of this information and how the system should react must take into consideration the ordering of these events and the time duration between them. Hence TR based rules may be modelled and used to discriminate between normal conditions and potentially hazardous situations when an alarm condition should be raised. Consideration of the types of support required from a person can lead to support in four different ways:

- *Prevention of dangerous situations*: The information gathered from the person's environment can be assessed and also take into consideration their previous actions. The purpose is to provide an alarm when a person's actions may lead to a potentially hazardous situation.
- *Comfort:* This addresses the issues of the surrounding environment and changes these by taking into consideration the person's whereabouts and their actions, for example, turning lights on or off as they move from different rooms, or adjusting the temperature when it becomes too hot or too cold.
- *Security:* Rules may be generated to provide an element of security. For example, rules can be generated to act accordingly if an intruder or unexpected visitor enters into the house.
- *Health:* Information can be accessed from self-operated medical devices. It is often necessary to monitor vital signs over a period of time e.g. monitoring of blood sugar level. For example, if this has an abnormally low or high value for more than three successive recordings then medical staff should be contacted.

The analysis of the information generated can be based on an Active Database (ADB) framework, an approach commonly used in TR. ABD are strongly based on the concept of ECA (Event-Condition-Action) rules. ECA rules have the following form:

ON *event* IF *condition* DO *action*

The ON clause assesses in a given system the series of events that take place. For example, it is possible to detect an event defined as "a person being in bed at a given period of time during the day for a long period of time." According to these events they may be matched against certain criteria (IF clause of the ECA rule). For example, the criteria may be that according to their profile it is not normal to be in bed during this period. The final component of the ECA rule is the DO clause. When an instance of the events defined in the ON clause occurs, and given that the condition has been matched according to the IF clause this will lead to the system acting accordingly such as the DO clause. For example, a form of action/intervention must be taken to assess why the person is still in bed. The following ECA rule schema informally represents this scenario:

> ON "person being in bed for a long period of time"
> IF "not expected to be in bed during this period"
> DO "contact relevant carer"

Another example may be representation of a condition where the system knows a person is in their house, but when the doorbell has been activated by a visitor, the person in the house has not responded. This may be represented by the following ECA rule schema:

> ON "doorbell ringing and person not answering the call for a long time"
> IF "person inside house and not known to have hearing problems"
> DO "apply possible emergency plan"

Here we have just provided schema rules. Notice that there are specific time-related concepts that are embedded in these rules, like the expression "a long period of time" in the first rule and the need to detect the temporal sequence between the doorbell ringing and the person not reacting to it. These concepts are domain dependant and should be further specified in a specific language that can be adequately interpreted by a computer, for example Prolog. To see one example of such a language, including temporal references like those depicted in Figure 3, and a specific application of the scenario described in this section to smart homes for elderly care, we invite the reader to make reference to Augusto and Nugent (2004).

Through an ECA rule set specification it is possible to model the activities of the person and based on different activities the system can interact and change their living environment to promote independent living in a number of ways. The core element of such an interactive system will be reliant on an ADB management system hosted by an Internet-based server with capabilities to distinguish between normal situations and abnormal situations where some form of human intervention is required to assist the person.

Example 2: Use of TR in Medication Management

Medication compliance indicates the precision with which a patient follows a medical prescription. This is a relatively new research domain that is currently receiving much attention (Farmer, 1999; Wertheimer & Santella, 2003). Non-compliance has a negative impact on both the individual patient and the healthcare system and wider economy. The impact is clear: if a patient has been prescribed correctly following diagnosis, non-compliance becomes an important barrier in optimizing treatment.

To assist with the issue compliance aids have been devised, tested and employed. Typically, compliance aids have been classified into three categories; pill holders (Corlett, 1996), alarm-based aids (Szeto & Giles, 1997) and monitoring devices (Wertheimer & Santella, 2003). The latter of these are the most elaborate and form part of an integrated care model whereby the patient's adherence to their medication regime can be monitored remotely (e.g., via an Internet-based care model) and alarms issued in instances of non-compliance as a means of reminder. These systems facilitate the collation of unlimited information regarding the patient's compliance and hence provide for the means of data analysis. For example, for a given patient, analysis of their data may show that due to their lifestyle they repeatedly miss their medication dosages on Friday afternoons and Saturday nights. For a different patient, analysis may show that they have high levels of non-compliance during periods at which they may be at work. The resultant information can be used to readjust the medication regimen in instances of non-compliance. This is a complex procedure and must take into consideration the patient's ailment, their entire medication regime and the impact that missing one medication may have on the remainder of medications and the point in time when the medication must be taken, for example, before meals.

Qualitative and quantitative TR (as previously outlined) may be employed in this instance whereby a set of formal rules can be established and used to control the underlying principles by which a patient's medication regime must be altered to accommodate for instances of non-compliance. For example, a decision support system running on the central Internet-based server monitoring patient's compliance at home may have the following (hypothetical) rule set:

Qualitative Rule Set to Manage Instances of Non-Compliance

+Warfarin must be taken each day at 5pm
+Brufen must be taken after meals
+Insulin must be taken when sugar levels drop below a predefined value
+Acepril must not be taken at the same time as potassium supplements
....

Example 3: Hospital Management of Bed Occupancy

The estimation of the length of stay of patients in hospitals is important for patient treatment management, bed occupancy management and cost effectiveness measures. Mathematical models have been produced as decision aids in this area (McClean, 1994) and in addition many trauma scores can be used to give an indication of patient survival probability (Boyd, Tolson, & Cope, 1987) and severity of trauma indices (Champion, Copes, & Sacco, 1990). All of this information along with demographic details, treatment, physical and mental scores can be taken into consideration and used as means upon which to generate a time to event measure, that is, prediction of length of stay in hospital.

This is a complex and multi-factorial problem, but nevertheless one worthy of further attention. Qualitative and quantitative TR techniques may be used as the basis to model the time to event measures. These techniques can easily be integrated into a hospital based administrative system operating on an intranet/Internet basis which collates information from all in-patients and wards within a hospital environment, in addition to having access to centralized patient information. It can thus accommodate for the complex temporal nature of the problem through formalized approaches, for example:

> **Quantitative TR Rule Set to Manage Bed Occupancy**
>
> +Vacant beds should be increased by 10% during vacation periods, predicted flu epidemics etc. (months 5,6,7 and 11, 12)
> +Movement of patients between beds/wards should only occur between 12.00 and 13.00 or 17.00 and 18.00
> +A patient cannot remain within a bed for more than 24 hours following indications that discharge has been approved
> ....

Future Trends

It is clear that the Internet will have a long-term role within healthcare applications. At present its application and ability to improve upon current and existing healthcare practices has shown its promise and already we are slowly witnessing its uptake. Many of the drivers that will inevitably lead to success will come from governmental perspectives. For example, in the United Kingdom the government has put at the forefront of its healthcare organization agenda development and increased usage of Intranet and Internet services (Tyrrell, 1999). Provision of access to clinical information via Internet-based services allows not only access via common desktop PCs, but also through the usage of mobile phones with Web browsers and personal digital assistants. Hence users will no longer be geographically constrained and will be able to access information whilst undertaking house calls or on a hospital ward. This concept is further extended through the new and innovative research areas of Ambient Intelligence, Context-Aware Computing, User Profiling and Pervasive and Ubiquitous Computing (ERCIM News, 2001) where networks and interoperable communication infrastructures, along with intelligent processing of information gathered have evolved as the next generation of services which could further support this area.

In general, the role of ICT within healthcare will change the way in which patients interact with healthcare professionals and also the way in which healthcare professionals interact with each other. It is not unreasonable to consider that patients may be able to access information via the Internet in the future and self-diagnose based on their symptoms (Tyrrell, 1999). In addition, the recent success of sensor and textile technologies, wearable vital sign monitors and home based medical devices for self assessment make self-recording a possibility and provide tangible clinical information to be used in self-diagnosis. For this to become reality, regional and national networks of high standing will have to be established to ensure the integrity of the information and processes to which the patient may have access. From the healthcare professional's point of view, the Internet offers the future role of high quality information provision. This will provide enhanced access to new and evolving techniques, medical conditions and online learning capabilities. All of which will greater inform the healthcare professionals resulting in an overall improved level in healthcare quality.

From a more general perspective, the Internet has the ability to connect, on a global scale patients and healthcare professionals during consultation processes. For example, a patient may request a diagnosis directly from any private based clinic in the world, or on a "pay-per-use" basis, exploit the e-commerce nature of the Internet and decrease waiting times for private examinations or treatments. Although these scenarios are not unrealistic to consider, it must be appreciated that the term "the Internet" as a means to provide them all is a large scale underestimation of the complexity of the underlying processes. A multitude of components such as communication bandwidth, mobile communication infrastructures, security, interoperability, transmission protocols, information exchange standards, and decision support modules will all have a role to play. Considering the latter, which has been the main thrust of the current work, establishes the need to further improve upon existing approaches.

It is envisaged that efforts will continue to provide solutions, from a TR perspective, which can accommodate for the temporal nature of healthcare based information. Recent studies (Shahar, 1999; Augusto, 2003a) have identified several key areas of TR to be explored to provide more accurate and useful automated assistance in healthcare. Here we have covered general aspects while more specific and technical coverage is given in Augusto & Nugent (2004). Some of the tools and the information required for the medical practice is shared between the traditional scenario and the Internet-based approach to diagnosis as discussed above. However, this article attempts to illustrate that while TR remains an important concept to be considered in Internet-based healthcare, the infrastructure where it is applied and the interaction with this technology is substantially different and will pose further unresolved challenges.

Conclusion

It is evident that the way healthcare will be delivered in the future will involve the utilization of ICT solutions, especially the Internet, to ensure a cost-effective means of high quality of service. Coupled with this will be embedded automated medical decision support systems offering the ability to analyze and interpret the plethora of clinical information that can be acquired through such infrastructures and their data repositories. This will inevitability result in both patients and healthcare professionals being presented more often with ICT healthcare solutions resulting in a change in the way they interact.

The role of TR has been established and proven from a computer science perspective and has attained applied success in many domains, including the assistance in medical diagnosis and treatment. The last decade has witnessed a fruitful interaction between the areas of TR and medicine and significant advances were made in time-related diagnosis and therapy support. However, these advances were made under a different knowledge management paradigm and although we think they may provide a good starting point they cannot be directly applied to Internet-based healthcare systems.

Given that there is a large "time" element associated with clinical information and the opportunity to embed TR into the next generation of Internet-based healthcare, we identify this as a potential prolific area for future developments and an opportunity to identify "how" and "where" it can be applied. Specifically, it is our view that in the short term TR will be a valuable asset in the provision of automated support during Internet-based patient to healthcare professional consultations.

References

Allen, J. (1984). Towards a general theory of action and time. *Artificial Intelligence, 23*, 123-154.

Augusto, J.C. (2001). The logical approach to temporal reasoning. *Artificial Intelligence Review, 16*(4), 301-333.

Augusto, J. C. (2003a). Temporal reasoning in decision support for medicine. *Artificial Intelligence in Medicine, 33*(1), 1-24.

Augusto, J.C. (2003b). A general framework for reasoning about change. *New Generation Computing, 21*(3), 209-247.

Augusto, J.C. & Nugent, C.D. (2004). *The use of temporal reasoning and management of complex events in smart homes*. Technical report, School of Computing and Mathematics, University of Ulster at Jordanstown, UK. Online *http://www.infj.ulst.ac.uk/~jcaug/TR-AugustoNugent2004a.pdf*

Beolchi, L. (2003). *Telemedicine glossary* (5th edition). Working Document, Glossary of concepts, technologies, standards and users. November. European Commission Information Society Directorate-General.

Beun, J.G. (2003). Electronic healthcare record: A way to empower the patient. *International Journal of Medical Informatics, 69*, 191-196.

Boyd, C.R., Tolson, M.A. & Cope W.S. (1987). Evaluating trauma care. The TRISS method. *The Journal of Trauma, 27*, 370-392.

Bushko, R.G. (2002). *Future of health technology*. Amsterdam: IOS Press.

Champion, H.R., Copes, W.S. & Sacco W.I. (1990). A new characterization of injury severity. *The Journal of Trauma, 30*, 539-545.

Combi, C. & Shahar Y. (Eds.) (1997). Time-oriented systems in medicine. *Computers in Biology and Medicine, 27*(5).

Corlett, A. J. (1996). Aids to compliance with medication. *BMJ, 313*, 926-929.

Cukierman, D. & Delgrande, J. (2000). A formalization of structured temporal objects and repetition. *Proceedings of the Seventh International Workshop on Temporal Representation and Reasoning* (pp. 13-20).

ERCIM News (2001). Ambient intelligence. *ERCIM News, 47*.

Etzioni, O., Jajodia, S. & Sripada S. (Eds.) (1998). *Temporal databases: Research and practice*. Springer-Verlag.

Farmer, K.C. (1999). Methods for measuring and monitoring medication regimen adherence in clinical trials and clinical practice. *Clinical Therapeutics, 21*(6), 1074-1090.

Freksa, C. (1992). Temporal reasoning based on semi-intervals. *Artificial Intelligence, 54*, 199-227.

Galton, A. (1995). Time and change. In D. Gabbay, C. Hogger & J. Robinson (Eds.), *Handbook of logic in artificial intelligence and logic programming (epistemic and temporal reasoning)* (Vol. 4, pp. 175-240). Clarendon Press.

Isechi, A., Oda, C., Shinkura, R., Akiba, S., Fujikawa, H. & Yamazaki, K. (2004). Experiment of Internet-based tele-medicine in Amami Rural Islands. *Proceedings of Workshop on Internet to Support Social Welfare (SAINT2004)*, Tokyo, Japan, January. IEEE Press.

Keravnou, E.T. (Ed.) (1991). Medical temporal reasoning. Special Issue from *Artificial Intelligence in Medicine, 3*(6).

Keravnou, E.T. (Ed.) (1996). Temporal reasoning in medicine. Special Issue from *Artificial Intelligence in Medicine, 8*(3).

Korea Chonnam Hospital (n.d.). Online *http://kr.bea.com/case_studies/chonnam_hospital_eng.pdf*

Krohn, R. (2003). Making e-Prescribing work: A fresh approach. *Journal of Healthcare Information Management, 17*(2), 17-19.

Malaysian Telemedicine Blueprint (2001).

McClean, S.I. (1994). Modelling and simulation for health applications. In *Modelling hospital resource use* (pp. 21-28). London: RSM Press.

Meiri, I. (1991). Combining qualitative and quantitative constraints in temporal reasoning. *Proceedings of AAAI 1991* (pp. 260-267).

Middleton, H. (2000). Electronically transmitted prescriptions: A good idea. *The Pharmaceutical Journal, 265*, 172-176.

Mundy, D.P. & Chadwick, D.W. (2003). Security issues in the electronic transmission of prescriptions. *Medical Informatics & The Internet in Medicine, 28*(4), 253-277.

Newton-Smith, W.H. (1980). *The structure of time*. London: Routledge and Kegan Paul.

Nugent, C.D., Finlay, D.D. & Black, N.D. (2001). The design of a care model and associated peripherals to assist with non-compliance of medication. *Proceedings of the Engineering in Medicine and Biology Society International Conference (IEEE EMBS)*, Istanbul, Turkey.

Physicians Online (n.d.). Online *http://www.po.com*

Policies (n.d.). Online *http://www.evidence-based-medicine.org/medical-practice-management.htm*

Provost, M., Perri, M., Baujard, V. & Noyer, C. (2003). Opinions and e-Health behaviors of patients and health professionals in the USA and Europe. *Proceedings of Medical Informatics Europe Annual Meeting (MIE 2003)*. St. Malo. IOS Press.

Sadri, F. (1987). Three recent approaches to temporal reasoning. In A. Galton (Ed.), *Temporal logics and their applications* (Chapter 4). Academic Press.

Sandewall, E. (1994). *Features and fluents*. Oxford: Oxford University Press.

Shahar, Y. (1999). Timing is everything: Temporal reasoning and temporal data maintenance in medicine. *Proceedings of Seventh Joint European Conference on Artificial Intelligence in Medicine and Medical Decision Making (AIMDM'99)* (pp. 30-46).

Shahar, Y. & Combi C. (Eds.). (1999). Intelligent temporal information systems in medicine. *Special Issue from Journal of Intelligent Information Systems*, *13*(1-2).

Shoham, S. (1987). *Reasoning about change*. Boston: Cambridge Press.

Stallman, F.W. & Pipberger, H.V. (1961). Automatic recognition of electrocardiographic waves by digital computer. *Circulation Research*, *9*, 1138-1143.

Szeto, A. & Giles, J. (1997). Improving oral medication compliance with an electronic aid. *Engineering in Medicine & Biology Magazine*, *16*, 48-58.

Tansel, A., Clifford, J., Gadia, S., Jajodia, S., Segev, A. & Snodgrass, R. (1993). *Temporal data bases (theory, design and implementation)*. The Benjamin Cummings Publisher Co.

Tyrrell, S. (1999). *Using the Internet in healthcare*. Abingdon, Oxford: Radcliffe Medical Press.

Van Bemmel, J. & Musen M. (Eds.). (1997). *Handbook of medical informatics*. Springer Verlag.

Wertheimer, A.I. & Santella, T. (2003). Medication compliance research: Still so far to go. *Journal of Applied Research in Clinical and Experimental Therapeutics*, *3*(3).

Willems, J., Abreu-Lima, C., Arnaud, P., Van Bemmel, J., et al. (1988). Effect of combining electrocardiographic interpretation results on diagnostic accuracy. *European Heart Journal*, *9*, 1348-1355.

Chapter IV

Interactive Information Retrieval as a Step Towards Effective Knowledge Management in Healthcare

Jörg Ontrup, Bielefeld University, Germany

Helge Ritter, Bielefeld University, Germany

Abstract

The chapter shows how modern information retrieval methodologies can open up new possibilities to support knowledge management in healthcare. Recent advances in hospital information systems lead to the acquisition of huge quantities of data, often characterized by a high proportion of free narrative text embedded in the electronic health record. We point out how text mining techniques augmented by novel algorithms that combine artificial neural networks for the semantic organization of non-crisp data and hyperbolic geometry for an intuitive navigation in huge data sets can offer efficient tools to make medical knowledge in such data collections more accessible to the medical expert by providing context information and links to knowledge buried in medical literature databases.

Introduction

During the last years the development of electronic products has lead to a steadily increasing penetration of computerized devices into our society. Personal computers, mobile computing platforms, and personal digital assistants are becoming omnipresent. With the advance of network aware medical devices and the wireless transmission of health monitoring systems pervasive computing has entered the healthcare domain (Bergeron, 2002).

Consequently, large amounts of electronic data are being gathered and become available online. In order to cope with the sheer quantity, data warehousing systems have been built to store and organize all imaginable kinds of clinical data (Smith & Nelson, 1999).

In addition to numerical or otherwise measurable information, clinical management involves administrative documents, technical documentation regarding diagnosis, surgical procedure, or care. Furthermore, vast amounts of documentation on drugs and their interactions, descriptions of clinical trials or practice guidelines plus a plethora of biomedical research literature are accumulated in various databases (Tange, Hasman, de Vries, Pieter, & Schouten, 1997; Mc Cray & Ide, 2000).

The storage of this exponentially growing amount of data is easy enough: From 1965—when Gordon Moore first formulated his law of exponential growth of transistors per integrated circuit—to today: "Moore's Law" is still valid. Since his law generalizes well to memory technologies, we up to now have been able to cope with the surge of data in terms of storage capacity quite well. The retrieval of data however, is inherently harder to solve. For structured data such as the computer-based patient record, tools for online analytical processing are becoming available. But when searching for medical information in freely formatted text documents, healthcare professionals easily drown in the wealth of information: When using standard search engine technologies to acquire new knowledge, thousands of "relevant" hits to the query string might turn up, whereas only a few ones are valuable within the individual context of the searcher. Efficient searching of literature is therefore a key skill for the practice of evidence-based medicine (Doig & Simpson, 2003).

Consequently, the main objective of this chapter is to discuss recent approaches and current trends of modern information retrieval, how to search very large databases effectively. To this end, we first take a look at the different sources of information a healthcare professional has access to. Since unstructured text documents introduce the most challenges for knowledge acquisition, we will go into more detail on properties of text databases considering MEDLINE as the premier example. We will show how machine learning techniques based on artificial neural networks with their inherent ability for dealing with vague data can be used to create structure on unstructured databases, therefore allowing a more natural way to interact with artificially context-enriched data.

Sources of Information in Healthcare

The advance of affordable mass storage devices encourages the accumulation of a vast amount of healthcare related information. From a technical point of view, medical data can be coarsely divided into structured and unstructured data, which will be illustrated in the following sections.

Data Warehouses and Clinical Information Systems

One major driving force for the development of information processing systems in healthcare is the goal to establish the computer-based patient record (CPR). Associated with the CPR are patient registration information, such as name, gender, age or lab reports such as blood cell counts, just to name a few. This kind of data is commonly stored in relational databases that impose a high degree of structure on the stored information. Therefore, we call such data "structured." For evidence based medicine the access to clinical information is vital. Therefore, a lot of effort has been put into the goal to establish the computer-based patient record (CPR) as a standard technology in healthcare. In the USA, the Institute of Medicine (1991) defined the CPR as "an electronic patient record that resides in a system specifically designed to support users by providing accessibility to complete and accurate data, alerts, reminders, clinical decision support systems, links to medical knowledge, and other aids." However, as Sensmeier (2003) states, "this goal remains a vision [...] and the high expectations of these visionaries remain largely unfulfilled." One of the main reasons why "we are not there yet" is a general lack of integration.

A recent study by Microsoft[1] identified as much as 300 isolated data islands within a medical center. The unique requirements of different departments such as laboratory, cardiology, or oncology and the historical context within the hospital often results in the implementation of independent, specialized clinical information systems (CIS) which are difficult to connect (McDonald, 1997). Consequently, main research thrusts are strategies to build data warehouses and clinical data repositories (CDR) from disjoint "islands of information" (Smith & Nelson, 1999). The integration of the separated data sources involves a number of central issues:

1. Clinical departments are often physically separated. A network infrastructure is necessary to connect the disparate sites. Internet and Intranet tools are regarded as standard solutions.
2. Due to the varying needs of different departments, they store their data in various formats such as numerical laboratory values, nominal patient data or higher dimensional CT data. Middleware solutions are required to translate these into a consistent format that permits interchange and sharing.
3. In the real world, many data records are not suitable for further processing. Missing values or inconsistencies have to be treated: Some fields such as blood cell counts are not always available or data was entered by many different persons, and is thus

often furnished with a "personal touch." Cleaning and consistency checks are most commonly handled by rule-based transformations.

4. Departmental workflow should not be strained by data copying tasks. Automated script utilities should take care of the data migration process and copy the data "silently in the background."

None of the steps above is trivial and certainly it requires a strong management to bring autonomous departments together and to motivate them to invest their working power for a centralized data warehouse. But once such a centralized repository of highly structured data is available it is a very valuable ground for retrospective analysis, aggregation and reporting. Online analytical processing (OLAP) tools can be used to perform customized queries on the data while providing "drill downs" at various levels of aggregation. A study with an interactive visualization framework has shown that it can massively support the identification of cost issues within a clinical environment (Bito, Kero, Matsuo, Shintani, & Silver, 2001). Silver, Sakata, Su, Herman, Dolins, and O'Shea (2001) have demonstrated that data mining "helped to turn data into knowledge." They applied a patient rule induction method and discovered a small subgroup of inpatients on the DRG level which was responsible for more than 50 percent of revenue loss for that specific DRG (diagnoses related group). The authors describe a "repeatable methodology" which they used to discover their findings: By using statistical tools they identified regions and patterns in the data, which significantly differentiated them from standard cases. In order to establish the *reason* for increased costs in those regions, they found it necessary that a human expert further analyzes the automatically encircled data patterns. Therefore, a closed loop between clinical data warehouse, OLAP tool and human expert is necessary to gain the maximum of knowledge from the available data.

A recently proposed data mart by Arnrich, Walter, Albert, Ennker, and Ritter (2004) targets in the same direction: Various data sources are integrated into a single data repository. Results from statistical procedures are then displayed in condensed form and can be further transformed into new medical knowledge by a human expert.

Unstructured Data in Biomedical Databases

As the discussion on hospital information systems shows, interactive data processing tools form a large potential to extract hidden knowledge from large structured data repositories. A yet more challenging domain for knowledge acquisition tasks is the field of unstructured data. In the following we will briefly present the two premier sources of unstructured data available to the healthcare professional.

Narrative Clinical Data

Many hospitals do not only store numerical patient data such as laboratory test results, but also highly unstructured data such as narrative text, surgical reports, admission notes, or anamneses. Narrative text is able to capture human-interpretable nuances that numbers and codes cannot (Fisk, Mutalik, Levin, Erdos, Taylor, & Nadkarni, 2003). However, due to the absence of a simple machine-parsable structure of free text, it is a challenging task to design information systems allowing the exchange of such kind of information.

In order to address the interoperability within and between healthcare organizations widely accepted standards for Electronic Health Records (EHR) are important. Health Level 7 (HL7)—an ANSI accredited organization, and the European CEN organization both develop standards addressing this problem. The HL7 Clinical Document Architecture (CDA) framework "stems from the desire to unlock a considerable clinical content currently stored in free-text clinical notes and to enable comparison of content from documents created on information systems of widely varying characteristics." (Dolin, Alschuler, Beebe, Biron, Boyer, Essin, Kimber, Lincoln, & Mattison,, 2001). The emerging new standard based on HL7 version 3 is purely based on XML and therefore greatly enhances the interoperability of clinical information systems by embedding the unstructured narrative text in a highly structured XML framework. Machine-readable semantic content is added by the so-called HL7 Reference Information Model (RIM). The RIM is an all-encompassing look at the entire scope of healthcare containing more than 100 classes with more than 800 attributes. Similar to HL7 v3, the European ENV 13606 standard as defined by the TC 251 group of the CEN addresses document architecture, object models and information messages. For a more detailed discussion on healthcare information standards see Spyrou, Bamidis, Chouvarda, Gogou, Tryfon, and Maglaveras (2002).

Another approach to imprint structure on free text is the application of ontologies to explicitly formalize logical and semantic relationships. The W3C consortium has recently (as of May 2004) recommended the Web Ontology Language (OWL) as a standard to represent semantic content of web information and can therefore be seen as a major step towards the "semantic Web" (Berners-Lee, Hendler, & Lassila, 2001). In the medical domain, the Unified Medical Language System (UMLS) uses a similar technology to represent biomedical knowledge within a semantic network (Kashyap, 2003). Once an ontology for a certain domain is available, it can be used as a template to transfer the therein contained explicit knowledge to corresponding terms of free text from that domain. For example, the term "blepharitis" is defined as "inflammation of the eyelids" in the UMLS. If a medic professional would search for the term "eyelid" within the hospital's EHRs, documents containing the word "blepharitis" would not show up. However with knowledge from the UMLS, these documents could be marked as relevant to the query. However, the tagging of free text with matching counterparts from an ontology is either a cost intensive human labor or requires advanced natural language processing methods. We shall see below, how computational approaches could address the problem of adding semantics to free narrative text.

Biomedical Publications

For most of the time in human history, the majority of knowledge acquired by mankind was passed on by the means of the written word. It is a natural scientific process to formulate new insights and findings and pass them to the community by publishing. Consequently, scientific literature is constantly expanding. It is not only expanding at growing speed, but also increasing in diversification, resulting in highly specialized domains of expertise.

The MEDLINE database at the U.S. National Library of Medicine (NLM) has become the standard bibliographic database covering the fields of medicine, nursing, veterinary medicine, dentistry, pharmacology, the healthcare system, and the preclinical sciences. The majority of publications in MEDLINE are regular journals and a small number of newspapers and magazines. The database is updated on a regular daily basis from Tuesdays to Saturdays and on an irregular basis in November and December. At the time of this writing, MEDLINE contains approximately 12 million references to articles from over 37,000 international journals dating back to 1966. In 1980 there were about 100,000 articles added to the database, and in 2002 the number increased to over 470,000 newly added entries. That means, currently over 50 new biomedical articles are published on an hourly basis. Clearly, no human is able to digest one scientific article per minute 24 hours a day, seven days a week. But the access to the knowledge within is lively important nonetheless: "The race to a new gene or drug is now increasingly dependent on how quickly a scientist can keep track of the voluminous information online to capture the relevant picture ... hidden within the latest research articles" (Ng & Wong, 1999).

Not surprisingly, a whole discipline has emerged which covers the field of "information retrieval" from unstructured document collections, which takes us to the next section.

Computational Approaches to Handle Text Databases

The area of information retrieval (IR) is as old as man uses libraries to store and retrieve books. Until recently, IR was seen as a narrow discipline mainly for librarians. With the advent of the World Wide Web, the desire to find relevant and useful information grew to a public need. Consequently, the difficulties to localize the desired information "have attracted renewed interest in IR and its techniques as promising solutions. As a result, almost overnight, IR has gained a place with other technologies at the center of the stage" (Baeza-Yates & Ribeiro-Neto, 1999).

Several approaches have shown that advanced information retrieval techniques can be used to significantly alleviate the cumbersome literature research and sometimes even discover previously unknown knowledge (Mack & Hehenberger, 2002). Swanson and Smalheiser (1997) for example designed the ARROWSMITH system which analyses journal titles on a keyword level to detect hidden links between literature from different specialized areas of research. Other approaches utilize MeSH headings (Srinivasan &

Rindflesch, 2002) or natural language parsing (Libbus & Rindflesch, 2002) to extract knowledge from literature. Rzhetsky, Iossifov, Koike, Krauthammer, Kra, Morris, Yu, Duboue, Weng, Wilbur, Hatzivassiloglou, and Friedman (2004) have recently proposed an integrated system which visualizes molecular interaction pathways and therefore has the ability to make published knowledge literally visible.

In the following sections we go into more detail and look how computational methods can deal with text.

Representing Text

The most common way to represent text data in numerical form such that it can be further processed by computational means is the so-called "bag of words" or "vector space" model. To this end a standard practice composed of two complementary steps has emerged (Baeza-Yates & Ribeiro-Neto, 1999):

1. **Indexing**: Each document is represented by a high-dimensional feature vector in which each component corresponds to a term from a dictionary and holds the occurrence count of that term in the document. The dictionary is compiled from all documents in the database and contains all unique words appearing throughout the collection. In order to reduce the dimensionality of the problem, generally only word stems are considered (i.e., the words "computer", "compute", and "computing" are reduced to "comput"). Additionally, very frequently appearing words (like "the" & "and") are regarded as *stop words* and are removed from the dictionary, because they do not contribute to the distinguishability of the documents.

2. **Weighting**: Similarities between documents an then be measured by the Euclidean distance or the angle between their corresponding feature vectors. However, this ignores the fact that certain words carry more information content than others. The so-called *tf-idf* weighting scheme (Salton & Buckley, 1988) has been found a suitable means to deal with this problem: Here the *term frequencies (tf)* in the document vector are weighted with the *inverse document frequency idfi* = $\log(N/s_i^d)$, where N is the number of documents in the collection, and s_i^d denotes the number of times the word stem s_i occurs in document d.

Although the "bag of words" model completely discards the order of the words in a document, it performs surprisingly well to capture its content. A major drawback is the very high dimensionality of the feature vectors from which some computational algorithms suffer. Additionally, the distance measure typically does not take *polysemy* (one word has several meanings) and *synonymy* (several words have the same meaning) into account.

Evaluating Retrieval Performance

In order to evaluate an IR system, we need some sort of quality measure for its retrieval performance. The most common and intuitively accessible measures are *precision* and *recall*. Consider the following situation: We have a collection of C documents and a query Q requesting some information from that set. Let R be the set of relevant documents in the database to that query, and A be the set of documents delivered by the IR system as answers to the given query Q. The *precision* and *recall* are then defined by

$$precision = \frac{|R \cap A|}{|R|} \quad \text{and} \quad recall = \frac{|R \cap A|}{|R|},$$

where $R \cap A$ is the intersection of the relevant and the answer set, that is, the set of found relevant documents. The ultimate goal is to achieve a precision and recall value of one, that is, to find only the relevant documents, and all of them. A user will generally not inspect the whole answer set A returned by an IR system—who would want to click through all of the thousands of hits typically returned by a web search? Therefore, a much better performance description is given by the so-called *precision-recall-curve*. Here, the documents in A are sorted according to some ranking criterion and the precision is plotted against the recall values obtained when truncating the sorted answer set after a given rank—which is varied between one and the full size of A. The particular point on the curve where the precision equals the recall is called *break even point* and is frequently used to characterize an IR system's performance with a single value.

Therefore, the quality of a search engine which generates hit lists is strongly dependent on its ranking algorithm. A famous example is PageRank, which is part of the algorithm used by Google. In essence, each document (or page) is weighted with the number of hyperlinks it receives from other web pages. At the same time the links themselves are weighed by the importance of the linking page (Brin & Page, 1998). The higher a page is weighted, the higher it gets within the hit list. The analogue to hyperlinks in the scientific literature are citations, that is, the more often an article is cited, the more important it is (Lawrence, Giles & Bollacker, 1999).

The Role of Context

So, we can handle unstructured text and measure how good an IR system works, but a critical question was not answered yet: How do we determine, which documents from a large collection are actually *relevant* to a given query? To answer this question, we have to address the role of *context*.

User Context

Consider the following example: A user enters the query "Give me all information about jaguars." The set of relevant documents will be heavily dependent on the context in which the user phrases this query: A motor sports enthusiast is probably interested in news about the "Jaguar Racing" Formula 1 racing stable, a biologist probably wants to know more about the animal. So, the context—in the sense of the situation in which the user is immersed—plays a major role (Johnson, 2003; Lawrence, 2000).

Budzik, Hammond and Birnbaum (2001) discuss two common approaches to handle different user contexts:

1. Relevance feedback: The user begins with a query and then evaluates the answer set. By providing positive or negative feedback to the IR system, this can modify the original query by adding positive or negative search terms to it. In an iterative dialogue with the IR system, the answer set is then gradually narrowed down to the relevant result set. However, as studies (Hearst, 1999) have shown, users are generally reluctant to give exhaustive feedback to the system.

2. Building user profiles: Similar to the relevance feedback, the IR system builds up a user profile across multiple retrieval sessions, that is, with each document the user selects for viewing, the profile is adapted. Unfortunately, such a system does not take account of "false positives", that is, when a user follows a link that turned out to be of no value when inspecting it closer. Additionally, such systems integrate short term user interests into accumulated context profiles, and tend to inhibit highly specialized queries which the user is currently interested in.

Budzik, Hammond & Birnbaum (2001) presented a system that tries to guess the user context from open documents currently edited or browsed on the work space. Their system constitutes an "Information Management Assistant [which] observes user interactions with everyday applications and anticipates information needs [...] in the context of the user's task." Another system developed by Finkelstein and Gabrilovich (2002) analyzes the context in the immediate vicinity of a user-selected text, therefore making the context more focused. In an evaluation of their system both achieved consistently better results than standard search engines are able to achieve without context.

Document Context

Context can also be seen from another point of view: Instead of worrying about the user's intent and the context in which a query is embedded, an information retrieval system could make the context in which retrieved documents are embedded more explicit. If each document's context is immediately visible to the information seeker, the relevant context might be quickly picked out. Additionally, the user might discover a context he had not in mind when formulating the query and thus find links between his intended and an unanticipated context. In the following we describe several approaches that aim in this direction.

Augmenting Document Sets with Context

Text Categorization

One way to add context to a document is by assigning a meaningful label to it (Le & Thoma, 2003). This constitutes a task of text categorization and there exist numerous algorithms that can be applied. The general approach is to select a training set of documents that are already labelled. Based on the "bag of words" representation, machine learning methods learn the association of category labels to documents. For an in depth review of statistical approaches (such as naives Bayes or decision trees) see Yang (1999). Computationally more advanced methods utilize artificial neural network architectures such as the support vector machine which have achieved break even values close to 0.9 for the labelling of news wire articles (Joachims, 1998; Lodhi, 2001).

However, in the medical domain, 100 categories are seldom adequate to describe the context of a text. In case of the MEDLINE database, the National Library of Medicine has developed a highly standardized vocabulary, the Medical Subject Headings (MeSH) (Lowe & Barnett, 1994). They consist of more than 35,000 categories that are hierarchically organized and constitute the basis for searching the database. To guarantee satisfactory search results of constant quality, reproducible labels are an important prerequisite. However, the cost of human indexing of the biomedical literature is high: according to Humphrey (1992) it takes one year to train an expert the task of document labelling. Additionally, the labelling process lacks a high degree of reproducibility. Funk, Reid, and McGoogan (1983) have reported a mean agreement in index terms ranging from 74 percent down to as low as 33 percent for different experts. Because the improvement of index consistency is such demanding, assistance systems are considered to be a substantial benefit. Recently, Aronson, Bodenreider, Chang, Humphrey, Mork, Nelson, Rindflesh, and Wilbur (2000) have presented a highly tuned and sophisticated system which yields very promising results. Additionally to the bag of words model their system utilizes a semantic network describing a rich ontology of biomedical knowledge (Kashyap, 2003).

Unfortunately, the high complexity of the MeSH terms makes it hard to incorporate a MeSH-based categorization into a user interface. When navigating the results of a hierarchically ordered answer set it can be a time-consuming and frustrating process: Items which are hidden deep within the hierarchy can often only be obtained by descending a tree with numerous mouse-clicks. Selection of a wrong branch requires backing up and trying a neighboring branch. Since screen space is a limited resource only a small area of context is visible and requires internal "recalibration" each time a new branch is selected. To overcome this problem, *focus and context* techniques are considered to be of high value. These are discussed in more detail below.

The Cluster Hypothesis

An often cited statement is the *cluster hypothesis,* which states that documents which are similar in their bag of words feature space tend to be relevant to the same request (van Rijsbergen, 1979). Leuski (2001) has conducted an experiment where an agglomerative clustering method was used to group the documents returned by a search engine query. He presented the user not with a ranked list of retrieved documents, but with a list of clusters, where each cluster in turn was arranged as a list of documents. The experiment showed, that this procedure "can be much more helpful in locating the relevant information than the traditional ranked list." He could even show that the clustering can be as effective as the relevance feedback methods based on query expansion. Other experiments also validated the cluster hypothesis on several occasions (Hearst & Pedersen, 1996; Zamir & Etzioni, 1999). Recently the *vivisimo*[2] search engine has drawn attention by utilizing an online clustering of retrieval sets, which also includes an interface to PubMed / MEDLINE.

Visualizing Context

Another way to create context in line with the spirit of the cluster hypothesis is by embedding the document space in a visual display. By making the relationship between documents visually more explicit, such that the user can actually *see* inter-document similarities, the user gets *(i)* an overview of the whole collection, and *(ii)* once a relevant document has been found, it is easier to locate others, as these tend to be grouped within the surrounding context of already identified valuable items. Fabrikant and Buttenfield (2001) provide a more theoretical framework for the concept of "spatialization" with relation to cognitive aspects and knowledge acquisitions: Research on the cognition of geographic information has been identified as being important in decision making, planning and other areas involving human-related activities in space.

In order to make use of "spatialization", that is, to use cognitive concepts such as "nearness" we need to apply some sort of transformation to project the documents from

Figure 1. A map of animals

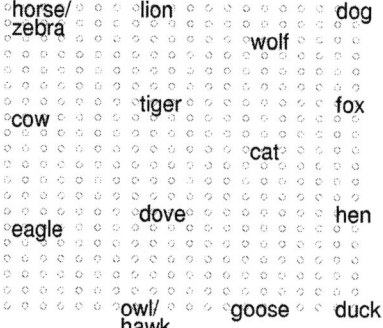

their high-dimensional (typically several thousands) bag-of-words space onto a two dimensional canvas suitable for familiar inspection and interaction. The class of algorithms performing such a projection is called *multi dimensional scaling* (MDS). A large number of MDS methods such as Sammon mapping, spring models, projection pursuit or local linear embeddings have been proposed over the years. Skuping and Fabrikant (2003) provide an excellent discussion of the most common MDS variants in relation to spatialized information visualization.

Self-Organizing Maps

The notion of the self-organizing map (SOM) has been introduced by Kohonen (1982) more than 20 years ago. Since then is has become a well-accepted tool for exploratory data analysis and classification. While applications of the SOM are extremely wide spread—ranging from medical imaging, classification of power consumption profiles, or bank fraud detection—the majority of uses still follows its original motivation: to use a deformable template to translate *data similarities* into *spatial relations*.

The following example uses a simple toy dataset to demonstrate the properties of the SOM algorithm. Consider a 13-dimensional dataset describing animals with properties such as s*mall, medium, big, two legs, hair, hooves, can fly, can swim, and* so on. When displaying the data as a large table, it is quite cumbersome to see the inter-relationship between the items, that is, animals. The map shown in Figure 1 depicts a trained SOM with neurons placed on a 20x20 regular grid. After the training process, each animal is "presented" to the map, and the neuron with the highest activity gets labelled with the corresponding name. As can be seen in the figure the SOM achieves a semantically reasonable mapping of the data to a two-dimensional "landscape": birds and non-birds are well separated and animals with identical features get mapped to identical nodes.

There have been several approaches to use the SOM to visualize large text databases and relations of documents therein. The most prominent example is the WEBSOM project. Kohonen, Kaski and Lagus, (2000) performed the mapping of 7 million patent abstracts and obtained a semantic landscape of the corresponding patent information. However, screen size is a very limited resource, and as the example with 400 neurons above suggests, a network with more than one million nodes—2500 times that size—becomes hard to visualize: When displaying the map as a whole, annotations will not be readable, and when displaying a legible subset, important surrounding context will be lost. Wise (1999) has impressively demonstrated the usefulness of compressed, map-like representations of large text collections: His *ThemeView* reflects major topics in a given area, and a zoom function provides a means to magnify selected portions of the map—unfortunately without a coarser view to the surrounding context.

Focus & Context Techniques

The limiting factor is the two dimensional Euclidean space we use as a data display: The neighborhood that "fits" around a point is rather restricted: namely by the square of the distance to that point. An interesting loophole is offered by *hyperbolic space*: it is

Figure 2. Navigation snapshot showing isometric transformation of HSOM tessellation. The three images were acquired while moving the focus from the center of the map to the highlighted region at the outer perimeter. Note the "fish-eye" effect: All triangles are congruent, but appear smaller as further they are away from the focus.

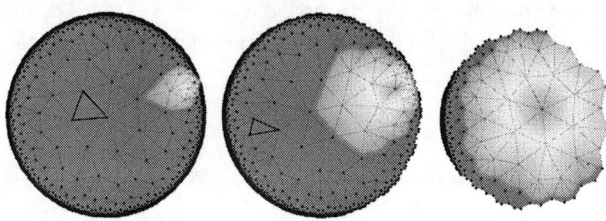

characterized by uniform negative curvature and results in a geometry, where the neighborhood around a point increases *exponentially* with the distance. This exponential behavior was firstly exploited (and patented) by the "hyperbolic tree browser" from Lamandpng & Rao (1994), followed by a Web content viewer by Munzner (1998). Studies by Pirolli, Card and van der Wege (2001) showed that the particular *focus & context* property offered by hyperbolic space can significantly accelerate "information foraging".

Naturally, it becomes apparent to combine the SOM algorithm with the favorable properties of hyperbolic space (Ritter, 1999). The core idea of the hyperbolic self-organizing map (HSOM) is to employ a grid of nodes in the hyperbolic plane H^2. For H^2 there exists an infinite number of tessellations with congruent polygons such that each grid point is surrounded by the same number of neighbors (Magnus, 1974). As stated above, an intuitive navigation and interaction methodology is a crucial element for a well-fitted visualization framework. By utilizing the *Poincaré* projection and the set of *Möbius* transformations (Coxeter, 1957) a "fish-eye" fovea can be positioned on the HSOM grid allowing an intuitive interaction methodology. Nodes within the fovea are displayed with high resolution, whereas the surrounding context is still visible in a coarser view. For further technical details of the HSOM see (Ritter, 1999; Ontrup & Ritter, 2001). An example for the application of the fish-eye view is given in Figure 2. It shows a navigation sequence where the focus was moved towards the highlighted region of interest. Note, that from the left to the right details in the target area get increasingly magnified, as the highlighted region occupies more and more display space. In contrast to standard zoom operations, the current surrounding context is not clipped, but remains gradually compressed at the periphery of the field of view. Since all operations are continuous, the focus can be positioned in a smooth and natural way.

Browsing MEDLINE in Hyperbolic Space

The following example presents a HSOM framework that combines the aspects of information retrieval, context enrichment and intuitive navigation into a single application. The following images show a two dimensional HSOM, where the nodes are arranged on a regular grid consisting of triangles similar to those in Figure 2. The HSOM's neurons can be regarded as "containers" holding those documents for which the corresponding neuron is "firing" with the highest rate. These containers are visualized as cylinders embedded in the hyperbolic plane. We can then use visual attributes of these containers and the plane to reflect document properties and make context information explicit:

- The container sizes reflect the number of documents situated in the corresponding neuron.

- A color scale reflects the number of documents which are labelled with terms from the *Anatomy* hierarchy of MeSH terms, i.e. the brighter a container is rendered, the more articles labelled with *Anatomy* MeSH entries reside within that node.

- The color of the ground plane reflects the average distance to neighboring nodes in the semantic bag-of-words space. This allows the identification of thematic clusters within the map.

Figure 3 shows a HSOM that was trained with approximately 25,000 abstracts from MEDLINE. The interface is split into two components: the graphical map depicted below and a user interface for formulating queries (not shown here). In the image below, the user has entered the search string "brain". Subsequent to the query submission, the system highlights all nodes belonging to abstracts containing the word "brain" by elevating their corresponding nodes. Additionally the HSOM prototype vectors—which were autonomously organized by the system during a training phase—are used to generate a key word list which annotate and semantically describe the node with the highest hit rate. To this end, the words corresponding to the five largest components of the reference vector are selected. In our example these are: *brain, perfusion, primari, neuron,* and *injuri*. In the top left the most prominent MeSH terms are displayed, that is, 34 articles

Figure 3. MEDLINE articles mapped with the HSOM. The user submitted the query string "brain" and the corresponding results are highlighted by the system.

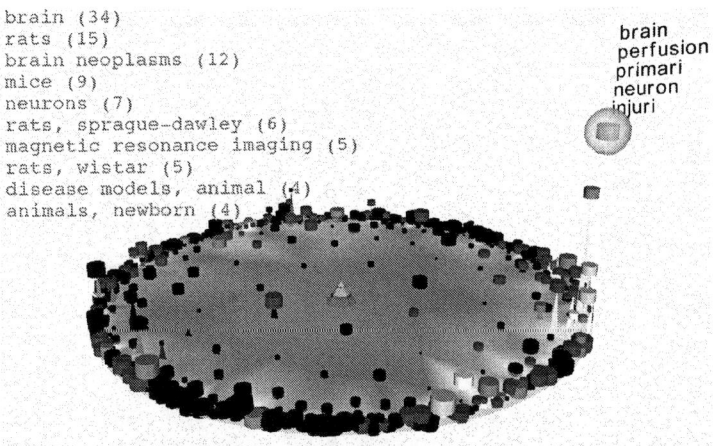

Figure 4. The user has moved the focus into the target area and can now inspect the context enriched data more closely.

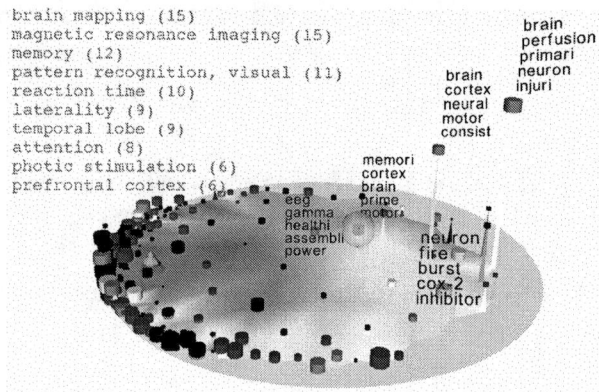

are tagged with the medical subject heading "brain", 15 are tagged with "rats", and so on.

The map immediately shows that there are several clusters of brain-related articles in the right part of the map. Moving the focus into this area leads to the situation shown in Figure 4. By selecting single neurons with the mouse, the user can selectively retrieve key words of individual containers. Additionally, the MeSH terms of the selected node are displayed. As can be seen from the keyword distribution, all nodes share semantically related characteristics. The two leftmost labelled nodes do not contain key words like "brain" or "neuron", but the labelling with "cord", "spinal", "eeg", and "gamma", respectively, indicates a close relation to the selected "brain-area".

The drill-down process to the document level combines the strengths of the graphical and the textual display: After zooming-in on a promising region, the titles of documents pertaining to a particular node can be displayed by a double mouse click. Note, that due to the ordering process of the SOM algorithm these documents are typically very similar to each other—and in the sense of the clustering hypothesis therefore might all provide an answer to the same query. A further double click on a document title then displays the selected abstract on the screen. In this way, the hyperbolic self-organizing map can provide a seamless interface offering rapid overview of large document collections while allowing at the same time the drill down to single texts.

Benefits for a Hospital Information System

In the framework of a hospital information system, such a context enriched interactive document retrieval system might be of great benefit to the clinical personal. According to Chu and Cesnik (2001) a hospital may generate up to five terabytes of data a year, where 20 to 30 percent of such data are stored as free text reports such as medical history, assessment and progress notes, surgical reports or discharge letters. Free text fields in medical records are considered of invaluable importance for medical relevance (Stein, Nadkarni, Erdos, & Miller, 2000). By using document standards like the CDA as briefly discussed in Section 2.2, the clinician is able to browse and retrieve the data according to the structure provided by the document framework. However, many existing information systems do not yet take advantage of the complex CDA or the even more substantial Reference Information Model—because it is a very time consuming task to transfer documents to the new standards and to fit the enormous mass of model objects to free text data.

Nevertheless, these systems store valuable information—just in a format which cannot be directly analyzed by structured methods. In order to provide access by content, the aforementioned Information Retrieval methodologies can significantly contribute to facilitate these tasks: a self-organizing map creates structure and context purely on the statistical word distributions of free text and thus allows a semantic browsing of clinical documents. This in turn offers the possibility to gain a problem orientated perspective on health records stored in information systems, that is, as Lovis et al (2000) have noted: "intelligent browsing of documents, together with natural language emerging techniques, are regarded as key points, as it appears to be the only pertinent way to link internal knowledge of the patient to general knowledge in medicine."

Conclusion

Recent advances in hospital information systems have shown that the creation of organization-wide clinical data repositories (CDR) plays a key role for the support of knowledge acquisition from raw data. Currently, many case studies are published which emphasize that data mining techniques provide efficient means to detect significant patterns in medical records. Interactive OLAP tools support best practice and allow for informed real time decision-making and therefore build the solid foundation for further improvements in patient care.

The next step to further broaden the knowledge base in healthcare environments will be the integration of unstructured data, predominantly free narrative text documents. The nature of this type of data will demand for new approaches to deal with the inherently vague information contained in these documents. We have shown how artificial neural networks with their ability for self-organization of non-crisp data provide an almost natural link between the hard computational world and the soft-computing of the human brain. We believe that interactive information retrieval methodologies that account for the rich context of natural language will significantly contribute to knowledge acquisition through text analysis.

Copyright © 2005, Idea Group Inc. Copying or distributing in print or electronic forms without written permission of Idea Group Inc. is prohibited.

References

Arnrich, B., Walter, J., Albert, A., Ennker, J. & Ritter, H. (2004). Data mart based research in heart surgery: Challenges and benefit. *Proceedings of the MEDINFO 2004,* San Francisco.

Aronson, A.R., Bodenreider, O., Chang, H.F., Humphrey, S.M., Mork, J.G., Nelson, S.J., Rindflesh, T.C. & Wilbur, W.J. (2000). The NLM indexing initiative. *Proceedings of the Annual AMIA Symposium* (pp. 17-21).

Baeza-Yates, R. & Ribeiro-Neto, B. (1999). *Modern information retrieval.* New York: ACM Press/Addison-Wesley.

Bergeron, B.P. (2002). Enterprise digital assistants: The progression of wireless clinical computing. *Journal of Medical Practice Management, 17*(5), 229-233.

Berners-Lee, T., Hendler, J. & Lassila, O. (2001). The Semantic Web. *Scientific American* (May), 35-43.

Bito, Y., Kero, R., Matsuo, H., Shintani, Y. & Silver, M. (2001). Interactively visualizing data warehouses. *Journal of Healthcare Information Management, 15*(2), 133-142.

Brin, S. & Page, L. (1998). The anatomy of a large-scale hypertextual web search engine. *Proceedings of the 7th International WWW Conference,* Brisbane, Australia.

Budzik, J., Hammond, K.J. & Birnbaum, L. (2001). Information access in context. *Knowledge-Based Systems, 14*(1-2), 37-53.

Chu, S. & Cesnik, B. (2001). Knowledge representation and retrieval using conceptual graphs and free text document self-organisation techniques. *International Journal of Medical Informatics, 62*(2-3), 121-133.

Coxeter, H.S.M. (1957). *Non Euclidean geometry.* Toronto: University of Toronto Press.

Doig, G.S. & Simpson, F. (2003). Efficient literature searching: A core skill for the practice of evidence-based medicine. *Intensive Care Medicine, 29*(12), 2109-11.

Dolin, R.H., Alschuler, L., Beebe, C., Biron, P.V., Boyer, S.L., Essin, D., Kimber, E., Lincoln, T. & Mattison, J.E. (2001). The HL7 clinical document architecture. *Journal of the American Medical Informatics Association, 8*(6), 552-69.

Fabrikant, S.I. & Buttenfield, B.P. (2001). Formalizing semantic spaces for information access. *Annals of the Association of American Geographers, 91,* 263-280.

Finkelstein, L. & Gabrilovich, E. (2002). Placing search in context: the concept revisited. *ACM Transactions on Information Systems, 20*(1), 116-131.

Fisk, J.M., Mutalik, P.G., Levin, F.W., Erdos, J., Taylor, C. & Nadkarni, P.M. (2003). Integrating query of relational and textual data in clinical databases: A case study. *Journal of the American Medical Informatics Association, 10*(1), 21-38.

Funk, M.E., Reid, C.A. & McGoogan, L.S. (1983). Indexing consistency in MEDLINE. *Bulletin of the Medical Library Association, 2*(71), 176-183.

Hearst, M.A. (1999). User interfaces and visualization. In R. Beaza-Yates & B. Ribeiro-Neto. *Modern information retrieval.* New York: ACM Press/Addison-Wesley.

Hearst, M.A. & Pedersen, J.O. (1996). Reexamining the cluster hypothesis: scatter/gather on retrieval results. *Proceedings of 19th ACM International Conference on Research and Development in Information Retrieval* (pp. 76-84).

Humphrey, S.M. (1992). Indexing biomedical documents: From thesaural to knowledge-based retrieval systems. *Artificial Intelligence in Medicine, 4*(5), 343-371.

Institute of Medicine. (1991). *The computer-based patient record: An essential technology for healthcare.* Washington, DC: National Academy Press.

Joachims, T. (1998). Text categorization with support vector machines: Learning with many relevant features. *Proceedings of 10th European Conference on Machine Learning (ECML)* (pp. 137-142).

Johnson, J.D. (2003). On contexts of information seeking. *Information Processing and Management, 39*(5), 735-760.

Kashyap, V. (2003). The UMLS semantic network and the Semantic Web. *Annual Conference of the American Medical Informatics Association (AMIA)* (pp. 351-355).

Kohonen, T. (1982). Self-organized formation of topologically correct feature maps. *Biological Cybernetics, 43,* 59-69.

Kohonen, T., Kaski, S. & Lagus, K. (2000). Organization of a massive document collection. *IEEE Transactions on Neural Networks. Special Issue on Neural Networks for Data Mining and Knowledge Discovery,* 574-585.

Lamping, J. & Rao, R. (1994). Laying out and visualizing large trees using a hyperbolic space. *ACM Symposium on User Interface Software and Technology,* 13-14.

Lawrence, S. (2000). Context in Web search. *IEEE Data Engineering Bulletin, 23*(2), 25-32.

Lawrence, S., Giles, C.L. & Bollacker, K. (1999). Digital libraries and autonomous citation indexing. *IEEE Computer, 32*(6), 67-71.

Le, D.X. & Thoma, G.R. (2003). Automated document labeling for Web-based online medical journals. *Proceedings of the 7th World Multiconference on Systemics, Cybernetics and Informatics,* July (pp. 411-415).

Leuski, A. (2001). Evaluating document clustering for interactive information retrieval. *Conference on Information and Knowledge Management (CIKM)* (pp. 33-40).

Libbus, B. & Rindflesch, T.C. (2002). NLP-based information extraction for managing the molecular biology literature. *Annual Conference of the American Medical Informatics Association (AMIA),* 445-449.

Lodhi, H., Shawe-Taylor, J., Christianini, N. & Watkins, C. (2001). Text classification using string kernels. *Advances in Neural Information Processing Systems (NIPS), 13,* 563-569.

Lovis, C., Baud, R.H., & Planche, P. (2000). Power of expression in the electronic patient record: Structured data or narrative text? *International Journal of Medical Informatics,* 101-110.

Lowe, H.J. & Barnett, G.O. (1994). Understanding and using the medical subject headings (MeSH) vocabulary to perform literature searches. *Journal of the American Medical Association, 271*(4), 1103-1108.

Mack, R. & Hehenberger, M. (2002). Text-based knowledge discovery: search and mining of life-sciences documents. *Drug Discovery Today, 7*(11), 89-98.

Magnus, W. (1974). *Noneuclidean tesselations and their groups.* London: Academic Press.

McCray, A.T. & Ide, N.C. (2000). Design and implementation of a national clinical trials registry. *Journal of the American Medical Informatics Association, 7,* 313-23.

McDonald, C. (1997). The barriers to electronic medical record systems and how to overcome them. *Journal of the American Medical Informatics Association, 4,* 213-221.

Munzner, T. (1998). Exploring large graphs in 3D hyperbolic space. *IEEE Computer Graphics and Applications, 18*(4), 18-23.

Ng, S.K. & Wong, M. (1999). Towards routine automatic pathway discovery from on-line scientific text abstracts. *Genome Informatics, 10,* 104-112.

Ontrup, J. & Ritter, H. (2001). Hyperbolic self-organizing maps for semantic navigation. *Advances in Neural Information Processing Systems (NIPS), 14,* 1417-1424.

Pirolli, P., Card, S.K. & van der Wege, M.M. (2001). Visual information foraging in a focus + context visualization. *ACM Conference on Human Factors in Computing Systems* (pp. 506-513).

Ritter, H. (1999). Self-organizing maps in non-euclidean spaces. In E. Oja & S. Kaski (Eds.), *Kohonen maps* (pp. 97-110). Amer Elsevir.

Rzhetsky, A., Iossifov, I., Koike, T., Krauthammer, M., Kra, P., Morris, M., Yu, H., Duboue, P.A., Weng, W., Wilbur, W.J., Hatzivassiloglou, V. & Friedman, C. (2004). GeneWays: A system for extracting, analyzing, visualizing, and integrating molecular pathway data. *Journal of Biomedical Informatics, 37*(1), 43-53.

Salton, G. & Buckley, C. (1988). Term-weighting approaches in automatic text retrieval. *Information Processing and Management, 24*(5), 513-523.

Sensmeier, J. (2003). Advancing the state of data integration in healthcare. *Journal of Healthcare Information Management, 17*(4), 58-61.

Silver, M., Sakata, T., Su, H.C., Herman, C., Dolins, S.B. & O'Shea, M.J. (2001). Case study: How to apply data mining techniques in a healthcare data warehouse. *Journal of Healthcare Information Management, 15*(2), 155-164.

Skupin, A. & Fabrikant, S.I. (2003). Spatialization methods: A cartographic research agenda for non-geographic information visualization. *Cartography and Geographic Information Science, 30*(2), 95-119.

Smith, A. & Nelson, M. (1999). Data warehouses and clinical data repositories. In M.J. Ball, J.V. Douglas & D. Garets (Eds.). *Strategies and technologies for healthcare information* (pp. 17-31). New York: Springer Verlag.

Spyrou, S.S., Bamidis, P., Chouvarda, I., Gogou, G., Tryfon, S.M. & Maglaveras, N. (2002). Healthcare information standards. *Health Informatics Journal, 8*(1), 14-19.

Srinivasan, P. & Rindflesch, T.C. (2002). Exploring text mining from MEDLINE. *Annual Conference of the American Medical Informatics Association (AMIA)*.

Stein, H.D., Nadkarni, P., Erdos, J. & Miller, P.L. (2000). Exploring the degree of concordance of coded and textual data in answering clinical queries from a clinical data repository. *Journal of the American Medical Informatics Association, 7,* 42-54.

Swanson, D.R., Smalheiser. N.R. & Sleeman, D.H. (1997). An interactive system for finding complementary literatures: A stimulus to scientific discovery. *Artificial Intelligence, 91*(2), 183-203.

Tange, H.J, Hasman, A., de Vries, R., Pieter, F. & Schouten, H. C. (1997). Medical narratives in electronic medical records. *International Journal of Medical Informatics, 46*(1), 7-29.

Van Rijsbergen, C.M. (1979). *Information retrieval.* London: Butterworths.

Wise, J. (1999). The ecological approach to text visualization. *Journal of the American Society for Information Science, 50*(13), 1224-1233.

Yang, Y. (1999). An evaluation of statistical approaches to text categorization. *Information Retrieval, 1*(2), 69-90.

Zamir, O. & Etzioni, O. (1999). Grouper: A dynamic clustering interface to Web search results. *Computer Networks, 31,* 1361-1374.

Endnotes

[1] *http://www.microsoft.com/resources/casestudies/CaseStudy.asp?CaseStudyID=14967*

[2] *http://vivisimo.com*

Chapter V

The Challenge of Privacy and Security and the Implementation of Health Knowledge Management Systems

Martin Orr, Waitemata District Health Board, New Zealand

Abstract

Health information privacy is one of the most important and contentious areas in the development of Health Knowledge Systems. This chapter provides an overview of some of the daily privacy and security issues currently faced by health services, as health knowledge system developments risk outpacing medico-legal and professional structures. The focus is a mixture of philosophy and pragmatics with regard to the key "privacy" and "security" issues that challenge stakeholders as they try to implement and maintain an increasing array of electronic health knowledge management systems. The chapter utilises a number of evolving simple visual and mnemonic models or concepts based on observations, reflections and understanding of the literature.

Introduction

The focus of this chapter is largely shaped by the common themes and thoughts expressed, and dilemmas experienced, within the environment in which the Author works. However many of these local opinions are shaped by more universal forces, media, and experiences, and common themes, concepts and challenges can be found internationally, both within health, and other complex systems that handle personal information (Anderson, 1996; Coiera & Clarke, 2003; Tang, 2000). Health Knowledge Management systems are assisted by processes that provide complete, accurate, and timely information. Issues of security and privacy have the capacity to facilitate or inhibit this process. However, there are a myriad of perspectives with regard to the meaning, significance, and interrelation of the terms privacy, security, and health knowledge system, which shall be discussed throughout the chapter.

A Health Knowledge system should aim to integrate and optimise stakeholders' "capacity to act" (Sveiby, 2001) or "capacity to C.A.R.E." (that is, the capacity to deliver in a coordinated fashion the integral Clinical, Administrative, Research and Educational functions of healthcare). The Electronic Patient Record term typically aims to describe the technology or software that stores the record of care or provides a degree of decision support. However, the term "Health Knowledge Management System" aims to better capture or identify the overall system changes required to implement decision support systems, such as changes in underlying processes and the development of a culture that values, respects and protects the acquisition, distribution, production and utilisation of available knowledge in order to achieve better outcomes for patients (Standards Australia, 2001; Wyatt, 2001).

A Health Knowledge Management System should facilitate closing the communication gaps on an ongoing basis, between all the key stakeholders involved in optimising care, GPs, Allied health services (including hospitals), and the often forgotten Patients and

Table 1. Health knowledge systems—Closing the C.A.R.E. G.A.P.S. F.I.R.S.T.

C.A.R.E.	Clinical Administration Research Education
G.A.P.S.	General Practitioner (primary and community care) Allied Health Services (including secondary and tertiary care) Patients Supports
F.I.R.S.T.	Fast Intuitive Robust Stable Trustworthy

their Supports, who all need and should benefit from an improved "capacity to C.A.R.E." The system should also aim to be fast, intuitive, robust, stable, and trustworthy (Orr, 2000).

Privacy and Security

Internationally there is a growing array of privacy and security codes, laws, and standards with many shared core themes (Office of the Privacy Commissioner, 2002; Standards Australia and Standards New Zealand, 2001). However, creating a shared understanding of the essential nature of "Privacy" continues to afford particular challenges, as many of its associated elements are contextual, perceptual and personal. Clarke (2004) and Anderson (2004) provide comprehensive resources exploring the dimensions and complexities of privacy, security, and related concepts.

There is the potential for conflict, as well as the need for balance between the hierarchy of needs (Maslow, 1943) or wants of an individual or a particular group, versus the needs, and wants and capabilities of the wider community or system. The original Latin origin of the word "Privacy" aims to convey a sense of something "set apart", owned by oneself rather than the state or public (Online Etymology Dictionary, 2001). Therefore we could view private information as something we feel ownership of or particularly connected to, information that should be "set apart" and out of the public domain, or at least over which we should have particular control and an understanding of the competence and motivation of those who may use it.

However with regard to our health (and accordingly our health information) we are rarely truly "set apart" from the public or our communities. Our health, our health management, and our behaviour are continually impacted on by the complex biopsychosocial (Engel, 1977) system in which we live. The need or desire to be "set apart" or have control can lead to conflict or at least opportunity costs with the needs of other aspects of the finely balanced system within which we function. Each stakeholder group may attribute different weights to different factors and believe the balance should lie to one side or the other.

The Latin origin of the word "Security" conveys the sense of something safe or without care (Online Etymology Dictionary, 2001). Security with respect to Information Systems typically focuses on the preservation of confidentiality integrity, and availability. Confidentially refers to the need to limit access only to those authorised to have access, integrity refers to ensuring the completeness and accuracy of information and associated system processes while availability refers to the need to ensure that information and associated system processes are accessible to authorised users when required (Standards Australia/New Zealand, 2001).

Perceptions of privacy and sub-optimal care risk according to degree of information flow can be depicted by way of a schematic (Figure 1).

As a background to a discussion on the challenges of privacy impact assessment and informed consent or "e-consent" systems, it is worth trying to stimulate a shared

Figure 1. Privacy versus sub-optimal care risk: Possible clinician view

conceptual understanding, of some of the perceptions of what the impacts or risks and benefits of a health knowledge management system might be. As a beginning, we might consider Figure 1, which illustrates how a clinician might perceive the relative decrease in sub-optimal care risk and increase in privacy risk as health information increasingly flows and is shared and integrated.

Privacy risk is defined as the perceived risk of an adverse outcome or sub-optimal care related to a privacy infringement. Sub-optimal care risk is defined as the perceived risk of non-optimal care related to information flow restriction or disintegration.

The focus of electronic health knowledge management system value analysis is often on preventing death or adverse events. However there is a whole range of care that, although it does not result in a reported adverse event or death, could be described as non- or sub-optimal. A health outcomes curve (Figure 2) may have optimal care at one end with everything else considered increasingly sub-optimal as it moves through various degrees of poorly co-ordinated and inefficient care, towards adverse events, permanent disability and death at the other end. This poorly co-ordinated or inefficient care may for example include duplication of assessment or investigation or the use of expensive or multiple interventions, without evidence of greater benefit over a cheaper or single intervention or simply failing to stop or review an intervention.

Figure 2. Health outcomes curve: Death to-optimal care

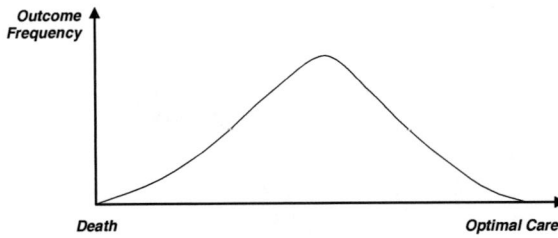

Figure 3. Privacy versus suboptimal care risk: Possible patient/consumer view

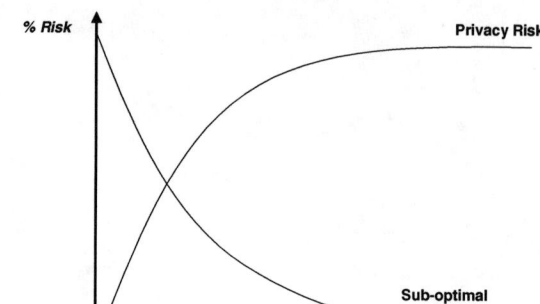

Therefore the hypothetical clinician's opinion (Figure 1) may be shaped by the perspective that, on balance, the greatest risk our communities face is a whole range of suboptimal care, that at least in part is attributable to the poor flow and integration of health information. Based on this hypothesis, better integrated health information leads to better integrated healthcare, the desired outcome from an effective health knowledge management system would be to shift the norm of the health outcomes curve towards the optimal end of the curve minimising the degree of sub-optimal care.

However while not disputing the desire to minimise sub-optimal care, a patient may perceive the relative risks differently (Figure 3). It is currently, and likely to remain, difficult to provide "evidence" based data on relative privacy versus sub-optimal care risks to an individual or indeed populations, as there will be strong perceptual differences as to what the scope of that evidence should be and how it should be weighed. However we could assume the relative risk graphs are unlikely to be reflected by straight lines. The relative risks are likely to be influenced by individual perception, respective roles, the nature of the data, and to change over time. The frequency, imminence, likelihood, and magnitude of risk would also need to be considered, as would the ability to identify, evaluate, manage, and review risk.

In the absence of definitive data, and with the recognition that the perceived risk arising from a privacy infringement has such a personal perceptual component, it is important to highlight some of the views and perceptions with regard to this matter.

It could be argued, that severe physical disability or death is unlikely except in the case of a privacy infringement leading to severe psychological distress resulting in attempted or completed suicide. However there are a number of counter arguments to this. These suggest that what we do now even with good intention could have future potentially unforeseen but not unforeseeable adverse impacts. These suggest that there is not only the potential for sub-optimal care, but also for active discrimination if not persecution and political or regime sanctioned killings.

Perceived risk of a privacy infringement may lead to a patient withholding information, or delaying presentation to health services. This may result in delayed investigation and treatment and potentially sub-optimal care including avoidable morbidity and death. This morbidity may extend to others particularly in the case of infectious diseases, not least partners and unborn children in the case of sexually transmitted diseases.

Current data integration, data mining, risk prediction, and genetic profiling developments have the potential for great benefit for our communities. However there is also the risk that without appropriate regulation, certain segments of the populations could be identified at high health risk and become unemployable, uninsurable and unable to obtain credit, a mortgage or appropriate affordable healthcare (Kennedy, 2004).

Insurers and employers are generally not innately bad. They are businesses charged with making a profit (or minimising loss) by their shareholders, and can be expected to use every legal means and loophole possible to identify and manage their risk particularly if it gives them an advantage over a competitor.

However it can be argued that the potential unfettered ability of insurance companies and health management organisations to be progressively able to segment or cut up the market, and discard those least profitable should be a concern for every community.

Unique Health Identification Numbers

Unique health identification numbers can facilitate the process of data integration across multiple electronic systems, and lead to better-integrated care. It can also be argued that within the context of a large integrated electronic system they may decrease privacy or security risk by minimising the amount of non secure activity around bringing a disintegrated paper record together, such as multiple phone calls to other services inquiring as to the availability of records, and the subsequent transfer of paper records.

However, it can be argued that if there is an infringement of an integrated unique identifier based electronic system, there is the opportunity for thousands of records to be accessed, with unique identifiers facilitating access to a wide range of comprehensive health information. An infringement could in theory be the result of a hacker gaining unauthorised access to the health network. However, of likely greater risk is a health service staff member, acting out of curiosity, malice or pecuniary gain, or having simply been misled or talked into it by someone posing as a patient or clinician on the telephone. This latter method of infringement has been highlighted by Anderson (1996). The risk of staff access, facilitating a range of crimes including murder has also entered the popular fictional literature (Gerritsen, 2001), also shaping community perceptions.

It could also be argued that significant parts of humanity's past and present has been dominated by war, invasion, totalitarian regimes, and state sanctioned terror, and even those of us currently living in relatively benign environments are naïve to believe that such horrors will not feature in our future. Such invading or totalitarian regimes have and would use every means possible to control and terrorise the population, including health data. The provision therefore of a national unique health identification number could facilitate this process, including the singling out of specific subgroups for discrimination, persecution or killing (Black, 2001).

There are also general privacy infringement concerns, including health information systems, being cross referenced, or matched with other government systems including social services, benefits and police, and the use of data for unauthorised research or commercial purposes.

Are patient privacy concerns sometimes just a smokescreen for clinician privacy concerns? Are clinicians sometimes just trying to minimize the transparency of a whole range of issues from limitations of their skill and efficiency to fraudulent practice? It is understandable why many clinicians would find these questions challenging if not insulting. However in terms of exploring the challenges that might impact on the effective implementation of health knowledge management systems, they are perspectives that cannot be ignored, and following is a brief limited discussion of some of the potential underlying issues.

Evidence Based Medicine (EBM) or the concept that treatment choices should be based on, or backed by, the best evidence available, has unfortunately in many situations, become perceived as an excuse for "evidence bound" medicine. That is the establishment of rigid regimes, or marked limitation of treatment choices that stifle innovation or new developments and do not recognise individual patient variation. This has particularly become associated with the term "managed care". Managed care is a concept that in its pure form has some worth, as it promotes the idea of effective clinical care pathways and efficient targeted integrated use of resources. However the form of "managed care" espoused by profit focused health organisations, and insurance companies is perceived as having little to do with effectively managing care in the patients interests, but everything to do with managing costs and subsequently maximising profits.

Accordingly clinicians may be rightly concerned with respect to whether, Health Information Management Systems that propose to support "evidence based medicine" and "quality management" are not just Trojan Horses, implemented with purely financial goals that will lead to a state of "evidence bound medicine" (Orr, 2000).

It can be argued that in a fee-for-service environment concerns for patient privacy may be a smokescreen to prevent monitoring of over-servicing and over-investigation and subsequent over-claiming. Unfortunately this argument might have some credence, but hopefully it applies to only a small minority of clinicians.

The clinician population, just like the patient population, will have a spectrum of views and needs and behaviours with regard to their own personal privacy. The view that those who are doing nothing wrong have nothing to worry about may have some validity, but it does not dismiss the fact that some of those who have done nothing wrong will worry, or certainly will not enjoy or embrace their employing, funding or regulatory body's ability to closely monitor or control their behaviour.

Clinicians may rightly question a health information management system's capacity for "evidence building medicine", the potential to support research and effectively improve clinical processes from their perspective and to augment their clinical skill and professionalism.

Coiera and Clark (2003) have comprehensively described some of the parameters, complexities, and potential solutions for the concept of "e-consent" with regard to the handling of electronic patient information. It may be useful to focus on some of the perceived potential limitations or challenges of the concept (or more specifically the concept of so called "informed consent") in practice, when it interacts with the complex system of health and clinical care information flows and utilisation. As a context to this discussion, a series of perspectives, and metaphors about how the current health system (and particularly clinical practice) functions is presented.

Informed consent implies some form of formal risk benefit analysis; or at least that a patient is freely consenting to a particular action having been appropriately informed of the potential risks and benefits of such an action. In relation to an electronic Health Information System, informed consent can be used to imply that a patient, based on perceived privacy risk, should have the right to suppress certain information, or to prevent certain caregivers from viewing specific information. Additionally, to protect privacy, it can be argued the caregiver should be given no indication that information has been suppressed or is deliberately being withheld from their view. One argument that can be used for this is that patients have always withheld information from caregivers, and caregivers have always been willing to make incomplete notes on patient's request.

A caregiver utilises a Health Knowledge System with a view to making more effective or optimal decisions about a patient's care. A clinician may recognise that, just like a paper record, the electronic record may not always be complete accurate and timely. However is it acceptable for an employer to provide systems that can deliberately deceive or mislead a clinician by suppressing information, while giving no indication that information is being withheld?

Biopsychosocial Healthcare (Engel, 1977) should be related to Biopsychosocial Pattern Recognition. Healthcare that seeks to address a patient's illness or disease, not just as biological pathology, but within a psychosocial context, needs to have an understanding of that context. From an information system perspective, assistance with pathology recognition and general treatment advice may be helpful, for example if a patient presents with certain symptoms and signs, the clinician is provided with list of potential diagnoses and general investigation and management advice.

However, for optimal biopsychosocial healthcare, it is important to know an individual's specific risk profile and psychosocial circumstances, their coping mechanisms and supports and how they may have progressed and responded in the past, with the aim of facilitating individualised chronic care and task management. Pattern recognition and appropriate interventions are often related to subtle changes for that specific patient. Longitudinal history and knowing not just how a patient is presenting now but how they have presented and responded over time is one of the major benefits of a longstanding relationship between a patient and a trusted clinician.

In an environment where patient and clinician relationships may be increasingly fragmentary, with multiple short term interventions by multiple caregivers, a health knowledge system should seek to act as an organisational glue and organisational memory, holding together and integrating both the shared collective and longitudinal knowledge of a patient and appropriately co-ordinated responses.

A Health Knowledge System should aim to bring together as soon as possible the required "pieces" that will increase the "capacity to act" (Sveiby, 2001), or capacity to C.A.R.E. (that is the capacity to deliver in a coordinated fashion the integral clinical, administrative, research and educational functions of healthcare). Like a jigsaw, the more pieces already in place, the quicker it is to start seeing the overall picture and what pieces (areas) best to solve next. Indeed, in terms of biopsychosocial pattern recognition, a clinician may sometimes think they are working on putting together the pieces to build one picture, but get more pieces and realise they are actually working on another.

Media reviews of adverse events or deaths using a "retrospectascope" often see simple clarity where none was evident to clinicians involved at the time. Using the "retrospectascope" often one can see all the pieces of the jigsaw at a single place and point in time when in reality the pieces will have emerged from multiple sources over time. In reality, a clinician often needs to solve the jigsaw, as if riding around it in a roller coaster through a dark tunnel with only glimpses from different perspectives of not only the index problem, but multiple problems for multiple patients they may be being asked to solve at the same time.

A machine or computerised jigsaw, with increasing levels of intelligence, could bring pieces together for you, sort and frame pieces into groups and provide decision support telling you a piece is missing or a piece cannot go here as it does not fit with another (e.g., preventing medication errors). If you switch on the privacy mode of this "intelligent" jigsaw should it place mittens on you, slowing the problem solving process or say "sorry you cannot see that piece" or even withhold pieces without even telling you? There will undoubtedly be situations where the withholding of specific information from specific caregivers can be justified. However the concept of informed consent should at least seek to convey an understanding of not just the perceived privacy risks and benefits of information flowing or not flowing, but also the clinical risks and benefits.

However there are different perceptions or perspectives of what those risks and benefits are, and these can differ for each disorder, patient, caregiver, and can vary over time. For example information typically considered sensitive includes that related to infectious and sexually transmitted diseases, alcohol, and drug and mental health history and obstetric and gynaecological history (particularly in relation to induced abortion).

However some patients may consider, for example a family or personal history of carcinoma as, or more, sensitive while one clinician suggested he would be more worried about an insurance company (via their GP) checking his lipids. We also need to consider the time and resource implications of information flow "informed consent" processes and how realistic it is at the time of information collection to make decisions on all information collected in terms of its current and future information flows; for example, that information item A can go to Doctor A but not Doctor B or Nurse C, but item C can only be viewed by Doctor D. In the common situation of resource limitation and prioritisation it can be argued there may be a pragmatic need for "implied consent", with the patient having to actively indicate they want a particularly piece of information limited in its flow.

Similarly it can be argued that it is misleading to even attempt to utilise the concept of informed consent to imply that a health service can truly offer a patient the option of control over their information flows. There may be clinical, statutory, regulatory or financial requirements that require a clinician or health service to pass certain information or partial information to other bodies. A clinician or health service can do their best to make a patient aware of the nature and purpose of the information they are collecting and how it may be utilised, but it is misleading to imply that the patient has total, un-coerced control over that information, for which they can freely offer or withdraw consent.

Accordingly a patient may be advised that if they do not agree to their information being stored or shared in a particular way, the clinician or health service may not be in a position to offer them a service. The Foundation for Information Policy Research (Anderson, 2004) has criticised the draft UK NHS patient information sharing charter (NHS 2002) for including a statement of this nature that makes specific reference to the potential refusal of treatment, if a patient's decision to restrict information sharing, is considered to make them untreatable or their treatment dangerous.

Challenges to Privacy and Security Processes

Healthcare, and indeed human physiology and disease progression are increasingly recognised for their complexity and non-linear dynamics, where there are limitations for reductionistic views and solutions that do not or can not recognise or adapt to that complexity (Plsek & Greenhalgh, 2001; Goldberger, 1996). This section will not attempt an exploration of the theory or to utilise the formal language of non-linear dynamics, but again, through the use of metaphor and perspectives, aims to convey some of that complexity, particularly as it relates to privacy and security and the implementation of electronic health knowledge management systems.

Knowledge Neurones and the Therapeutic Knowledge Alliance

From a distance and at a fixed point in time clinical information may appear to be exchanged in a simple linear chain like fashion. For example, organisation or person A communicates with organisation or person B who communicates with organisation or person C, and all may be expected to exchange the same information in a standardised format.

Therefore conceptually, one might hope to control the flow of information by mapping these connections and asking a patient who trusted connections they consent or agree to have in their own particular chain and under what conditions information should flow along it. However in reality there are often a series of intermingling web-like connections, both within and between multiple organisations and their constituent individuals that are continually being realigned, reshaped and restructured over time. Additionally, as opposed to information being passed in a single standardised format, there are often different variants or segments of the information passed between each connection.

Knowledge Neurones

In addition to electronic health information systems storing text-based records clinicians utilise a multitude of other channels to transfer and manage health information. These include paper records, phones, personal digital assistants, pagers, voicemail, fax, e-mail, and not least conversation or personal and group interaction. Therefore, rather than being considered a simple link in a linear knowledge chain, a clinician can be considered as being at the centre of a multi-pronged spherical neurone like structure, with each prong representing a different channel through which they may connect with others, to push, pull, and produce knowledge.

Electronic health knowledge management systems may assist the clinician in the continuous process of integrating these multiple channels and adding or discerning intelligence. The clinician may strengthen or utilise more channels perceived to add value and ignore or weaken channels perceived to add less value. "Value" may be a function of perceived ability to integrate and add intelligence, and can vary by individual, context, and time.

Historically the "therapeutic alliance" concept has had an evolving meaning, but carries the sense of clinician and patient working together (Goldstein, 1999). Similarly, the earlier definition of a health knowledge management system attempted to convey the sense of GPs, allied health services, and patients and their supports all working together. We could consider each of these groups as forming their own knowledge neurones and conceptualise a "therapeutic knowledge alliance" comprising of a web of these interweaving knowledge neurones, all supporting each other and working together to make a healthy difference.

Healthcare may involve a spectrum of clinical presentations or scenarios that have to be managed, which can be described as simple certainty to complex ambiguity (Figure 4). The range from simple to complex relates to the number of caregivers and interventions that may have to be coordinated in a patient's care in relation to their condition or multiple conditions. Certainty to ambiguity relates to the degree of perceived accuracy or certainty with regard to the diagnosis or diagnoses and the potential effectiveness of interventions or treatments.

Traditionally, as clinicians have moved from simple certainty to complex ambiguity, they have utilised an increasing array of resources within their therapeutic knowledge

Figure 4. Resources and uncertainty to be managed with increasing complexity and ambiguity

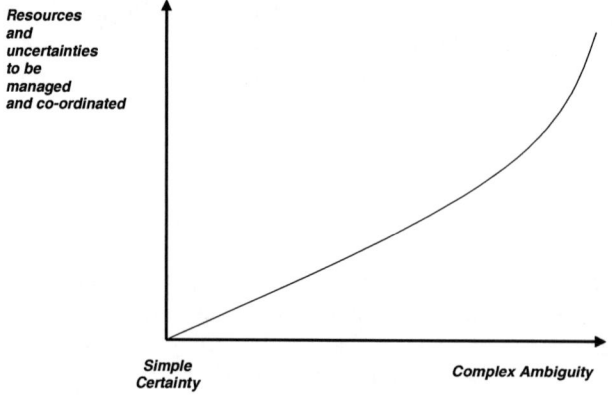

alliance. This typically includes their multidisciplinary team colleagues directly involved in the patient's care in the form of ward rounds or case conferences, but can also involve de-identified case presentations, to non-involved peers or senior colleagues for support and advice on how to best manage this complexity and ambiguity.

The quest for increasingly highly structured data runs the risk of adopting a cloak of scientific certainty that loses the art of dealing with this often-prevalent complex ambiguity in clinical practice. A specialist may have more factual knowledge or skill in a particular clinical area than a generalist or someone in training, but it is often their ability to recognise, adapt and cope with what they do not know, or what is not initially clear that provides their particular value. A patient's presentation may not fit neatly into a specific textbook or diagnostic category, but a specialist may say to a referrer: "I have a sense this patient may be suffering from condition A or maybe B; however, I do not know the exact answer either, but let's monitor and manage the situation this way." Similarly, the value in sharing a case with colleagues is often not so much that they provide an immediate diagnosis or recommend you dramatically change the management, but more that they provide support and contain anxiety that you have not missed anything obvious, and that you are taking a course that would best manage a range of possibilities that may become evident over time (Smith, 1996).

Intuition and learning from experience to cope with complex ambiguity is an aspect of the art of medicine that could arguably be just sophisticated learned pattern recognition and risk management that is and will be increasingly codifiable in electronic form. Indeed, a central aim of an electronic health knowledge management system should be to enhance and develop the capability of the health system and the individuals within it, to adapt and cope with varying degrees of complexity and ambiguity. However, at present there is a need to recognise the relative limitations of structured electronic data and guidelines, for dealing with complex ambiguous clinical scenarios, and to recognise the current reality of how clinicians cope with that complex ambiguity (Figure 5).

Figure 5. Therapeutic knowledge alliance versus structured data and guidelines

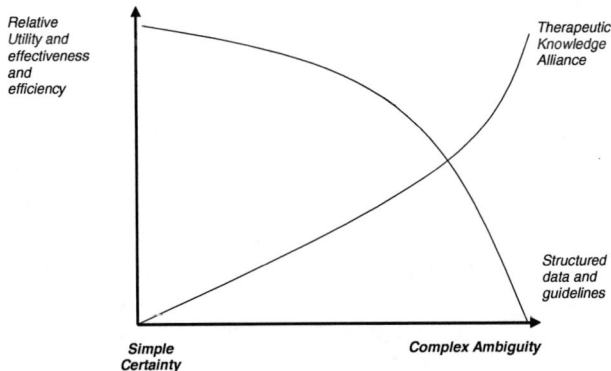

Figure 6. Utility of an electronic health knowledge management system in clinical scenarios of increasing complexity and ambiguity

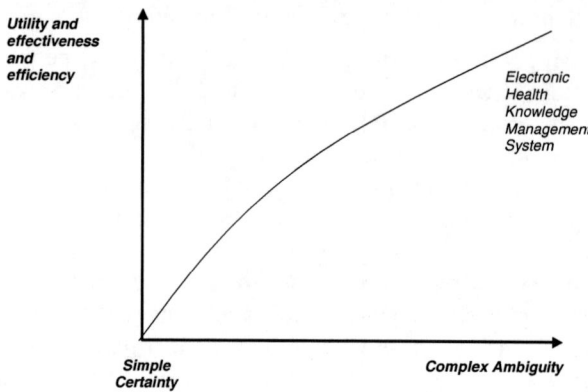

The above section and Figure 5 refer to the increasing relative utility of the wider therapeutic knowledge alliance versus structured data and guidelines when dealing with increasing levels of complex ambiguity. However, this of course does not infer no utility. An electronic health knowledge management system, particularly one focused on care coordination management, could be of increasing utility, effectiveness, and efficiency in helping manage complex ambiguous health scenarios (Figure 6).

Channel Management and the Therapeutic Knowledge Alliance

Many Information System developments may just add additional channels to manage, diluting limited resource and consuming more time and may not be of immediate perceived benefit. The clinician may have to utilise multiple poorly integrated channels, with the concern that increasing levels of non-productive channel management activity may lead to decreased performance across the range of clinical, administrative, research, and educational activities. Privacy and security interventions should aim to optimise performance and minimise the level of associated channel management activities that may lead to decreased performance (Figure 7).

Senior clinicians' failure to use electronic information systems to the same degree as their junior colleagues is often attributed to their reluctance or fear of change or technology, when the reality may be they have established channel management methods that are more productive and efficient in terms of their limited time. The clinician, rather than seeking or entering data on a computer, may perceive the most efficient use of their time may be to seek information from a trusted colleague who can integrate, and quickly add intelligence or make sense of the data, or may communicate with a trusted individual verbally to carry out some recommended course of action. A telephone or face-to-face interaction may be perceived as a more effective way to clarify ambiguity and convey knowledge not easily structured or codified, and come to a shared understanding on the course forward.

Figure 7. Performance versus channel management activity

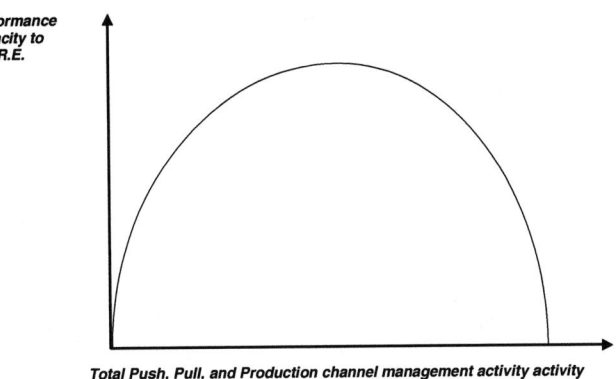

Elevator, Corridor and Coffee Table Conversations

There have been longstanding concerns with regard to clinicians discussing patient information in public places, where they can be overheard (Vigod, Bell & Bohnen, 2003). A lack of respect or appreciation of patient confidentiality may be one facet of this behaviour. However within the concept of the therapeutic knowledge alliance we also need to consider some of the other drivers. These include the increasing mobility and fragmentation of the workforce and use of mobile communications and the decrease in segregated or dedicated space for clinicians. The clinician may be answering an urgent cell phone or pager call, or these locations may be one of the few contexts in which they perceive they can access a colleague or group of colleagues in a timely fashion to discuss or seek advice or support on a patient's care.

In terms of advising a patient on the risks and benefits of limiting their health information flow, there are limitations as to how much we can identify what will be the high value information, result, channel or knowledge neurone within the caregiver's therapeutic knowledge alliance that may facilitate the patient's healthcare at a future stage. Rather than denying or trying to close down the therapeutic knowledge alliance, we should increasingly recognise the centrality of patients and their supports within it, in terms of building trust and recognising and managing the complexity and ambiguity of their healthcare.

The Shift from Passive Decision Support to Active Autonomous Intervention

Many health systems worldwide are currently implementing electronic health knowledge management systems introducing various levels of electronic passive to active decision support. Oncken (1999) has described the concept of levels of initiative as applied to people. Similarly to the Oncken concept, we can expect to see the development of increasing levels of initiative and autonomous intervention by the computer or machine:

1. Online patient data and online textbooks which user has to actively seek out;
2. Computer generated prompts asking whether you would like to go and look at more information on a particular subject, for example, in the form of a hypertext link;
3. General suggestions automatically provided as information, or guidelines, for example, general drug or evidence based medicine disease management information;
4. Specific optional recommendations automatically provided. These may outline a specific course of action, but can be over ridden by various levels of clinician justification with clinician still effectively making decision;
5. Mandatory recommendations automatically provided. These, for example, may relate to preventing life threatening drug interactions, or allergic reactions. Here the system or machine can be viewed as moving from realm of decision support into realm of decision making, (if only under mandate of an overarching organisational protocol);
6. The machine, only if specifically told or requested, will utilise data to make a decision, which the machine rather than the clinician then acts on;
7. The machine, automatically utilises data to make a decision, which it then acts on, but advises clinician on all actions immediately;
8. The machine, automatically utilises data to make a decision, which it then acts on, but clinician only advised immediately if exceeds certain parameters;
9. The machine automatically utilises data to make a decision which it then acts on, and reports to clinician at certain agreed periods; and
10. The machine automatically utilises data to make a decision, which it then acts on, with no active feedback to clinician.

Moving along this spectrum, there is a progressive increase in machine control and decrease in immediate human control (Figure 8). The clinical user is given increasing levels of active advice and overview by the machine, in terms of patient data and specificity of medical evidence and treatment recommendations, while the machine gets decreasing levels of active immediate advice and overview of its use of patient data by the individual clinician user. In general relatively few Health Knowledge Systems have reached level 3, but there is the capacity for this to change rapidly with significant privacy and security implications. For example, what are the privacy and security implications of a "booking system" that could draw data from several sources and act on its own to decide on suitability or eligibility, and then instigate certain forms of care?

Pragmatically this increase in the machine's capacity to act will be limited by the difficulties in coding for the complexities of the health environment. However, as the machine becomes less dependent on immediate human overview, there will be an

Figure 8. Machine versus human control

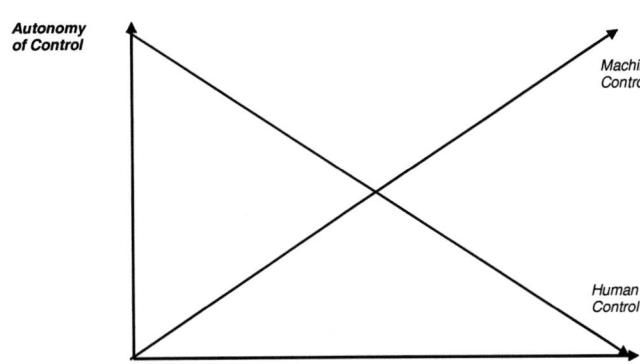

increasing need to consider the risks and benefits in respect of privacy and security. The lack of overview, and integrity of the data and processes by which decisions are made, will be a significant issue. However possibly of greater moment, will be how we perceive, value, or balance the impartiality versus heartlessness of a machine accessing our health information and making decisions based on protocols.

Internationally, there is a growing awareness of the need for information system developments to undergo a Privacy Impact Assessment, underpinned by a state sanctioned privacy code (Slane, 2002). For example the New Zealand Privacy Impact assessment recommendations and underlying code sets out a number of principles in relation to unique identifiers, the purpose, source, and manner of collection, storage, security, access, and correction, and limits on storage, use, and disclosure of personal information.

Internationally similar principles or rules are common to a myriad of legal, professional, and research codes and acts that relate to the collection, storage, and utilisation of health information.

Assessing the impact, or future consequences of current actions, presents a particular challenge for health knowledge systems. The risk-benefit equation, used to assess how certain information is handled may change over time and we have to question how far forward we can see, or how far forward we are expected to see.

Privacy and Security Objectives

Privacy and security developments may be focused on addressing a number of key objectives or concerns including medico-legal or patient trust or confidence concerns, within the context of a belief that better integrated information will lead to better integrated care for our communities. However what is perceived best by or for an individual may be in conflict with what is perceived as best for the community.

Each clinician and health service, depending on their location may be conceptually subject to a multitude of privacy and security codes and laws, for which in reality at the clinical coal face there is only limited compliance due to various combinations of lack of knowledge, attention, priority, will, ability, or perception of unacceptable costs and burden. There may be broad agreement that patient privacy should be protected, but a range of views as to what that pragmatically can or should mean in practice and how much any law or code may achieve this. For example, Marwick (2003) outlines similar views and responses as having greeted the introduction in the USA of the privacy rule of the Health Insurance Portability and Accountability Act.

However health organisations need to increasingly strive to create a culture that respects and protects health information, and seek to demonstrate and reinforce that culture through a number of basic or initial communication, human resource or technical steps. These include creating with, and communicating to, their communities clear open policies around the nature and purpose of health information flows and utilisation.

These include the risks and benefits of information flowing or not flowing and respective privacy versus sub-optimal care risks .In pragmatic terms this may take the form of conversations, leaflets, posters or web-sites.

Human resource processes may include training and development and professionalisation of all healthcare workers in terms of their attitude to health information and clear disciplinary procedures for malicious use. Anderson (1996) has highlighted the importance of training and procedures for the high-risk area of providing patient information on the telephone. Davis, Domm, Konikoff and Miller (1999) have suggested the need for specific medical education on the ethical and legal aspects of the use of computerised patient records.

Technical processes may include ensuring that an electronic information system has at least an audit trail that allows who has viewed or accessed a particular piece of health data to be monitored, providing some degree of psychological reassurance to patients and psychological deterrence against malicious use.

While having highlighted some of the concerns around restriction of information flows, particularly if the clinician is not advised of the suppression there is of course a place for restricted access for sensitive information. This may include allocating graded access levels to certain categories of information and graded access levels for providers or users, with the user only able to access information for which they have an appropriate level of clearance. The system may also include a "break glass" or override facility for emergencies, which allows access to restricted information, but triggers a formal audit or justification process. Denley & Smith (1999) discuss the use of access controls as proposed by Anderson (1996).

However for all these processes we can predict an inverse relationship between complexity and utility (and subsequent uptake or compliance). When planning privacy or security developments, we should strive to make it easier to do the right thing. This can include making login processes as fast and intuitive as possible, so as to decrease the behavioural drivers for clinicians to leave themselves logged in, or the sharing of

Table 2. SAFE-diffusibility factors

SAFE-Diffusibility factors	
Scalable	Retaining implementability / usability (Fast, Intuitive, Robust, Stable, Trustworthy)
Affordable	Resource/Time/Risk
Flexible	Individual/Local/National needs
Equitable	Perceived Equity/Relative Advantage

personal or generic logins or passwords. With unlimited resource or the passage of time and decreasing costs, this may mean installing the latest proximity login or biometric authentication device that can log a clinician in or out as they move towards or away from a information access point, with instantaneous fingerprint or retinal scan verification. However initial steps may involve configuring systems so they minimise the login time, and developing fast, intuitive, and clearly understood administration systems for the issuing (and terminating) of logins or passwords so that new or locum clinicians can immediately access systems without having to utilise generic logins or "borrowing" other clinicians logins.

Our vision may be to make a healthy difference by facilitating the development of Health Knowledge Systems that help us provide safe and effective integrated care, within a culture that respects and protects both the value and privacy of health information. However recognising the difficulties of implementing an information system within the complex health environment (Heeks, Salazar & Mundy 1999), each step or building block towards attaining that vision, including privacy and security developments, needs to be **SAFE**: **Scalable** (while retaining usability and implementability), **Affordable** (in terms of resource time and risk); **Flexible** (enough to meet individual, local and national needs) and **Equitable** (in that potential stakeholders perceive a relative advantage for them in terms of adopting the change or development) (Table 2).

Future Trends

As we look to the future we can expect to see both increasing perceived benefits and privacy and security concerns with respect to data mining and risk profiling particularly genetic and geographic profiling, and increasing attention to the related actions of insurance, financial, and health organisations and government. We can expect to see greater use of technology in the provision of healthcare and broadening of the therapeutic knowledge alliance, both at the triage stage via call centres and so on to the chronic care management stage with the increasing use of texting, email and web broadcast reminders, as well as web based patient self evaluation and shared or self management.

With an increasingly elderly population, and corresponding ageing and diminishing healthcare provider population, we can also expect to see more healthcare workers working from home. While minimising the importance of travel distance to work and retaining the ability to for example support their own children or elderly parents at home, healthcare teleworkers will have an increasing technology based capacity to provide triage functions, telemonitoring of essential functions, parameters or progress, or telepresence while patients for example take medication or monitor their blood glucose level.

There will be ongoing debate around the issue of anonymity and the correlation and matching of data across databases. While there may be a current public focus on privacy or confidentiality issues, as health services become more dependent on electronic systems, we can expect a greater media, public, and clinician appreciation for the integrity and availability aspects of information system security. This would be true particularly if there were a major system availability failure causing at best major inconvenience or disruption; or if a data integrity error were to lead to a significant adverse event or sub-optimal care.

There is a need for ongoing iterative research into the identification and minimisation of privacy and security risk and the effective implementation of a local and national culture that respects, protects, and values health information. There is also an ongoing need for research into the public's views on health information privacy and security, and the priortisation of limited health resource. In the UK the NHS has published some work in this area (NHS Information Authority, 2002) but the methodology and results reporting have been criticised as unbalanced and misleading (Anderson, 2004).

There has been an underlying argument or assumption throughout this chapter that electronic health knowledge management systems providing better integrated information will lead to better integrated care and outcomes for our communities. However it has to be recognised that this is another area that requires ongoing research and development as there are still significant limitations and challenges to the evidence base of widespread successful implementations of electronic systems that make a cost effective positive impact on patient care and outcomes (Heeks et al, 1999; Littlejohns, Wyatt & Garvican, 2003; Ash, Berg & Coiera, 2004; Ash, Gorman, Seshdri & Hersh, 2004).

Unique identifiers will continue to be a cause for concern, and resistance in some countries, however whether through incremental stealth or necessity in combination with increased implementability, we can expect to see their wider use. Also we can expect the increasing use of unique clinician identifiers, which although facilitating the development of electronic ordering and task management systems may also be perceived negatively as a method to monitor, micromanage and restrain the clinical workforce. We can also envisage the increasing use of both patient and clinician unique identifiers linked to global positioning systems or tracking technologies, which although having potential benefits for both patient and clinician safety and resource management could also be perceived similarly negatively.

While we may initially see increasing integration between General Practitioners or primary care and allied healthcare or hospital services, over time we should expect increasing integration of patients and their supports into the health knowledge management system. These developments will increasingly highlight data availability and integrity, including who should have the right to access or view versus who should have the right to make entries or change data.

We may also see a movement from electronic decision support to decision making to autonomous active intervention, which will have significant implications for privacy and security.

We will also increasingly have to recognise that within our communities there will be a spectrum of different needs and ability to participate in and benefit from these technological developments together with an increasing knowledge gap between those that have and those that do not have the required technology to access the therapeutic knowledge alliance.

Should a health service spend limited resource on creating and maintaining complex email or web based assessment or self management channels that the most deprived or in need are least able to access? These complex systems may clearly be of benefit, but are potentially of greatest benefit to those already most able to advocate for themselves and access the resource. Subsequently we will need to continuously revisit the question of, what are the needs of our community and how can we best met those needs with limited resource? Should the more complex, resource intensive systems be provided on a user pays or targeted basis? Should a health service focus largely on the channels most accessible to most people, or the channels most appropriate for those sub groups with the greatest levels of deprivation and identified need? When considering the Therapeutic Knowledge Alliance, as a whole and not just electronic channels, the wider security issues of "integrity" and "availability" still need to be considered. Getting clinicians out into deprived communities, making "available" an "interpersonal" rather than "electronic" channel in the therapeutic knowledge alliance may be the most effective way to ensure data "integrity" and identify and address unmet needs.

Conclusions

Health information privacy and security is an area of at times, strongly held and diverse perspectives. The focus of this chapter has not been on supporting or refuting a single view, but introducing a range of perspectives (even if they challenge my innate clinician bias) and argue for balance and shared understanding. Privacy risk is not a single simple definable entity, but instead complex and multidimensional with quantitative and qualitative perceptual aspects, dynamically changing over time as various factors along the knowledge chain interact, change and possibly compete. Clinicians and patients and the wider community of healthcare stakeholders may vary markedly in their perceptions of privacy risk, particularly when compared to the sub-optimal care risk that may arise from the non flow of health information.

Knowledge management and data mining techniques, provide an opportunity to utilise large clinical data repositories to identify trends and patterns and "risk". This ability to identify risk can be perceived as a double edged sword. It can be used positively to promote or support further research and lead to changes in clinical processes and effective targeting of efficacious and cost efficient interventions; or could be used negatively to discriminate with potential perceived consequences ranging from increased insurance premiums, to refusal of insurance and health cover, mortgages or employment and the creation of an uninsurable, unemployable "underclass", to fears with regards to persecution from totalitarian regimes.

Sporadic incidents or reports of illegal "hacking", or privacy infringement of individual patient records by healthcare workers out of carelessness, malice or pecuniary gain will remain of concern. However as a community looking to the future, we should be possibly more concerned for potential frequent legal and systematic uses of health information that lead to discrimination by insurers, employers or indeed government agencies.

Before designing "privacy" solutions it is important to be clear on privacy risks and priorities (whether medico-legal concerns, deterring malicious access or use, or the increase in patient trust) and how any solution will contribute to or deter from providing better integrated information for better integrated care. Technical processes, such as firewalls, restricted access and audit trails, can provide important physical and psychological deterrents to malicious use and psychological reassurance to patients. However significant risk is likely to lie beyond the protection of our technological armour and within the behaviour of end users.

Pragmatic "privacy" solutions focused on end user behaviour can include making login processes as fast and intuitive as possible, (decreasing behavioural drivers to leave self-logged in or to share logins), privacy training and clear censure processes for malicious use.

There is a need for shared zones or models of acceptance within our communities, with regard to privacy risk, sub-optimal care risk and information flow. In building a shared zone of acceptance, we need to appreciate the complex ambiguity of many aspects of healthcare, and the non linear nature of the therapeutic knowledge alliance, and the need to integrate patients and their supports within that alliance. Recognising the difficulties of implementing change within the complex health system, we need to adopt an incremental iterative S.A.F.E. (Table 2) stepping stone ,or building block approach ,reflecting on and learning from each intervention, and progressively adapting to our changing context or understanding of that context.

Within an environment of limited resource, we need to recognise the diminishing returns or additional benefit we may receive for each unit of expenditure, and how individually worthy needs may have to be balanced against each other for an optimal outcome for our communities. Electronic developments may certainly assist with data confidentiality, integrity and availability, but education and the "professionalisation" of all health knowledge system users, (clinical, non-clinical staff, patients and supports), is possibly more important.

Clinicians and health services will increasingly need to "respect, value and protect" health information if they are to maintain patient confidence, and the comprehensiveness and integrity and validity of information volunteered by patients for entry into electronic systems. However potentially of greater risk is not clinicians or health services' use of patient data or an electronic health knowledge management system , to carry out their integral clinical, administrative, research and educational roles, but how much that data is respected, valued and protected within wider society. If the fears of discrimination and creation of an uninsurable, unemployable underclass come to pass it will be the result of individual and collective actions of all us. The simple individual action of for example accepting cheaper insurance or health premiums as we are in a highly identifiable low risk group, without at least recognising that our gain is someone's loss; the collective action of governments failing to legislate to protect against discrimination, and to provide a safety net for those most vulnerable.

Assessing or predicting the impact or future consequences of current actions will remain an ongoing challenge for health knowledge management system developments. We may certainly carry out Privacy Impact Assessments and risk benefit analysis, based on our current knowledge and context. However, as a global community, it is likely that the challenges of grasping the double-headed sword of electronic health knowledge management systems have only just begun.

References

Anderson, R. (1996). Clinical system security: Interim guidelines. *British Medical Journal 312,* 109-111.

Ash, J.A., Berg, M. & Coiera E. (2004). Some unintended consequences of information technology in health care: The nature of patient care information system-related errors. *JAMIA. 11:104-112.* First published online as doi: 10.1197/jamia.M1471.

Ash, J.A., Gorman, P.N., Seshadri, V. & Hersh, W.R. (2004). Computerized physician order entry in U.S. hospitals: Results of a 2002 survey. *JAMIA, 11,* 95-99. First published online as doi:10.1197/jamia.M1427.

Berger, R.G. & Kichak, J.P. (2004). Viewpoint paper: Computerized physician order entry: Helpful or harmful? *JAMIA 2004, 11,* 100-103. Pre-Print published November 21, 2003; doi:10.1197/jamia.M1411.

Black, E. (2001). *IBM and the Holocaust: The strategic alliance between Nazi Germany and America's most powerful corporation.* New York: Crown Publishers.

Coiera, E. & Clarke, R. (2003) "e-Consent": The design and implementation of consumer consent mechanisms in an electronic environment. *JAMIA, 11*(2), 129-140. Pre-Print published March 1, 2004; doi:10.1197/jamia.M1480.

Davis, L., Domm, J.A., Konikoff, M.R. & Miller, R.A. (1999). Attitudes of first year medical students toward the confidentiality of computerised patient records. *JAMIA, 6*(1).

Denley, I. & Weston Smith, S. (1999). Privacy in clinical information systems in secondary care. *British Medical Journal, 318,* 1328-1331.

Engel, G.L. (1977). The need for a new medical model: A challenge for biomedicine. *Science, 196* (4286).

Etymology Dictionary (2001). Retrieved March 2004 from *http://www.etymonline.com*

Gerritsen, T. (2001). *The surgeon.* London: Bantam Books.

Goldberger, A.L. (1996). Non-linear dynamics for clinicians: Chaos Ttheory, fractals, and complexity at the bedside. *The Lancet, 347*(9011).

Goldstein, W. (1999). Dynamically based psychotherapy: A contemporary overview. *Psychiatric Times, XVI*(7).

Heeks, R., Mundy, D. & Salazar, A. (1999). Why health care information systems succeed or fail. Information systems for public sector management. *Working Paper Series, paper no. 9.* Institute for Development Policy and Management, Manchester, UK.

Kennedy, H. (2004). Stop taking uncivil liberties with DNA. *New Scientist, 2439.*

Littlejohns, P., Wyatt, J.C. & Garvican, L. (2003). Evaluating computerised health information systems: Hard lessons still to be learnt. *British Medical Journal, 326,* 860-863.

Maslow, A.H. (1943). A theory of human motivation. *Psychological Review, 50,* 370-396.

NHS Confidentiality Consultation (2003). FIPR (Foundation for Information Policy Research). Retrieved March 2004 from *http://www.cl.cam.ac.uk/users/rja14/*

NHS Information Authority (2002) Draft: National patient information sharing charter. Retrieved March 2004 from *http://www.nhsia.nhs.uk/confidentiality*

NHS Information Authority (2002). Share with care! People's views on consent and confidentiality of patient information.

Oncken III, W. (1999). Having initiative, executive excellence. *Provo, 16*(9), 19.

Orr, M. (2000). *Implementation of health information systems: Background literature review.* MBA research dissertation, Southern Cross University.

Plesk, P.E. & Greenhalgh, T. (2001). The challenge of complexity in health care. *British Medical Journal, 323,* 625-628.

Roger Clarke's Web site (n.d.). Online *http://www.anu.edu.au/people/Roger.Clarke/*

Ross Anderson's Web site (2004). Online *http://www.cl.cam.ac.uk/users/rja14/*

Slane, B.H. (2002). In Forward to *Privacy impact assessment handbook.* Wellington, New Zealand: Office of the Privacy Commissioner.

Smith, R. (1996). What clinical information do doctors need? *British Medical Journal, 313,* 1062-1068.

Standards Australia (2001). Knowledge management: A framework for succeeding in the knowledge era. HB275.

Standards Australia/Standards New Zealand (2001). Information technology – Code of practice for information security management, AS/NZS ISO/IEC 17799:2001.

Sveiby, K-E. (2001). Knowledge management – Lessons from the pioneers. Online *http://www.sveiby.com/KM-lessons.doc*

Tang, P.C. (2000). An AMIA perspective on proposed regulation of privacy of health information. *JAMIA, 7*(2).

Wyatt, J. (2001). Top tips on knowledge management. *Clinical Governance Bulletin, 2*(3).

Section II

Organisational, Cultural and Regulatory Aspects of Clinical Knowledge Management

Chapter VI

Knowledge Cycles and Sharing:
Considerations for Healthcare Management

Maurice Yolles, John Moores University, UK

Abstract

Healthcare organizations have the same problem as any other organization that is run by sentient but mentally isolated beings. It is a problem that comes out of constructivist thinking and relates to the ability of people, once they start to communicate, to share knowledge. The popular knowledge management paradigm argues the importance of knowledge to management processes and organizational health. It may be said that it is likely that this paradigm will in due course give way to the "intelligent organization" paradigm that addresses how knowledge can be used intelligently for the viability of the organization. Part of the knowledge management paradigm centers on the use of knowledge sharing. This takes the view that while knowledge is necessary for people to do their jobs competently, there is also a need to have the potential for easy access to the knowledge of others. This chapter centers on the capacity of organizations to know what knowledge they have and to coordinate this knowledge.

Introduction

The incapacity of healthcare organizations to coordinate such knowledge is typified by the old joke[1] about a hospital asking its consultant doctors to provide some guidance in coming to a decision about the construction of a new wing at the hospital. The allergists voted to scratch it; the dermatologists preferred no rash moves; the gastroenterologists had a gut feeling about it; the neurologists thought the administration had a lot of nerve; the obstetricians stated they were laboring under a misconception; the ophthalmologists considered the idea short-sighted; the orthopedists issued a joint resolution; the pathologists yelled, "over my dead body"; the pediatricians said, "grow up"; the proctologists said, "we are in arrears"; the psychiatrists thought it was madness; the surgeons decided to wash their hands of the whole thing; the radiologists could see right through it; the internists thought it was a hard pill to swallow; the plastic surgeons said, "this puts a whole new face on the matter"; the podiatrists thought it was a big step forward; the urologists felt the scheme wouldn't hold water; and the cardiologists didn't have the heart to say no. The message that this joke gives is that people working together in an organization see things from their own perspectives, which are formed by the knowledge that they have. The minimum requirement for an organization to work as a single system is for perspectives to be coordinated, and this can only occur through knowledge sharing: one can only coordinate perspective when one knows what perspectives there are to coordinate.

Positivists normally see knowledge as a commodity that has value to individuals within a social context. It can be identified, coded, transferred through communications, decoded, and then used. The message that is provided in this chapter is that this commodity model is not only inadequate, but is actually dangerous for organizations because it allows them to assume that no work has to be put into the process of knowledge sharing. The constructivist view of sharing knowledge centers on the notion that knowledge is fundamentally a property of individuals and shaped by their experiences and worldviews. As such, it cannot be transferred, and all communications carrying knowledge are seen as catalysts that simply initiate the creation of new local knowledge. Effective knowledge migration occurs when there is a strong relationship between the local semantic patterns of a message source and sink (referred to as semantic entanglement), and this requires human interaction and knowledge validation processes.

Healthcare and Information and Knowledge

Healthcare provision is a knowledge-intensive activity and the consequences of an organization failing to make best use of the knowledge assets at its disposal can be severe (Lelic, 2002). Knowledge and knowledge processes (including sharing) in healthcare have both an individual and an organizational dimension. These dimensions are defined as:

1. Individual, involving:
    a. Patient attributes for whose benefit healthcare establishments are established, where knowledge and information can assist patients to appreciate their condition and help them to maintain their treatments, and
    b. Staff members of a healthcare organization that can only properly satisfy an employment role if they have relevant knowledge.
2. Organizational, where healthcare is benefited from knowledge and knowledge processes by enabling them to understand their own organizational capacity to maintain and improve quality patient services and to respond to the need to coherently create new knowledge by becoming a learning organization.

Part of the knowledge process in the UK National Health Service (NHS) centers on a need to involve patients more in their own healthcare; and there are sound financial and medical arguments for this that satisfy the needs of both consultant practitioners and management. In traditional positivist culture that still operates in so many healthcare establishments, the patients are viewed as a commodity input to the healthcare system represented as objectivated[2] "cases" rather than subjectivated individuals with their own learning needs. As a consequence, it is not unknown for patients to become invisible as their "cases" are discussed with a third party in their presence. Baldwin, Clarke, Eldabi, & Jones (2002) note that there is a call for healthcare professionals to engage more fully with their patients, and to see them more as some kind of partner in their healthcare rather than as a paternal authority. While Baldwin et al. are primarily interested in information, without knowledge this has no value or significance to a recipient. It is knowledge that provides the capacity for patients to understand their own conditions, recognize what constitutes relevant information, and contribute to the decision making process both in regard to primary and secondary care.

There is also a need in healthcare organizations to ensure that staff are provided not only with the information and knowledge that enables them to effectively perform their tasks, but that they are also included within the organizational processes that enables them to become motivated and participate in organizational improvement. This human resource management approach is normal to techniques of Organization Development (Yolles, 1999).

In healthcare organizations the nature of the knowledge processes that are undertaken can be expressed in terms of organizational quality. Stahr (2001), in a study on quality in UK healthcare establishments, uses the definition by Joss, Kogan, & Henkel (1994) to identify three levels of quality: *technical*, *generic* and *systemic*. While the word technical is often used to mean "control and predication", for Joss et al. it is taken to mean the employment of specialist knowledge and expertise to solve a problem. The word generic is expressed in terms of normative organizational healthcare standards. The word systemic is concerned with making sure that the whole organization works as an integrated whole in order to ensure long term success. For Stahr (2001), if quality approaches are to be useful they need to affect the culture of an organization, and to do this they need to be systemic.

The systems approach to quality is more than just "joined up governance", intended to convey the impression of organizational cohesion through policy and processes of coherent group behavior. Rather, it is characterized by full integration of all aspects of its activities into focused action on continuous improvement and patient needs (though Stahr does not consider whether these needs should be considered from an objectivistic or subjectivistic perspective). Systemic approaches are more likely to be successful, it is reasoned, than generic and technical approaches, because they impact on everything that managers and clinicians do. Stahr also suggests that systemic approaches become the culture of the organization. However, they should instead be seen to be distinct but intimately connected with that culture (Yolles & Guo, 2003).

While knowledge is important to healthcare organizations, there is also a current tendency to explore it in terms of knowledge management (KM). *Wickramasinghe (2003, p. 295) offers what seems to amount to an* information systems (IS) conceptualization of the nature of KM:

> *"Knowledge management deals with the process of creating value from an organization's intangible assets (Wigg, 1993). It is an amalgamation of concepts borrowed from the artificial intelligence/knowledge-based systems, software engineering, business process re-engineering (BPR), human resources management, and organizational behavior (Liebowitz, 1999). In essence then, knowledge management not only involves the production of information, but also the capture of data at the source, the transmission and analysis of this data, as well as the communication of information based on, or derived from, the data, to those who can act on it (Davenport & Prusak, 1998)."*

This provides little access to a proper understanding of the nature of KM, nor in particular, or the distinction between knowledge processes and data/information processing. The conceptualization of knowledge in the IS view limits ones understanding of knowledge processes, and dilutes the understanding that KM is about knowledge and human understanding rather than about information technology (IT).

It is clear from *Wickramasinghe (2003)* that the subjective dimension of KM is recognized, but the IS approach tends to diminish the barriers to appreciating KM and its implementation. What is the objective nature of knowledge referred to here, and how does it differ from subjective knowledge? It is enough to distinguish between tacit and explicit knowledge? How does the subjective component of knowledge differ from tacit knowledge? More questions, it seems, are raised than responded to. Unsurprisingly, by exploring a number of case studies *Wickramasinghe (2003)* found that knowledge based systems did not support the subjective aspect of knowledge, and by not doing so, their function is reduced to that of an explicated organizational memory.

We have said that KM links with organizational learning. This is an interest of Grieves and Mathews (1997) for healthcare establishments. For them:

> "Learning is produced by organizational members themselves by actively creating the conditions for individual and organizational development. If organizational learning is viewed this way then it is both tacit/ informal and explicit/ formal. Learning requires the assimilation of knowledge and skills by individuals and groups who take responsibility for their own activity and come to 'own' what they learn. This type of learning is non-directive since its purpose is not to transmit information through a trainer.
>
> All human learning requires the ability to name, classify, construct and communicate cognitive imagery conveying both spatial and temporal characteristics (Bateson, 1979). Mental maps can be considered as tacit, in the head methods, for making sense of and for performing tasks. Such methods are acquired by individuals either through contextually-tied trial-and-error techniques, or through imaginative thinking that is essentially abstract and not tied to an immediate context. By contrast, it is possible to argue that organizations create mental maps through methods that formally articulate rules and procedures to guide the activities of their members."

The term "organizational learning" is attributed to Argyris and Schon (1978) intended to cover the notion that organizations, like organisms, adapt to a changing environment. Learning in organizations can be said to require the development of both systems and processes in order that changes in the external (and internal) environments filter through to attitudes, procedures and practices in a way that facilitates constant review of operating norms at a variety of levels throughout the organization. Interestingly, this implies a relationship to Stahr's (2001) adopted notions of systemic and generic quality. Since the concept of learning also relates directly to the acquisition of knowledge, it entails a fundamental link to quality.

Having discussed some notions of information, quality and the learning organization in terms of knowledge and knowledge management, we are now in a position to consider some knowledge conceptualizations that are appropriate to both individual and organizations dimensions of healthcare.

The Theoretical Approach

The development of a cohesive approach to the coordination of perspectives through knowledge sharing in healthcare organization requires a clear frame of reference to be provided. This section provides such a frame of reference.

Organizations and Complexity

Organizations are seen by many to exist in social environments that entertain rapid changes, an interdependence between different organizations, and complexity. When we talk of complexity, we really mean a situation composed of structures (and their associated processes) that have considerable variety in their microscopic distinction (i.e., microstructures). When a variation in microstructure is perceived, then it may be referred to as an event in time or space. Classes of space are *physical* or *conceptual*, and related to the *social*, *emotional* or *cultural* context. Each of these can involve a relatively large number of elements and associated relationships and processes. For example, a situation may involve socially differentiable organizations. These have conceptual forms that are represented physically through legal paperwork and the presence of individual workers. If there is seen to be considerable variety in the way these organizations are differentiated, then the situation is complex; otherwise, it is simple. The paradigm supported in this chapter is that of viable systems theory. It is sufficiently rich to enable organizations to be modeled in such a way as to explore how they might "successfully" operate in complex dynamic environments. The notion of success may be seen as related to an organization being viable in relation to what it does, and today there is a view that this can be facilitated through the creation and use of knowledge.

Viable Systems

A viable system is an active, purposeful, and adaptive organization that can operate in complex situations and survive. Since complex situations entail variety differentiation, in surviving a viable system responds to changing situations by generating sufficient variety through self-organization to deal with the situational variety it encounters (called requisite variety). It is often said in the cybernetic literature that variety is a measure of complexity.

In developing on from and relating the work of Checkland and Scholes (1990) and Kuhn (1970), two types of worldviews may be defined, informal (weltanschauung), and formal (paradigm). By formal we are referring to the expression of ideas through language. A formalization enables a set of explicit statements (propositions and their corollaries) to be made about the beliefs and other attributes that enable (more or less) everything that must be expressed, to be expressed in a self-consistent way. Informal worldviews are more or less composed of a set of undeclared assumptions and propositions, while formal ones are more or less declared. Both are by their very nature bounded, and thus constrain the way in which perceived situations can be described. Now paradigms can change (Kuhn, 1970; Yolles, 1999), so that the nature of the constraint is subject to a degree of change - however bounded it might be. Consequently, the generation of knowledge is also constrained by the capacities and belief systems of the worldviews.

The idea of a worldview (Yolles, 1999) is that it:

a. is culture centered,
b. has cognitive organization (beliefs, values, and attitudes) are its attributes,
c. has normative and cognitive control of behavior or action that can be differentiated from each other, and
d. it has a cognitive space of concepts, knowledge and meaning that is strongly linked to culture.

Figure 1. Relationship between the phenomenal, virtual and existential domains in a viable system

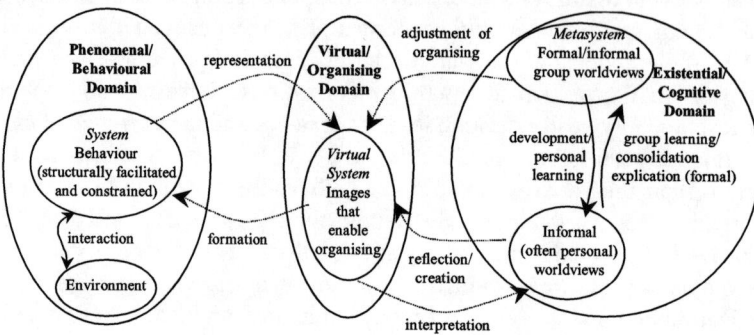

Worldviews interact, and following the cybernetic tradition, this interaction can be placed in a cognitive domain that drives the purposeful adaptive activity system. The system has form, thus has structure, process and associated behavior. It is assigned to an energetic behavioral domain. The knowledge related cognitive domain is the "cognitive consciousness" of the system that it drives. According to Yolles (1999), the two domains are connected across a gap that we refer to as the transformational or organizing domain, and that may be subject to surprises. It is strategic in nature, and operates through information (Figure 1). The three existential/cognitive, virtual/organizing, and existential/behavioral domains are analytically and empirically independent. This model can be applied to any purposeful adaptive activity system by distinguishing between cognitive, strategic, and behavioral aspects of a situation.

This defines the basis of viable systems (as defined by Yolles) that, through transformational self-organizing processes, are able to support adaptability and change while maintaining stability in their behavior. In a plastic organization the nature of that behavior may change, and in so doing a viable system will maintain behavioral stability.

The Cognitive Domain Commodities

Relationships exist between the cognitive domain commodities of Table 1. These are often poorly defined however (Roszak, 1986). For instance, systems analysts frequently say that information is data, and information theorists that knowledge is information. Our definition is as follows:

- Data are a set or string of symbols that can be associated with structures and behaviors. They are meaningful only when related to a given context. They can be stored, and if storage is to be meaningful or coherent, then within that context storage will occur according to a set of criteria that are worldview derived. Data are able to reflect variety differentiation in complexity. Stored data are also retrievable

Table 1. Relationship between human cognitive interests, purpose, and influences

	Sociality Properties		
Cognitive properties	**Kinematics** (through energetic motion)	**Orientation** (determining trajectory)	**Possibilities/potential** (through variety development)
Cognitive interests	Technical	Practical	Critical Deconstraining
Phenomenal (conscious) domain — Activities, Energy	Work. This enables people to achieve goals and generate material well-being. It involves technical ability to undertake action in the environment, and the ability to make prediction and establish control.	Interaction. This requires that people as individuals and groups in a social system gain and develop the possibilities of an understanding of each others subjective views. It is consistent with a practical interest in mutual understanding that can address disagreements, which can be a threat to the social form of life.	Degree of emancipation. For organizational viability, the realizing of individual potential is most effective when people: (i) liberate themselves from the constraints imposed by power structures (ii) learn through precipitation in social and political processes to control their own destinies.
Cognitive purposes	Cybernetical	Rational/Appreciative	Ideological/Moral
Virtual or organizing (subconscious) domain — Organizing, Information	Intention. This is through the creation and strategic pursuit of goals and aims that may change over time, enables people through control and communications processes to redirect their futures.	Formative organizing. Enables missions, goals, and aims to be defined and approached through planning. It may involve logical, and/or relational abilities to organize thought and action and thus to define sets of possible systematic, systemic and behavior possibilities. It can also involve the (appreciative) use of tacit standards by which experience can be ordered and valued, and may involve reflection.	Manner of thinking. An intellectual framework through which policy makers observe and interpret reality. This has an aesthetical or politically correct ethical orientation. It provides an image of the future that enables action through politically correct strategic policy. It gives a politically correct view of stages of historical development, in respect of interaction with the external environment.
Cognitive influences	Social	Cultural	Political
Existential or cognitive (unconscious) domain — Worldviews, Knowledge	Formation. Enables individuals/groups to be influenced by knowledge that relate to our social environment. This has a consequence for our social structures and processes that define our social forms that are related to our intentions and behaviors.	Belief. Influences occur from knowledge that derives from the cognitive organization (the set of beliefs, attitudes, values) of other worldviews. It ultimately determines how we interact and influences our understanding of formative organizing.	Freedom. Influences occur from knowledge that affect our polity determined, in part, by how we think about the constraints on group and individual freedoms, and in connection with this to organize and behave. It ultimately has impact on our ideology and morality, and our degree of organizational emancipation.

according to the pattern of meaning created for it within the context (e.g., by defining a set of entities that meaningfully relate to each other)

- Information is a sign or set of signs or signals that predisposes an actor (that is, a person, or group of persons acting as a unit or organization) to action. This appears to be consistent with the notion of Luhmann (1995), who considers information to be a set of coded events. It can also be defined as, that which enables a viewer to perceive greater variety differentiation in a complex situation
- All knowledge is worldview local, and belief related. It can be defined as patterns of meaning that can promote a theoretical or practical understanding that enables the recognition of variety in complexity. These patterns are often developed

Figure 2. Connection between data, information and knowledge through the relationship between understanding and context, based on Bellinger (1996)

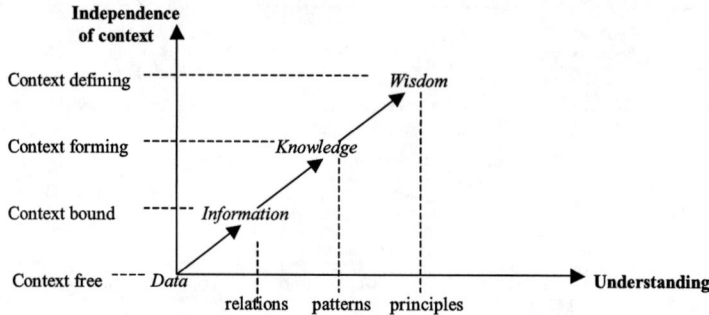

through a coalescing of information. If information is seen as a set of coded events, then consistency with Nonaka and Takeuchi occurs when they say that explicit knowledge is codified.

The second part of the definition for information derives from Information Theory. It supports the idea that if the entropy of a situation is increased, structures become less differentiated. Entropy may usefully be thought of as a *lack of information* (Brillouin, 1967, p. 160). In a well-ordered situation there is a high probability of finding differentiation. If this is expressed in terms of distinct microstructures (that is, microscopically distinct structures), then they are differentiated through the boundaries or frames of reference that distinguish them. If a viewer is to be able to recognize that these boundaries or frames of reference are differentiable, then that viewer must be able to adopt concepts (that is, characteristics of knowledge) that enable differentiation.

A view about the relationship between knowledge and information is based on Bellinger (1996), that provides an interconnection between data, information, data and wisdom derived through the relationship between contexts and understanding (Figure 2).

According to this construction, data, as an unattached symbol or mark, are context free, and with no reference to time any point in space and time. As such it is without a meaningful relation to anything else. Meaning is attributed to data by associating it with other things, that is, defining a context. Bellinger refers to wisdom as an understanding of the foundational principles responsible for the patterns representing knowledge being what they are, and it creates its own context totally. These foundational principles are completely context independent, and have been referred to here as context determining because the context is bound into the wisdom.

A traditional view in finding information is to seek data, and this leads us to seek an appreciation of the relationship between data and information, and indeed between information and knowledge. Such a relationship is proposed in Figure 3. Here, data can be processed into information (called data information) through the application of patterns of meaning that relate to organizational purpose. Data processing is also constrained by criteria of what constitutes a processing need. Information also exists phenomenally, through the very microstructural variety differentiation that exists in a structured situation. The model given in Figure 3 leads to questions about our understanding of knowledge creation, and has consequences for the way in which we see knowledge development in organizations. For instance, how and through what means are the patterns of meaning formed that enable data to be processed, and information to be coalesced. Further exploration of knowledge processes within organizations can be developed within the context of knowledge management.

There is a perhaps a better way than that of Figure 3 to describe the relationship between data, information and knowledge, that comes from an ontological model of viable systems that originates from Schwarz (1997).

While data is not information, data classifications or classes can be described as entities that, when woven into a relational pattern can become information when conditioned by knowledge within an action setting. Just as data is not information, information is not knowledge. However information can contribute to the creation of knowledge. This occurs when information is analyzed, interconnected to other information within a thematic context, and compared to what is already known.

A relationship between data, information, and knowledge cannot be considered independently of an agent that is involved in creating that relationship. Our interest lies in the generic relationship, rather than local detailed relationships between commodity elements that will be different for each agent. The generic relationship is defined in Figure 4. It presupposes that the agent has a purpose for inquiry and is involved in the process of either quantitative or qualitative measurement. Qualitative measurement involves conceptual assessment brought together with some form of mapping agent that is capable of generating a possibly complex scale of values that can be assessed as though they are quantitative measurements.

Figure 3. Relationship between data, information and knowledge

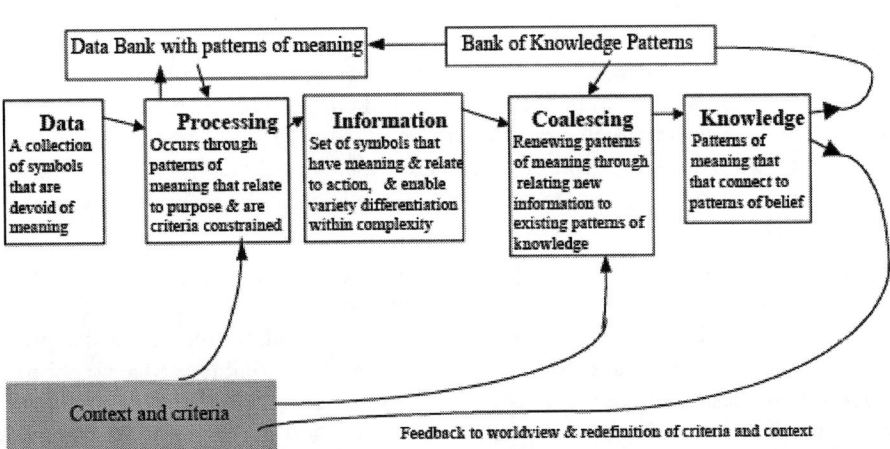

Figure 4. The ontological relationship between data, information, and knowledge

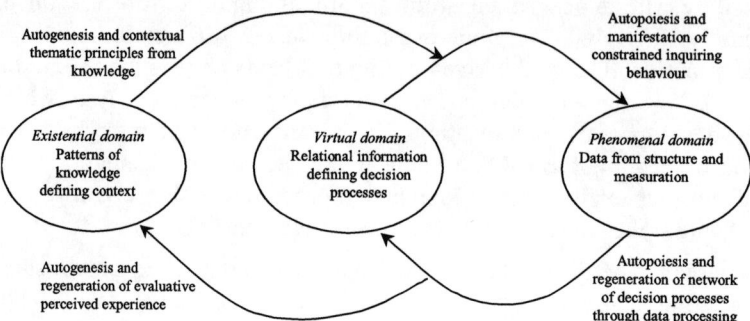

Knowledge Management

The management of knowledge is becoming an important area of interest. However, the question of what constitutes knowledge management may be posed in different ways (Allee, 1997). A traditional meaning approach discusses questions of ownership, control, and value, with an emphasis on planning. Another view is that knowledge is organic, and has a flow, a self-organizing process, and patterns. This latter approach explores how knowledge emerges, and how the patterns change. Also, in an organizational context, what happens to an organization's knowledge if a knowledgeable and wise employee leaves, and how can an organization manage to capture that. Knowledge is increasingly recognized as an important organizational asset (Iles, 1999). Its creation, dissemination and application are often now seen as a critical source of competitive advantage (Allee, 1997; Lester, 1996).

Following a tradition supported by Ravetz (1971), Nonaka and Takeuchi (1995, p. 8) distinguish between two types of knowledge: explicit and tacit (Table 2). Tacit knowledge includes cognitive and technical elements. Cognitive elements operate through mental models that are working worldviews that develop through the creation and manipulation of mental analogies. Mental models like schemata, paradigms, perspectives, beliefs and viewpoints, according to Nonaka and Takeuchi, help individuals perceive and define their world. The technical element of tacit knowledge includes concrete know-how, crafts, and skills. However, explicit knowledge is about past events or objects "there and then," and is seen to be created sequentially by "digital" activity that is theory progressive.

Nonaka and Takeuchi (Ibid.) develop on this to create what has become perhaps the most well known model for knowledge creation, referred to as the SECI model of knowledge creation and illustrated in Figure 5. It derives from their model of the conversion process between tacit and explicit knowledge, and results in a cycle of knowledge creation. The conversion process involves four processes: socialization, externalization, combination, and internalization, all of which convert between tacit and/or explicit knowledge.

Table 2. Typology of knowledge

Expression of knowledge type	Explicit Knowledge	Tacit Knowledge
Nonaka and Takeuchi	Objective Rationality (mind) Sequential (there and then) Drawn from theory (digital) Codified, formally transmittable in systematic language. Relates to past	Subjective Experiential (body) Simultaneous (here and now) Practice related (analogue) Personal, context specific, hard to formalize and communicate. Cognitive (mental models), technical (concrete know-how), vision of the future, mobilization process
Alternative	Formal and transferable, deriving in part from context related information established into definable patterns. The context is therefore part of the patterns.	Informal, determined through contextual experience. It will be unique to the viewer having the experience. Not transferable except through recreating the experiences that engendered the knowledge for others, and then the knowledge gained will be different.

Figure 5. The SECI cycle of knowledge creation (Nonaka & Takeuchi, 1995)

	Tacit Knowledge	**Explicit Knowledge**
Tacit Knowledge	*Socialisation* Existential; face-to-face Creates sympathised knowledge through sharing experiences, and development of mental models and technical skills. Language unnecessary	*Externalisation* Reflective; peer-to-peer Creates *conceptual* knowledge through knowledge articulation using language. Dialogue and collective reflection needed
Explicit Knowledge	*Internalisation* Collective: on-site Creates *operational* knowledge through learning by doing. Explicit knowledge like manuals or verbal stories helpful	*Combination* Systemic; collaborative Creates systemic knowledge through the systemising of ideas. May involve many media, and can lead to new knowledge through adding, combining & categorising

Socialization is the processes by which synthesized knowledge is created through the sharing of experiences that people have as they develop shared mental models and technical skills. Since it is fundamentally experiential, it connects people through their tacit knowledge. Externalization comes next and occurs as tacit knowledge is made explicit. Here, the creation of conceptual knowledge occurs through knowledge articulation, in a communication process that uses language in dialogue and with collective reflection.

The uses of expressions of communication are often inadequate, inconsistent, and insufficient. They leave gaps between images and expression while promoting reflection and interaction. It therefore triggers dialogue. The next process is combination, when explicit knowledge is transformed through its integration by adding, combining and categorizing knowledge. This integration of knowledge is also seen as a systemizing process. Finally, in the next process explicit knowledge is made tacit, by its internalization. This is a learning process, which occurs through the behavioral development of *operational* knowledge. It uses explicit knowledge like manuals or verbal stories where appropriate.

The different types of knowledge process are seen as phases in a knowledge creation cycle. This is driven by *intention*, seen as an aspiration to a set of goals. *Autonomy* is another requirement that enables the knowledge cycle to be driven. This increases the possibility of motivation to create new knowledge. There are three other conditioning factors. One is the need of *fluctuation and creative chaos*. This can generate signals of ambiguity and redundancy that inhibits the "improvement" of knowledge. The sharing of *redundant* information promotes a sharing of tacit knowledge when individuals sense what others are trying to articulate. Finally, *requisite variety* is needed if an actor is to deal with complexity. Thus, five factors condition the knowledge cycle enabling it to maintain developmental motion.

A Viable Approach to Knowledge Creation

The structured spiral of knowledge creation offered by Nonaka and Takeuchi adopts a positivist perspective. An alternative critical approach is possible that links closely with the viable system model of Figure 1. In addressing this, we note that each of the three domains have associated with them its own knowledge process, one connected with cognition, one with organizing, and one with behavior. This notion is consistent with Marshall (1995), whose interest lies in knowledge schema. Schema have four categories: (1) they are the mental organization of individual's knowledge and experience that allows him/her to recognize experiences that are similar; (2) they access a generic framework that contains the essential elements of all these similar experiences; (3) they use of this framework to plan solutions; and (4) they have the ability to utilize skills and procedures to execute the solution. For this purpose, Marshall identifies three types of knowledge:

- Identification knowledge—the facts and concepts making up the knowledge domain
- Elaboration knowledge—the relationships between the individual knowledge components and the way they are organized
- Execution knowledge—the conceptual skills and procedures required to execute an activity

Marshall does not attempt to address knowledge creation, though we shall do so through our own model. We consider that in social group situations, knowledge creation occurs through a process of knowledge migration from one worldview to another. It is an identification knowledge process. The basic knowledge management model is as given in Figure 6. It links to Table 1, and depicts the three fundamental phases of the knowledge process: migration, accommodation, and action. Migration is associated with the cognitive domain, accommodation with the organizing domain, and action with the behav-

Figure 6. The knowledge cycle

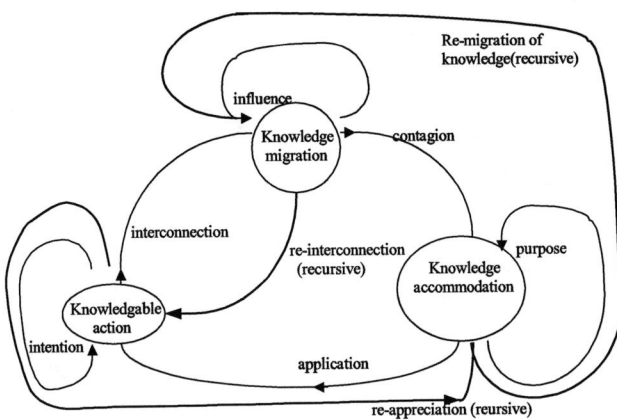

ioral domain. The way that migration occurs is conditioned by cognitive influence, accommodation though cognitive purpose, and action through cognitive intention. Each phase process has an input and an output. A feedback control process is able to condition each phase process directly, or through its input. The way that each phase process is conditioned by the feedback control is represented symbolically in Figure 6 by a loop around the process bubble, and we shall return to this in a moment.

The structured spiral of knowledge creation offered by Nonaka and Takeuchi adopts a positivist perspective. An alternative critical approach is possible that links closely with the 3-domain model of Figure 7. In addressing this, we note that each of the three domains

Figure 7. Ontological expression for knowledge migration cycle

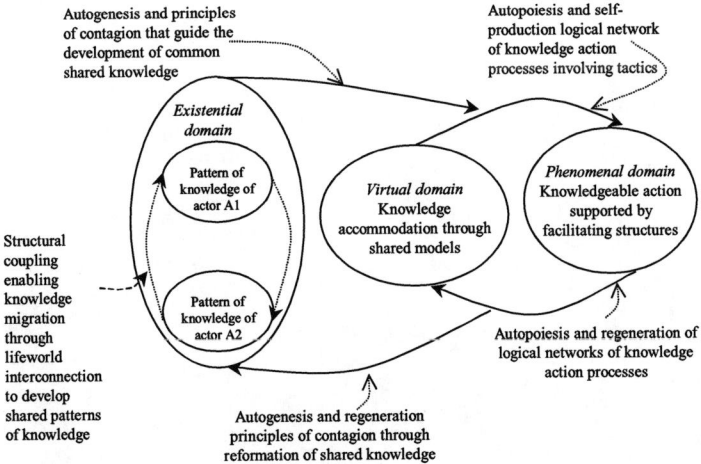

have associated with them its own knowledge process, one connected with cognition, one with organizing, and one with behavior. This notion is consistent with that of Marshall (1995) in connection with knowledge schema.

Marshall applies her ideas on knowledge schema to decision-making by people rather than by social groups. However, a link can be made between them by applying the typology of knowledge to the viable knowledge cycle of Figure 7. We consider that in social group situations, knowledge creation occurs through a process of knowledge migration from one worldview to another, and it is an identification knowledge process. The basic knowledge management model depicts the three fundamental phases of the knowledge process: migration, accommodation, and action. Migration is associated with the existential domain, accommodation with the virtual domain, and action with the behavioral domain. The way that migration occurs is conditioned by cognitive influence, accommodation though cognitive purpose, and action through cognitive interest. Each phase process has an input and an output. A feedback control process is able to condition each phase process directly, or through its input. The way that each phase process is conditioned by the feedback control is represented symbolically by a simple loop around the process bubble.

Returning now to the control process we referred to earlier, we show the explicit meaning of each return loop in Figure 8.

Control processes not only condition phase processes. They can also be responsible for re-scheduling them in the overall knowledge cycle. Within perspectives of traditional positivism, it is normal to consider controls in simple terms; but they may also be susceptible to complexity and chaos as illustrated in Table 3 (Yolles, 1999). This has implication for the development of a chaotic activation of a phase that is not sequentially ordered, and that can occur when complexity is affective.

Control processes, while often considered in terms of positivist or postpositivist paradigms, may also be seen from a critical theory perspective. To do this we invoke the propositions that:

a. knowledge that enables the nature of a control process to be understood is local and worldview dependent;

Figure 8. Basic form of the control model

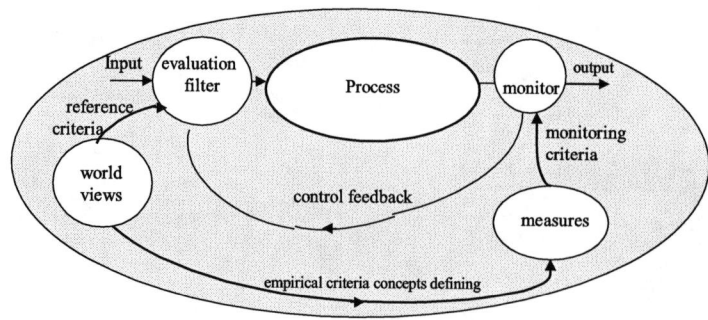

Table 3. Distinguishing between simple and complex feedback control loops

Simple Terms of Control	Complex Terms of Control
Likely to be linear and have a steady state behavior with clear relationships between the inputs and outputs of a process. Indications of instability will probably be predeterminable and boundable. The actuator that can take action to regulate the process will be deterministic or involve rational expectation.	Likely to be non-linear and far from a steady state behavior. Instability may appear without prior indication. The relationships between inputs and outputs will in general not be strictly causal, but unclear. The effects of the actuator will be uncertain. It is not always the case that standards, norms, and objectives that drive a control process will be well defined. It is not uncommon for them to be fuzzy whether or not it is *believed* that they are well defined, and it may be that such a belief can only be validated retrospectively. Even if they *are* well defined, it may be that their definition entails some level of unrealized flexibility. Measures of performance may be inadequate to indicate the nature of the output.

b. empirical and reference control criteria are worldview dependent, value laden, and will be susceptible to ideology and ethics; and

c. conditioning control processes are implemented in a local inquirer-relative way.

These propositions have implications for the way in which the social group, subject to the phase process conditioning: (i) responds to the control situations, and (ii) appreciates the need of semantic communications that make it broadly meaningful.

The conditioning processes of the knowledge cycle are illustrated in Figures 9-11. In Figure 6, we consider the control process involved with knowledge migration. This occurs through the development of (lifeworld) interconnections between the worldviews of the actors in a given suprasystem, and is the result of semantic communication. As part of the process of knowledge migration, new knowledge is locally generated for an actor. While this may be seen as part of a socialization process, it may also be seen as an actor local spontaneous thing when the process of knowledge migration operates as knowledge creation trigger.

The process of knowledge accommodation (Figure 10) can follow knowledge migration. An appreciation of how migrated knowledge can be of use to a relevant other is essential if they are to be able to harness it within a behavioral world. Knowledge accommodation by relevant others is dependent upon knowledge contagion to these others. However,

Figure 9. Control model for knowledge migration

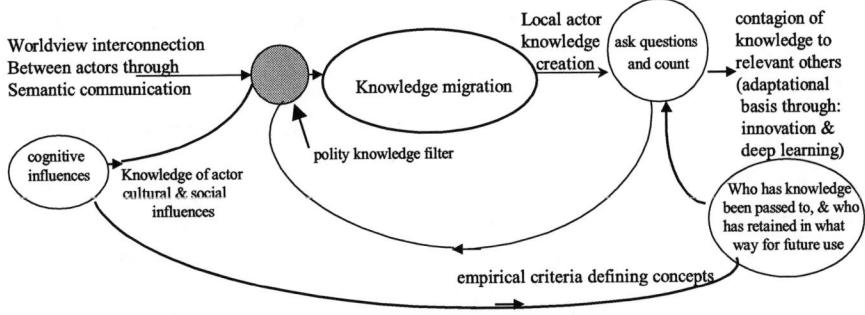

Figure 10. Control for knowledge accommodation

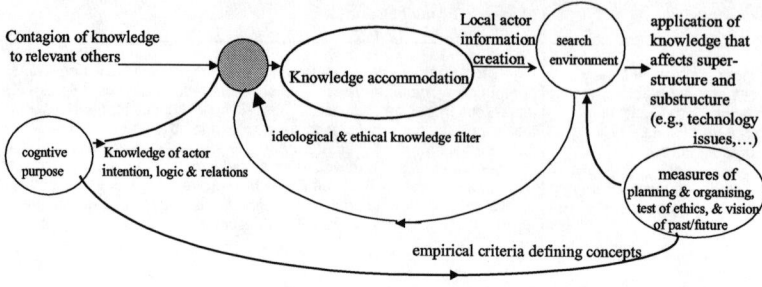

Figure 11. Control model for knowledgeable action

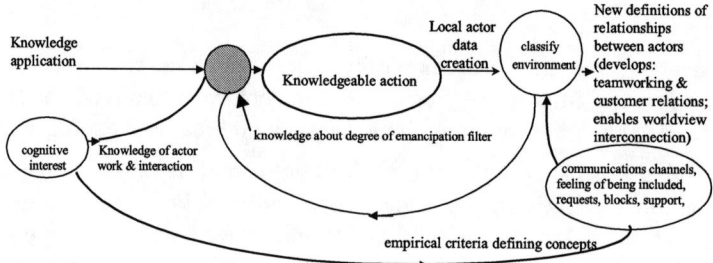

this is filtered through knowledge that activates weltanschauung derived ideology and ethics. In addition, the evaluation reference criteria derive from knowledge about intention and logico-relational cognitive purposes. Interestingly, this connects with the Marshall (1995) idea of planning knowledge—the knowledge of which pathways to select in order to achieve a solution.

A consequence of the process of knowledge accommodation is its intelligent application. We say intelligent, because its obverse, rote application may not require knowledge accommodation or even migration. Knowledge application can occur behaviorally within a superstructure and a substructure. Superstructure identifies the institutionalized political and cultural aspects of a situation, and is also issue related. Substructure is task orientated, and relates to the mode and means of production (e.g., technology) and the social relations (e.g., roles and their relationships) that accompany them.

The process of knowledgeable action (Figure 11) is dependent upon the application of knowledge. Knowledgeable action is action that occurs with awareness of what is being done within a behavioral world. Knowledgeable action in a situation is dependent upon knowledge application to the tasks that are perceived to require to be addressed within the situation. This is filtered through knowledge that activates weltanschauung derived emancipative capabilities, that enable knowledgeable action to occur. The evaluation reference criteria derive from knowledge about actor interests through work and interaction. It relates to the Marshall (1995) idea of execution knowledge, that is seen as the computational skills and procedures required to execute a behavior.

When the above control loops operate to make process changes, morphogenic changes occur in the knowledge phases of our knowledge cycle. When the control processes are complex and control action fails, knowledge process metamorphosis can occur (Yolles, 1999). As an example of a metamorphic change, a new concept may be born during the process of knowledge migration.

Conclusions

This is a good point to return to the hospital wing joke that we introduced this chapter with. We have argued that for healthcare perspectives to be coordinated there is a need to share knowledge, and we have explored the knowledge sharing possibilities for healthcare staff. However, the capability for organizations to share knowledge requires that healthcare organizations need to develop a capacity to recognize and use knowledge for patients and staff as well as organizationally. Organizational knowledge exists by virtue of the individuals associated with it, and there is a need to recognize that knowledge creation and sharing involves processes of knowledge migration, where knowledge transmitted in a communication from one individual to another also may not be the knowledge that is assembled.

There is a difference in the way knowledge creation is structured, whether one adopts a positivist or another epistemology. The ideas of Nonaka and Takeuchi would appear quite influential in the development of a theory of knowledge creation. While they are constructivist in their perception of each phase process, they are overall structurally positivist. It is not uncommon to have this type of usually benign methodological schizophrenia, though it may well be more aesthetic not to. An alternative approach that is fundamentally critical (even though it entertains the notion of control) and that does not suffer from the above problem derives from viable systems theory. This does not see knowledge creation as a set of sequential steps, but rather as a set of phases that are constantly tested and examined through possibly complex feedback. Shifts from one phase to another may occur according to the control phenomena that drive particular perspectives.

There are parallels between our proposed knowledge cycle (Figure 4) and that of Nonaka and Takeuchi (Figure 3). In the former knowledge can be created spontaneously within a migration process, and any socialization process that occurs is through communication that may be seen to act as a trigger for new knowledge. Unlike that of Nonaka and Takeuchi, our cycle is not required to be sequential continuous relative to a conditioning process. Rather, the process of continuity is transferred to the communication process, and knowledge creation is cybernetic, passing through feedback processes that can change the very nature of the patterns of meanings that were initiated through the semantic communications.

References

Allee, V. (1997). *The knowledge evolution.* Oxford: Butterworth-Heinemann.

Argyris, C. & Schon, D. (1978). *Organizational learning.* Reading, MA: Addison-Wesley.

Baldwin, L.P., Clarke, M., Eldabi, T. & Jones, R.W. (2002). Telemedicine and its role in improving communication in healthcare, *Logistics Information Management, 15*(4), 309-319.

Bateson, G. (1979). *Mind and nature.* New York: Dutton.

Bellinger, G. (1996). Knowledge management—Emerging perspectives. Online *http://www.outsights.com/systems/kmgmt/kmgmt.htm*

Brillouin, L. (1967). *Science and information theory.* New York: Academic Press.

Checkland, P.B. & Scholes, J. (1990). *Soft systems methodology in action.* Chichester, UK: John Wiley & Son.

Davenport, T. & Prusak, L. (1998). *Working knowledge.* Boston: Harvard Business School Press.

Dimbleby, R. & Burton, G. (1985), *More than words: An introduction to communication.* London: Routledge.

Foucault, M. (1982). The subject and power. *Critical Inquiry, 8,* 777-795. Also in Dreyfus, H., Rabinow, P. & Foucault, M. (2000). *Beyond structuralism and hermeneutics.* Chicago: University of Chicago Press. Also available in Faubion, F. (Ed.) (2000). *Power.* New York: New Press (translated by Robert Hurley).

Friedan, R. (1999). *Physics from Fisher information: A unification.* Cambridge: Cambridge University Press.

Grieves, J. & Mathews, B.P. (1997). Healthcare and the learning service. *The Learning Organization, 4*(3), 88-98.

Guba, E.G. & Lincoln, Y.S. (1994). Competing paradigms in qualitative research. In Denzin, N.K. & Lincoln, Y.S. (Eds.), *Handbook of qualitative research* (pp. 105-117). Thousand Oaks, CA: Sage.

Holsti, K.J. (1967). *International politics: A framework for analysis.* Prentice Hall.

Iles, P. (1999). Knowledge management, organizational learning and HR strategy: A model and case studies. In J. Kaluza (Ed.), *Strategic management and its support by information systems.* Technical University of Ostrava Faculty of Economics, Czech Republic.

Joss, R., Kogan, M. & Henkel, M. (1994). *Total quality management in the national health service: Final report of an evaluation.* London: Brunel University.

Jung, C.G. (1936). The archetypes and the collective unconscious, collected works, Vol. 9.i. Online *http://www.geocities.com/Athens/Acropolis/3976/Jung2.html*

Kuhn, S.T. (1970). *The structure of scientific revolutions.* Chicago: University of Chicago Press.

Lelic, S. (2002). Your say: KM in the healthcare industry. *Knowledge Management, 6*(4). Online *http://kmmagazine.com*

Lester, T. (1996). Minding your organization's knowledge base. *Human Resources.*

Liebowittz, J. (1999). *Knowledge management handbook.* London: CRC Press.

Lin, N. (1973). *The study of communication.* Indianapolis, NY: The Bobs-Merril Company.

Luhmann, N. (1995). *Social systems.* CA: Stanford University Press.

Marshall, S.P. (1995). *Schemes in problem solving.* Cambridge, UK: Cambridge University Press.

Nonaka, I. & Takeuchi, H. (1995). *The knowledge-creating company: How Japanese companies create the dynamics of innovation.* New York: Oxford University Press.

Ravetz, J.R. (1971). *Scientific knowledge and its social problems.* Oxford University Press. Reprinted by Penguin Books in 1973, Middlesex, UK.

Roszak, T. (1986). *The cult of information: A Neo-Ludite treatise on high-tech, artificial intelligence, and the true art of thinking.* Berkeley, CA: University of California Press.

Sakaiya, T. (1991). *The knowledge-value revolution.* New York: Kodansha.

Schwarz, E. (1997). Towards a holistic cybernetics: From science through epistemology to being. *Cybernetics and Human Knowing, 4*(1), 17-50.

Stahr, H. (2001). Developing a culture of quality within the United Kingdom healthcare system. *International Journal of Health Care Quality Assurance, 14*(4), 174-180.

Toffler, A. (1990). *Powershift: Knowledge, wealth and violence at the edge of the 21st century.* New York: Bantom.

Wickramasinghe, N. (2003). Do we practise what we preach: Are knowledge management systems in practice truly reflective of knowledge management systems in theory? *Business Process Management Journal, 9*(3), 295-316.

Wigg, K. (1993). *Knowledge management foundations.* Arlington, VA: Schema Press.

Yolles, M.I. (1999). *Management systems: A viable approach.* London: Financial Times Pitman.

Yolles, M.I. & Guo, K. (2003). Paradigmatic metamorphosis and organizational development. *Sys. Res., 20,* 177-199.

Endnotes

[1] This joke is taken from *http://www.med-psych.net/humor/joke0011.html*

[2] In the sense of Foucault (1982)

[3] The term joined up governance is reflected in various sources like *http://news.bbc.co.uk/1/hi/special_report/1998/11/98/e-cyclopedia/211553.stm*

Chapter VII

Key Performance Indicators and Information Flow:
The Cornerstones of Effective Knowledge Management for Managed Care

Alexander Berler, National Technical University of Athens, Greece

Sotiris Pavlopoulos, National Technical University of Athens, Greece

Dimitris Koutsouris, National Technical University of Athens, Greece

Abstract

It is paradoxical that, although several major technological discoveries such as Magnetic Resonance Imaging and Nuclear Medicine and Digital Radiology, which facilitate improvement in patient care, have been satisfactorily embraced by the medical community, this has not been the case with Healthcare Informatics. Thus, it can be argued that issues such as Data Management, Data Modeling, and Knowledge Management have a long way to go before reaching the maturity level that other technologies have achieved in the medical sector. This chapter proposes to explore trends and best practices regarding knowledge management from the viewpoint of performance management, based upon the use of Key Performance Indicators in healthcare systems. By assessing both balanced scorecards and quality assurance techniques in healthcare, it is possible to foresee an electronic healthcare record centered approach which drives information flow at all levels of the day-to-day process of delivering effective and managed care, and which finally moves towards information assessment and knowledge discovery.

Introduction

The advantages of the introduction of Information and Communication Technologies (ICT) in the complex Healthcare sector have already been depicted and analyzed in the Healthcare Informatics bibliography (Eder, 2000; Englebardt & Nelson, 2002; Harmoni, 2002; Norris, Fuller, Goldberg, & Tarczy-Hornoch, 2002; Shortliffe, Perreault, Wiederhold, & Fagan, 2001; Stegwee & Spil, 2001). It is nevertheless paradoxical that, although several major technological discoveries such as Magnetic Resonance Imaging, Nuclear Medicine and Digital Radiology, which facilitate improvement in patient care, have been satisfactorily embraced by the medical community, this has not been the case with Healthcare Informatics. Thus, it can be argued that issues such as Data Management, Data Modeling, and Knowledge Management have a long way to go before reaching the maturity level that other technologies have achieved in the medical sector.

A variety of reasons could be proposed for this issue, though with a short analysis it becomes rather clear that modern ICT present integration problems within the healthcare sector because of the way the latter is organized. Healthcare is a strongly people-centered sector in which ICT has been considered more as an intruder, as a "spy" to the healthcare professionals' way of doing things and as a competitor to this people-centered model. Thus, if ICT intend to prove its advantages towards establishing an information society, or even more a knowledge society, it has to focus on providing service-oriented solutions. In other words, it has to focus on people and this has not been the case in most of the circumstances. It is common knowledge that in order to install any type of information system in healthcare, especially if it involves knowledge management, six main groups of issues have to be dealt with (Iakovidis, 1998, 2000):

1. The organizational and cultural matters related to healthcare. This issue is rather important, regardless of any information system, since organizational models and culture do not endorse the continuity of care or any type of structured data collection. Issues such as mistrust between different specialists, between the different healthcare structures or between doctors and nurses prevent in many cases the effective sharing of information. Health reforms are currently under way in many countries stressing the will to deal with this problem.

2. The technological gap between healthcare professionals and information science experts. Doctors are often reluctant to use information systems that they believe are not designed for them. From another point of view, Healthcare Informatics have been introduced in healthcare institutions mostly on pilot-based projects aiming

at addressing specific issues and have proposed solutions addressing a small number of healthcare practitioners, resulting in establishing a complex map of information niches. This approach is the consequence of applying information technology to procedures that where not designed for it, thus creating a panspermia of information models which are neither compatible nor interoperable, even within a single institution's environment. Efforts in creating interoperability standards and protocols such as HL7 are proposing solutions to address this issue, thus enabling data manipulation and knowledge management.

3. The legal requirements on the confidentiality of personal and patient related data and on data privacy. It is clear that if this issue is not addressed at a managerial and procedural level by imposing suitable policies to meet these requirements, there is little chance that medical data will be kept digitally in a structured manner (thus allowing the transition from digital islands of clinical data towards a structured electronic healthcare record). The implementation of an information system, where the electronic healthcare record is considered to be the core of the system (patient-centered model), is the only way to drive data management towards creating new knowledge. The complexity of the problem can be explained if one just observes the course of implementation of both the Health Information Privacy and Accountability Act (HIPAA) in the US and Directive 95/46/EC in the EU. The issues seem to have been dealt with at the strategic level, but still a lot has to be done in the implementation and setup of those strategies.

4. The industrial and market position of Healthcare Informatics. In general, the healthcare market is seen by the industry as large in size but not highly profitable, mainly due to the lack of standards in implementing and interoperating healthcare informatics products. As a consequence, the industry has focused on creating mostly small-scale products (i.e., Laboratory Information Systems, Radiology Information Systems, Clinical Information Systems) and not on evangelizing the production of information system that are dealing with healthcare as a whole. The lack of end-to-end solutions is dealt with by interconnecting heterogeneous information systems (a rather complex task with constant change management issues) and by introducing solutions from other business sectors (i.e., ERP, SCM, CRM) that have often been rejected by "key users" as non-compliant with their job description. Nevertheless, the new Web technology approaches (Web services, XML, etc.) and the new information technology strategies (i.e., service oriented architecture) could be the drivers towards merging information technology and healthcare services and thus enabling the establishment of knowledge management products.

5. The lack of vision and leadership of healthcare managers and health authorities, and the lack of willingness to re-engineer healthcare processes for the benefits of efficiency and quality of care delivery. Some countries are in the process of introducing or implementing such Business Process Reengineering projects in order to address healthcare delivery in a more information flow conformant way. This is a key point in reaching knowledge management, knowledge re-use and sharing, and finally proposing a solution for the knowledge-based society of tomorrow. This issue should be dealt with by proposing strategies that focus on processes and by establishing key performance indicators, balanced scorecards,

or other metrics that are the upper level of a structured information flow-based model.

6. User acceptability and usability of the proposed information systems. This issue is the one most strongly related to the problem of dealing with the people-centered approach of the healthcare sector. This issue deals with information systems' user friendliness, with usability issues such as the time to reach a data entry point, the speed of information retrieval, the quality of information retrieval, the complex security procedures, and so on. In order to implement information systems and knowledge management systems, education and training must be addressed with high priority since user acceptability is strongly related to them. Service oriented models and patient-centered information systems have a higher chance of passing the user acceptability test. A system that is not accepted by the user is often a system with poor data quality (or no data at all) and knowledge management, business intelligence or data warehousing solutions are consequently inoperable and unsuccessful.

Taking all of the above issues into consideration, this chapter proposes to explore trends and best practices regarding knowledge management from the viewpoint of performance management, based upon the use of Key Performance Indicators (KPI) in healthcare systems. By assessing both balanced scorecards (Kaplan/Norton) and quality assurance techniques in healthcare (Donabedian), it is possible to foresee an electronic healthcare record centered approach which drives information flow at all levels of the day-to-day process of delivering effective and managed care, and which finally moves towards information assessment and knowledge discovery (both with administrative and medical data). KPIs should be regarded as the strategic assessment tool, for both the executives and the clinical decision-makers, that will lead healthcare delivery to excellence and to knowledge discovery and assessment.

Background

Knowledge Management as a Transformation Driver in Healthcare

Today, Knowledge Management (KM) is on everyone's mind. Healthcare organizations are no exception and are accepting the challenge to more effectively share knowledge both internally and externally (Strawser, 2000). The growth of KM projects (i.e., decision support systems, data mining tools, business intelligence solutions) signals a growing conviction that managing institutional knowledge is crucial to business success and possibly business survival. When the hype and confusion are stripped away, it is apparent that KM initiatives can profoundly change a healthcare enterprise for the better, and bring numerous advantages to Healthcare Information Management (HIM) professionals. For HIM professionals, KM is worthy of special attention because it informs them not only on how to do things, but also on how they might do them better. In order for this to happen, data should be provided in specific patterns and should be based upon a strategy that will empower a healthcare system by gaining knowledge of its processes, its outcomes, and its structures.

Despite the obvious advantages, many healthcare decision makers view the idea of a KM initiative with scepticism, possibly because of an incomplete or incorrect understanding of the tools needed to achieve it. Many of the tools and strategies associated with implementing KM are not new; what is new is a cohesive approach to KM design and implementation. Certainly there are pitfalls and limitations in using information technology for KM—trying to force fluid knowledge into rigid data structures, for example, or focusing too much on the tools and not enough on the content. But networks and computers, with their ability to connect people and store and retrieve virtually unlimited amounts of information, can dramatically improve departmental efficiencies. Some examples of knowledge management applications are listed below:

- **Data Mining tools** enable decision makers to search and analyze large sets of data by using specific querying methods and tools (Standard Query Language, Rough Data set, On Line Analytical Processing).

- **Document and Content Management systems** are widely used to store and archive files of any type (text, images, video, etc.) and correlated them with keywords that have a business meaning to the end user.

- **Knowledge Maps** are graphical or other representations of how and by whom a specific set of information is created, distributed and assessed. Knowledge Maps are very important tools in Total Quality Management projects.

- **Intelligent Agents** use a combination of profiling techniques, search tools, and recognition algorithms to provide up to date specific information to the end user. For example, intelligent agents could be used to forward completed test results to the corresponding physicians of a patient.

- **Web Browsers** are the most commonly used tools for searching information in an intranet or the Internet. As such, Web browsers are increasingly becoming the most common graphical user interface, even for specific software products such as financial accounting and patient order entry systems.

- **Business Intelligence tools and Data Warehouses** enable the decision maker to have predefined access to specific information of interest regardless of the physical location of the data. Such systems are ideal for performance management and executive reporting and serve as the technological base for supporting the idea of a digital dashboard of indicators.

- **Workflow applications** play a very important role in KM since knowledge is created during the process-based operations that take place in a healthcare institution. A computerised patient order entry system is a classic example of a process-based operation in healthcare that requires the constant monitoring of the workflow status.

- **E-learning and collaboration tools** are part of the knowledge distribution process, which is extremely important in healthcare, since continuous education is a key factor in effective practice of care.

The essence of effective knowledge management does not rely on the use of one or more existing or forthcoming information technology tools. It is mostly about people, about processes and about capturing the results of people following processes, about transforming information into knowledge (explicit or tacit) and reusing it within a healthcare framework.

Performance Management: Monitor and Manage Healthcare

In order to persuade healthcare decision makers to assess the added value of KM tools, the latter should initially be used to propose new performance measurement and performance management techniques at all levels of a healthcare system (Hurst & Jee-Hughes, 2001). In that sense, performance management has long been considered as a tool for controlling spending and for increasing the efficiency of healthcare systems (Oxley & MacFarlan, 1994). There are three broad goals that governments generally pursue in the healthcare area:

- Equity: where citizens should have access to some incompressible minimum level of healthcare and treatment based on the need for care rather that solely on income.
- Micro-economic efficiency: where quality of care and consumer satisfaction should be maximized at minimum cost.
- Macroeconomic cost control: where the healthcare sector should consume an "appropriate" share of GDP.

In addition, healthcare systems are often facing factors that put pressure on the system. As a consequence, an effective performance management framework is the only solution towards controlling factors such as:

- Population aging
- Increased income and higher demand for healthcare services;
- Increased access to healthcare services; and
- Increase of high technology usage which in turn increases the healthcare services usage creating sometime unnecessary demand (from a medical point of view).

Most existing policies for controlling the performance of healthcare systems were based upon financial assessment of past results (macroeconomic control of spending and micro-efficiency improvements) by giving incentives to payers and providers of a healthcare system. By its very nature the financial measurements are not forward looking and are exclusionary to non-financial measures. In addition, the emergence of the Information Society in the late 90s rendered many of the fundamental performance management assumptions obsolete. Information Society has brought a new set of assumptions that institutions have to include into their strategy. Amongst others we could refer to:

- The cross functional aspects of processes based upon specialisation, increased skills and high technology
- The integration of processes in the healthcare sector from the suppliers to the patient and the ability to manage and monitor materials upon requests and needs
- The ability to offer specific services to patients in accordance to their needs while being able to constraint costs of this customised care. In fact this is the essence of managed care: providing satisfactory and high quality of care at a reasonable cost.
- The global scale of healthcare: This sector is no exception to any other global marketplace, thus making healthcare delivery more comparable at a national or regional level and within accepted standards (e.g., clinical procedure guidelines)
- Innovation, which has been a key driver towards quality of care and quality of life for many years in the healthcare sector
- Knowledge workers: the increasing complexity of medicine and technology has created a need for highly skilled personnel at every level of a healthcare institution. Employee empowerment is driven by knowledge as it is created in the daily process of delivering healthcare services.

Nowadays, performance measurement is moving towards the adoption of a set of objectives for a healthcare system. To our knowledge, there is no complete agreement on what is meant by "performance" of health systems and many sets of objectives are generally proposed. The use of Key Performance Indicators (KPI) helps to establish this set of objectives more thoroughly by focusing on the real needs.

Using Balanced Scorecards

In order to address the issue of creating a set of KPIs, the Balanced Scorecard (BSC) framework (Kaplan, 2001; Norton & Kaplan, 1996) initially proposed by Kaplan and Norton for the strategic management of financial organizations in the mid 1990s, is one of the most suited approaches in that direction. This model has now proven its value and since it is a generic framework it is applicable to the healthcare sector. The BSC concept involves creating a set of measurements and objectives for four strategic perspectives:

- Financial: financial performance measures typically indicate whether a proposed strategy implementation and execution is contributing to bottom line improvements, based upon an accurate summary of economic consequences of actions already taken.

- Customer (i.e., the Patient): this perspective is a set of objectives that focuses on identifying the patient's needs, the targeted market (this could be the case of a specialised institution) and on measuring the performance of each specific business unit (i.e., a department) that has some influence on how the patient sees the healthcare organization.

- Internal Business Process. This perspective should gather all objectives related to processes and the way these are monitored and fine-tuned in order to achieve both excellence in financial accomplishments and patient satisfaction. In that sense, the BSC approach is a constant business process reengineering process based upon specific goals to meet and not on improving de facto established processes.

- Learning and Growth (innovation and vision). This is an important perspective of a BSC implementation because it focuses on the objectives and goals to achieve the incorporation of business innovation (e.g., installing a Positron Emission Tomography Device) and the continuous education of medical and administrative staff. In that sense, this perspective identifies measurable tasks in order to build long-term growth and improvement. In the healthcare sector, improvement is also measured by assessing the outcome of treatment and care.

These four perspectives comprise a framework, meaning that they must be assessed and populated in accordance to each business case. In order to achieve this, one has to set for each category a list of objectives that can be feasibly measured. To each objective, a specific target should be set and the initiative to reach that target well described.

The BSC approach retains the "traditional" financial perspective, which mostly focuses on reporting about past events. Financial figures, even if prospected in the future by statistical means, are inadequate for guiding and evaluating the journey that healthcare institutions must make to create future value through innovation, knowledge, patient satisfaction, employee empowerment, process refinement, technology assessment and material management. In that sense, the three other cited perspectives are completing the puzzle in order to create a more valid and future-oriented performance management strategy. In recent years, there have been some implementations of BSC frameworks in the healthcare sector mostly on a single institution base (i.e., a hospital, a clinic, an information technology component). Readers are encouraged to read and assess the best practices, cited as references to this chapter (Aidermark, 2001; Castaneda-Mendez, Mangan, & Lavery, 1998; Forgione, 1997; Freeman, 2002; Gordon & Geiger, 1999; Inamdar, Kaplan, & Bower, 2002; Oliveira J, 2001; Pink, Mc Killop, Schraa, Preyra, Montgomery, & Baker, 2001; Tarantino, 2003; Weber, 1999)

The model in Figure 1 enables decision makers to also value how the latter three perspectives have worked and thus it enables the measurement of intangible aspects of a system. The measurements focus on a single strategy that can be easily broken down to the various levels of depth, depending of the organization type. Thus, by using BSC

Figure 1. A healthcare oriented BSC framework

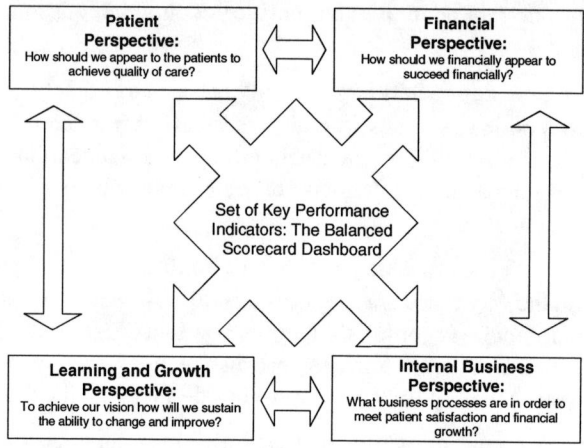

one can create a top-to-bottom design of a system, starting up from the needed strategy (what is the market, who are the customers, what are the critical processes, what is required by the stakeholders) going down to design metrics, processes, structures and finally the needed technical and functional specification to create an information system and a knowledge management framework capable of producing the right data and serve the strategy.

Quality Assurance in Healthcare and the Role of Information Technology

Known Problems and Issues of the Healthcare Sector

It is rather of common knowledge that the healthcare sector is not a sector without problems. Especially in Europe, the great majority of healthcare organizations are state funded. This means that institutions have in some cases very restricted budgets to satisfy their needs thus making performance management a critical issue in their daily routine. Other common problems are increased bed coverage (some institutions are almost at 100 percent), long length of stay, increased waiting lists, poor facilities, and so on. Taking all this into account and adding the life-related processes, one can clearly see that healthcare organizations are difficult to manage even for experienced managers of other sectors. One could list some of the major issues that healthcare stakeholders are confronted with:

- Diversity of cases
- Need of high technology medical devices
- Public Policy Restrictions (e.g., payment by per day quotas that do not cover the inpatient treatment costs)
- High lengths of stay
- Increased waiting times
- Obsolete facilities
- Restricted funds for training
- Restricted funds for maintenance
- Increased number of medical errors (sometimes fatal)
- Erroneous drug prescription and intake (sometime fatal)
- Geographical issues (too many patients, too few nurses and medical staff)
- Large lists of materials to be managed (more than 3000 on average)
- Excessive waste management (10 m^3 per day on average)

Quality Assurance and Performance Management

The application of a BSC framework will not by itself solve any of the aforementioned problems and issues. One could even say that BSC projects often fail as a consequence of misunderstanding or of not using a BSC strategy. BSC projects also fail because the variables of the scorecard are incorrectly identified as the primary drivers and because the improvement goals (the targeted objectives) are negotiated across an institution instead of being based upon stakeholders' requirements, fundamental processes and improvement process capabilities. They also fail because there is no deployment system installed to disseminate, maintain and promote the BSC framework, or because some very important KPIs are not used, or metrics are poorly defined.

Furthermore in order to create a structure that can be monitored the KPIs should not be more that 10 for each perspective, while non financial metrics should overcome the financial metrics by approximately six to one (Schneiderman, 1999). In order to support any BSC framework, a deployment and maintenance system based upon quality assurance specially designed for Healthcare should be established. Traditionally in healthcare, Quality Assurance (QA) has been meant to apply predominantly to healthcare itself as provided directly to patients by legitimate healthcare practitioners. We also include other services that directly affect the ability of practitioners to perform well, meaning services such as radiology, pharmaceutical, laboratory, and patient admission. The basic quality assurance terms (Donabedian, 2003) are:

- Efficacy: the ability of the science and technology of healthcare to bring about improvements in health when used under the most favorable circumstances.
- Effectiveness: the degree to which attainable improvements in health are in fact attained.
- Efficiency: the ability to lower the cost of care without diminishing attainable improvements in health.
- Optimality: the balancing of improvements in health against the costs of such improvements.
- Acceptability: conformity to wishes, desires, and expectations of patients and their families.
- Legitimacy: Conformity to social preferences as expressed in ethical principles, values, norms, mores, laws, and regulations.
- Equity: Conformity to a principle that determines what is just and fair in the distribution of healthcare and its benefits among the members of the population.

One can clearly see the benefit of applying quality assurance components in the development of a BSC strategy. A BSC framework that meets quality assurance in healthcare is most probable that will meet patient needs, practitioners' feelings, patient-practitioner relationship, amenities of care (e.g., confidentiality, privacy, comfort, cleanliness, convenience), as well as financial and organizational aspects required.

The most popular quality assurance model in healthcare is based, to the best of our knowledge, upon the Donadebian approach (Donabedian, 1980, 1982, 1985, 1993, 2003) where a healthcare organization (i.e., a hospital) is a system formed by the interaction of structures, processes, and outcomes. Structures are used to establish processes in order to create healthcare outcomes that have an effect on structures that need to change or adjust processes to meet the required outcomes. Strongly believing that healthcare outcomes are more important than financial outcomes in a healthcare system, we are confronted with a model where intangible assets are more important than tangible assets. This last statement makes a healthcare system very difficult to manage and a straightforward strategy hard to define.

The Need for a Specific Implementation Plan

In order to implement a viable performance management strategy (i.e., a BSC framework) the steps one needs to take include:

- Determining what to monitor
- Determining priorities in monitoring
- Selecting an approach to assessing performance

- Formulating criteria and standards (i.e., Key Performance Indicators)
- Obtaining the necessary information
- Determining how and when to monitor
- Constructing a monitoring system
- Managing changes and improvements

Building a Patient-Centered Healthcare Model

Finally, the proposed KPIs are directly or indirectly driven from healthcare processes. As an example, we propose to analyze a standard patient journey of a citizen in a healthcare institution. Figure 2 shows how the patient journey for a hospital is conceived (the patient journey in a primary care setting may be simpler). This workflow is the heart of the healthcare system and a prerequisite for any patient-centered information system to properly manage the information flow. This workflow is nowadays extended to include new processes such as emergency pre-hospital care and home care monitoring in order to create the hospital without walls of the 21st century.

Based on Figure 2 one can create a table where quality assurance and balanced scorecard features are confronted, analyzed, assessed and finally set. Table 1 is an example (non exhaustive) of the initial process.

From the above, it becomes apparent that the design and proposal of KPIs is not an easy task. KPI selection can vary upon specific measurement needs, upon goal set, and so on. In order to manage and validate the proposed KPIs by each BSC strategy, a set of KPI dashboards for each management entity, department or any other region of interest should be created.

Use Case: Regional Healthcare Authorities in Greece

In 2001, a reform of the Greek National Healthcare System was introduced in order to enhance the performance and control of healthcare provision in Greece (Greek National Healthcare System Reform Act, 2001; Vagelatos & Sarivougioukas, 2003). One of the main changes was the division of the country in 17 autonomous healthcare regions where the Regional Healthcare Authority (RHA) is responsible for the regional healthcare strategy. This reform introduced the need to establish a three-level decision-making and performance management mechanism as described in Figure 3.

The proposed methodology was used to reach an initial set of KPIs by assessing existing knowledge and future needs (Decision No 1400/97/EC, 1997; McKee & Healy, 2002; Polyzos, 1999). Those KPIs were processed especially to serve the new strategy introduced in Greece.

Figure 2. The patient journey centered model for a hospital institution

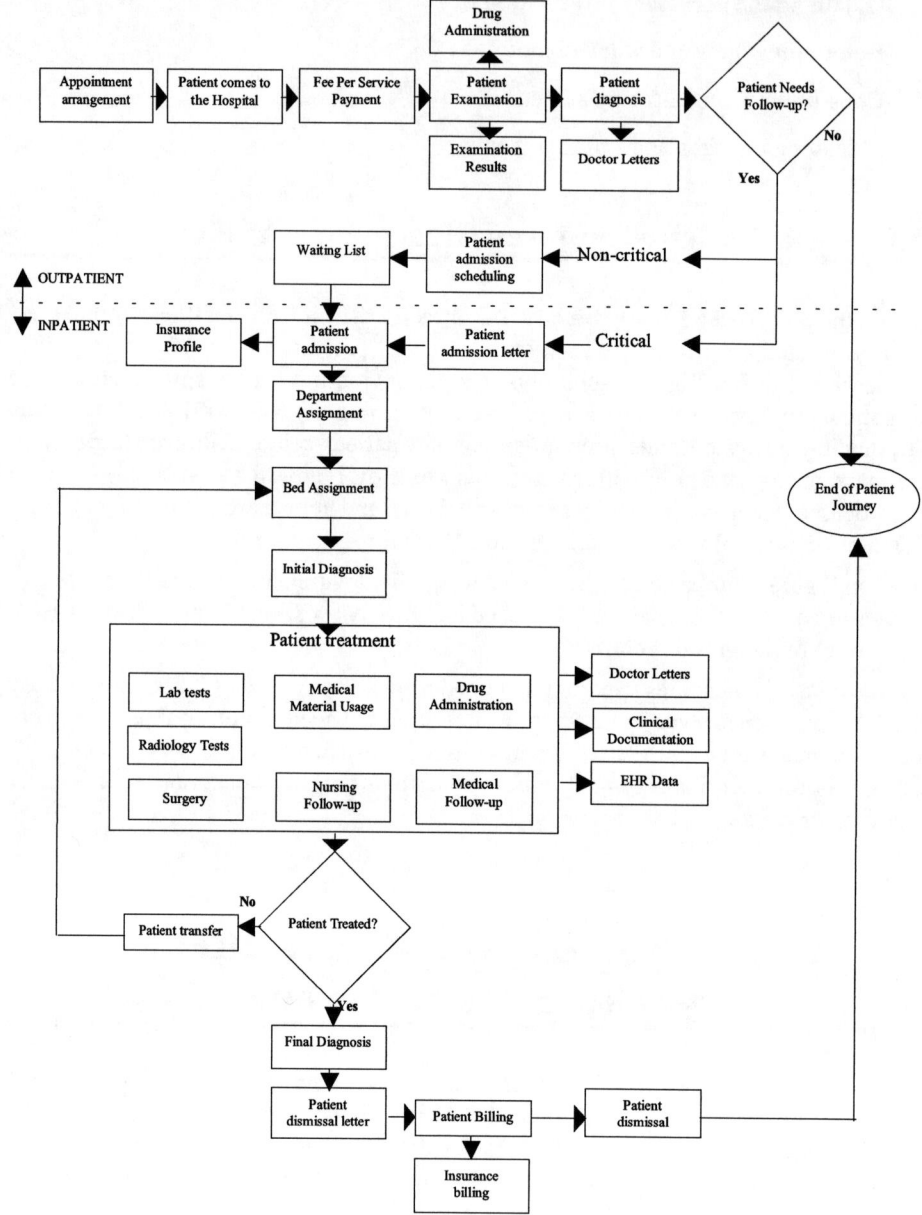

Table 1. Defining KPIs from processes, an example

Workflow process	BSC Perspective	KPI	QA approach	Source of information
Appointment arrangement	Patient	Appointments/day	Process	Hospital Information System (HIS), Scheduling S/W
Patient comes to the Hospital	Process	Number of outpatients	Process	Hospital Information System
Fee Per Service Payment	Financial	Mean cost per examination	Process	Billing, ERP S/W
Drug Administration	Process	Number of prescription/ drug	Outcome	Patient Order Entry S/W, HIS
Patient Examination	Patient	Patient satisfaction	Process	HIS, surveys
Examination Results	Growth	Number of patient with re-examination	Outcome	HIS, Electronic Healthcare Record (EHR)
Doctor Letters	Process	Number examinations/practitioner	Outcome	HIS, Electronic Healthcare Record (EHR)
Patient diagnosis	Growth	Visits/ICD codes	Outcome	HIS, Electronic Healthcare Record (EHR)
Patient Needs Follow-up?	Process	Number of inpatient from outpatient clinic	Process	HIS
Waiting List	Patient	Waiting time in days	Structure	Hospital Information System (HIS), Scheduling S/W
Patient admission scheduling	Patient	Equity of delivered care	Process	Hospital Information System (HIS), Scheduling S/W
Patient admission letter	Process	Number of emergency cases/day	Process	HIS
Patient admission	Process	Number of inpatients	Process	HIS
Insurance Profile	Financial	Net cash flow per insurance company	Process	HIS, ERP
Department Assignment	Financial	Mean operational cost per department	Structure	HIS
Bed Assignment	Process	Bed coverage rate	Structure	HIS
Initial Diagnosis	Process	Admission per case type (ICD 10)	Outcome	HIS, Electronic Healthcare Record (EHR)
Lab tests	Process	Mean value of lab test per doctor, per patient	Outcome	Laboratory Information System (LIS)
Radiology Tests	Process	Mean value of radiology test per doctor, per patient	Outcome	Radiology Information System (RIS)
Surgery	Financial	Mean cost of surgical procedure	Structure	HIS, Electronic Healthcare Record (EHR)
Medical Material Usage	Financial	Mean cost of medical material consumption	Outcome	HIS, ERP
Nursing Follow-up	Growth	Number of Nurses per bed	Process	HIS, Electronic Healthcare Record (EHR)
Medical Follow-up	Growth	Number of practitioners per bed	Process	HIS, Electronic Healthcare Record (EHR)
Clinical Documentation	Process	Number of medical procedures per day	Outcome	HIS, Electronic Healthcare Record (EHR)
EHR Data	Patient	Number of cases with EHR	Outcome	HIS, Electronic Healthcare Record (EHR)
Patient transfer	Process	Number of patient transfers/ patient or /day	Process	HIS
Patient Treated?	Growth	Number of patients treated under a specific critical pathway	Process	HIS, Electronic Healthcare Record (EHR)
Final Diagnosis	Growth	Cases per final diagnosis	Outcome	HIS, Electronic Healthcare Record (EHR)
Patient dismissal letter	Patient	Inpatient Satisfaction	Process	HIS, surveys
Patient Billing	Financial	Mean treatment cost per day	Process	Billing, ERP S/W
Patient dismissal	Process	Mean length of stay /per dept. per ICD code	Process	HIS
Insurance billing	Financial	Return of Capital Employed (ROCE)	Process	Billing, ERP S/W

As described in Figure 3, the regional healthcare system is comprised of a series of information systems covering the whole structures existing at any level, the processes required to meet the administrative and medical needs and finally, the outcomes that must come out from the implementation of such a complex interpolation of informatics infrastructure. The above information model was introduced to establish a community of networked healthcare organizations (hospitals, primary care) that are interoperating in order to support and implement the new healthcare strategy: to provide integrated and high quality healthcare services to the citizens based upon equal access to the resources (Information Society SA, 2003). In order to achieve this goal, two main issues were raised:

- How and when will information systems interoperate?
- What is the minimum required dataset to achieve the proposed strategy?

Figure 3. Regional healthcare information systems framework and interoperability

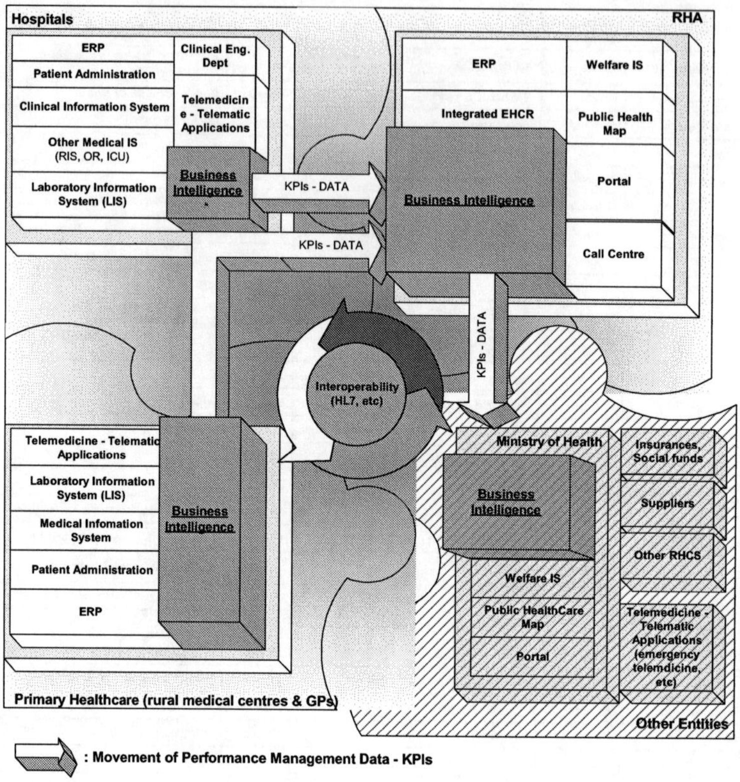

The first issue can be answered by using standards and protocols such as HL7 to meet with interoperability issues in healthcare (Spyrou, Berler & Bamidis, 2003). The second issue is partially addressed by the proposed initial set of KPIs presented below. The proposed KPIs are forming a complete set of metrics that enable the performance management of a regional healthcare system. In addition, the performance framework established is technically applied by the use of state-of-the art knowledge management tools such as data warehouses and business intelligence information systems.

The proposed KPIs are categorised into the four perspective stated by Norton & Kaplan, having also taken into account the raised quality assurance issues stated earlier. The performance indicators in a regional healthcare setting are depicted in Table 2 as below.

Each of the above KPIs is the result of analysis based upon the needs of a standard regional healthcare authority. The proposition of a set of KPIs is nevertheless not the complete solution of the problem. Implementing KPIs is a constant process based upon specific metrics that each regional healthcare authority and each department or institution under its control must periodically assess and reengineer. Assessment should be based upon specific goals met and reengineering is often required due to administrative, demographic or other important changes that must occur. Focusing for example in the KPI marked as "mean treatment cost per day" one should notice that the KPI does not mean much without a metric. In Greece, most of inpatient treatments are based upon fixed prices per day and do not follow the pay per service model which is financially more viable.

Healthcare financials are part of a national policy aiming at providing high quality healthcare services to all citizens regardless of their income, social status or other characteristic. As a consequence, the use of the fixed price model (per day quotas) in Greece serves that purpose albeit with its advantages and disadvantages. Current treatment cost per day (for an inpatient) has been fixed to about 135 Euros and this value could be used as an initial metric. If this value is exceeded this would mean that the RHA budget will cover the difference or transfer the cost to insurance companies and social welfare. In addition, a regional healthcare authority will then have credible proof that national standards are outdated and require revision in order to support the system. As a result, this KPI has now a specific meaning linked to regional strategy and budgetary needs. Following that example, all financial KPIs are therefore an important perspective of BSC since they are the measurement of the financial viability of the regional healthcare authority. In fact, all KPIs should be associated with adequate metrics in order to be

Table 2. KPIs in a regional healthcare setting

No	Financial KPI description	No	Process KPI description
F1	Mean treatment cost per day	P1	Length of stay
F2	Mean cost of medical treatment per patient	P2	Patient admission rate per medical unit
F3	Mean cost of drugs consumption	P3	Percentage of bed coverage
F4	Mean cost of radiology testing	P4	Vaccination rate
F5	Mean cost of laboratory testing	P5	Mean value of performed test per patient, per doctor
F6	Mean cost of material consumption	P6	Number of inpatients
F7	Mean cost of surgical procedure	P7	Number of outpatients
F8	Mean operational cost per dept./clinic	P8	Number of drug prescription
F9	Mean cost of vaccination procedures	P9	Number of laboratory tests
F10	Mean cost per medical examination	P10	Number of surgery procedures
F11	Return of capital employed (ROCE)	P11	Number of radiology tests
F12	Net Cash flow	P12	Number of visit in outpatient clinics
F13	Income per employee	P13	Number of visits in primary care institutions
F14	Payroll rate versus operational costs.	P14	Number of dental care processes
		P15	Number of processed emergency cases
		P16	Number of unprocessed order entries on the same day
		P17	Number of preventive care visits
		P18	Number of home care monitored patients
		P19	Assessment of patient satisfaction
No	Customer (patient) KPI description	No	Learning and Growth KPI description
C1	Mortality rate	L1	Medical device usage growth
C2	Morbidity rate	L2	Healthcare professionals training rate
C3	Number of medical staff per 1000 inhabitants	L3	Employee Satisfaction rate
C4	Number of beds per 1000 inhabitants	L4	Number of doctors per bed
C5	Accessibility of patients to the medical units	L5	Number of nurses per bed
C6	Time in a waiting list	L6	Number of existing healthcare professionals versus expected job positions
C7	Equity of delivered care	L7	Personnel productivity rate
C8	Number of readmission per patient	L8	Number of medical interventions per doctor
C9	Mean length of stay	L9	Number of patient with re-examinations
C10	Patient Satisfaction rate	L10	Visits/ICD codes
		L11	Admissions per case type (ICD 10)
		L12	Dismissals per case type (ICD 10)

assessed, thus driving the RHA towards the right strategic decisions.

In order to meet and populate the above-mentioned KPIs a regional healthcare authority has to implement a complex information technology system in order to gather up all needed information. Then the information collected through the use of enterprise resource planning software, hospital information systems, clinical information systems, radiology information system, and laboratory information systems has to be processed and interpolated to produce the final metadata set from which the KPIs are driven. In other words, the model is complemented by a business intelligence solution similar to the one depicted in Figure 4.

Figure 4 shows how data are collected from the various data sources, cleansed and homogenised, and finally redistributed to the knowledge workers and decision makers of the regional healthcare authority (Extraction, Transformation and Load—ETL).

The data collection process is extremely important since it is a basic feature of successfully populating the KPIs. In that sense both organizational and technological issues to achieve data quality should be considered. In the proposed setting, the regional healthcare authority has imposed on its healthcare units the use of specific classifications, codifications and taxonomies such as the 10th edition of the International Classification of diseases (ICD10). In addition, the proposed KPIs can be seen as attributes of structure, process or outcome (based upon the Donabedian approach) so that they can be used to draw an inference about quality. As such the KPIs are proposed, designed, tested and assessed by a panel of experts (executive officers of the RHA, practitioners). By implementing this organizational structure the quality level of the proposed KPIs is such that technological issues are greatly reduced.

Figure 4. Regional healthcare authority business intelligence framework

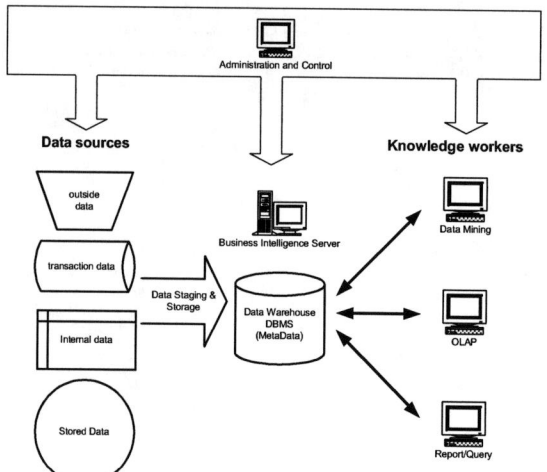

Future Trends

Up to this point, this chapter has mostly dealt with organizational and strategic features of knowledge management in healthcare. In the proposed use case we have shown an ongoing implementation of a Balanced Scorecard Framework in a regional healthcare authority environment. This has been done intentionally wishing to state that the technological part on such an implementation is probably the less important issue. If one regards the future trends in knowledge management, one can see that a multitude of new tools are already proposed for use. This chapter will briefly set the focus on the trends that to our knowledge are the most promising and present more opportunities to healthcare organizations in creating effective performance management facilities.

Service Oriented Architecture and Patient-Centered Architecture (Based on the Electronic Health Record)

The term of patient-centered architecture has been already in the literature for some years. Many techniques have been used in the past such as Corbamed (Object Management Group, 2001) and Distributed Healthcare environment (DHE). The introduction of web technologies such as the Extensible Markup Language (XML), the Simple Object Access Protocol (SOAP), the Web Services Description Language (WDSL), and more precisely the concept of Web Services (Deitel, Deitel, DuWaldt, & Trees, 2003; W3C, 2003) are driving information technologies towards a Service Oriented Architecture (SOA).

A service is a software component that is suitable for cross-application access. A service is never a complete application or transaction. It is always a building block. SOA is the architecture of an application that uses services. Services define reusable business functions; SOA binds services into applications. Logically, services are defined by their interfaces. Technically, services are defined by their implementations (sometimes complex integrated flows, other times a single simple program). SOA is a logical concept, and its design is focused on the definition of service interfaces and interactions between service interfaces. Fundamental to SOA is the loose coupling between its components. At the logical level, this translates to the ability to add a new service for the end-user unobtrusively to the service provider. At the technical level, this translates to the ability of software developers to deploy a new application that calls a service without the need to redeploy or change the service. The use of SOA will allow the creation of process-based components of applications that will manipulate knowledge and information based upon the processes and the required or designed outcomes.

The Semantic Web

The Semantic Web goes beyond the World Wide Web by encoding knowledge using a structured, logically connected representation, and by providing sets of inference rules that can be used to conduct automated reasoning. Whilst the idea of knowledge representation languages is not new, existing languages generally use their own set of ontologies and inference rules to identify and eliminate logical contradictions and inconsistencies. The Resource Description Framework—RDF (W3C, 2003) and XML Topic Maps (TopicMaps.Org, 2001) are the most promising tools towards the implementation of the Semantic Web in practice. Nevertheless, a long way towards maturity has still to be covered since issues such as specific metadata frameworks and data quality are not yet solved. In any case, the Semantic Web should enhance the promotion of clinical practice guidelines and evidence based medicine. They can be seen as taxonomies of medical cases that could be both used for performance monitoring (in respect to commonly agreed levels of delivered care) and decision support for the healthcare practitioners.

Critical Pathways

Critical Pathways (Wall & Proyect, 1998) are mechanisms for transforming a reactive bureaucratic ritual to a dynamic, indispensable, clinical improvement process. A critical pathway when established is a mechanism for:

- Integrating continuous quality improvements with traditional patients' care review
- Managing and impacting of clinical and financial outcomes for a specific treatment procedure
- Proactively addressing economic and regulatory changes
- Improving clinical outcomes through reduction in variation
- Controlling unnecessary cost and resource usage without jeopardising quality of care
- Fostering multi-disciplinary approach to patient care
- Linking quality management to staff education
- Managing limited financial resources
- Making efficient use of scarce organizational resources
- Increasing readiness for anticipated changed in healthcare
- Applying and using clinical practice guidelines and other taxonomies set up by different professional societies

Critical Pathways can be seen as "specialized" performance management tools that would provide a BSC framework with very specific performance indicators for each treatment or clinical process.

Conclusions

Performance management is a key issue in the continuous process of delivering high quality healthcare services. The use of KPIs has proved that the design of a Balance Scorecard acts as the "cockpit" of a regional (or national) healthcare authority where all metrics are the flight instruments enabling the provision of healthcare based upon equity, financial control, continuous process and structure refinement, and outcome measurements. In that sense, the proposed infrastructure is, technologically speaking, an important knowledge management tool that enables knowledge sharing amongst various healthcare stakeholders and between different healthcare groups. The use of BSC is an enabling framework towards a knowledge management strategy in healthcare since KM is about discovering knowledge from existing information, about creating new knowledge and about implementing processes and taxonomies that enable the reuse and assessment of information as part and bits of knowledge.

Knowledge can be seen as a performance management tool both for administrative purposes and clinical improvements.

During the implementation process of deploying a technological platform for performance management or any other type of knowledge management infrastructure, one must have in mind that:

- The six issues described in the introduction section must being taken into strong consideration from day one.
- It is important to focus on people, processes and outcomes, and to set-up a straight-forward strategy to plan, manage, assess, educate, disseminate, and maintain the developed BSC framework.
- Any type of knowledge management project bases its success on continuous improvement and assessment. Metrics and processes are meant to change in order to reflect improvements towards quality of healthcare.
- The knowledge and technology is there, but still very limited best practices have been successfully implemented.
- The implementation of BSC framework is a time consuming process that has to involve all stakeholders' representatives.

References

Aidermark, L.G. (2001). The meaning of balanced scorecards in the health care organization. *Financial Accountability & Management, 17*(1), 23-40.

Castaneda-Mendez, K., Mangan, K. & Lavery A.M. (1998, January / February). The role and application of the balanced scorecard in healthcare quality management. *Journal of Healthcare Quality, 20*(1), 10-3.

Decision No 1400/97/EC of the European Parliament and of the Council of 30th June 1997 adopting a program of Community action on health monitoring within the framework for action in the field of public health, *Official Journal L193*, 22/07/1997, 0001-0010.

Deitel, H.M., Deitel, P.J., DuWaldt, B. & Trees, L.K. (2003). *Web services: A technical introduction.* Prentice Hall Editions.

Donabedian, A. (1980). *Explorations in quality assessment and monitoring, Volume I: The definition of quality approaches to its measurement.* Ann Arbor, MI: Health Administration Press.

Donabedian, A. (1982). *Explorations in quality assessment and monitoring, Volume II: The criteria and standards of quality.* Ann Arbor, MI: Health Administration Press.

Donabedian, A. (1985). *Explorations in quality assessment and monitoring, Volume III: the methods and findings of quality assessment and monitoring – An illustrated analysis.* Ann Arbor, MI: Health Administration Press.

Donabedian, A. (1993). Continuity and change in the quest for quality. *Clinical Performance and Quality in Healthcare, 1*, 9-16.

Donabedian, A. & Bashshur, R. (2003). *An introduction to quality assurance in health care.* Oxford: University Press.

Eder, L. (2000). *Managing healthcare information systems with Web enabled technologies.* Hershey, PA: Idea Group Publishing.

Englebardt, S. & Nelson, R. (2002). *Healthcare Informatics: An interdisciplinary approach.* Mosby Edition.

Forgione, D.A. (1997). Health care financial and quality measures: International call for a balanced scorecard approach. *Journal of Health Care Finance, 24*(1), 55-58.

Freeman, T. (2002). Using performance indicators to improve health care quality in the public sector, a review of the literature. *Health Services Management Research, 15*(2), 126-137.

Gordon, D. & Geiger, G. (1999). Strategic management of an electronic patient record project using the balanced scorecard. *Journal of Healthcare Information Management, 13*(3), 113-123.

Greek National Healthcare System Reform Act (2001), *Greek Government Printing Office*, N2889/2001 / FEK-˜/37/02.03.2001 [In Greek].

Harmoni, A. (2002). *Effective healthcare information system.* Hershey, PA: IRM Press.

Hurst J. & Jee-Hughes, M. (2001). Performance measurement and performance management in OECD health systems. *Labour Market and Social Policy, Occasional Papers No. 47*, DEELSA/ELSA/WD(2000)8, OECD.

Iakovidis, I. (1998). Towards personal health record: Current situation, obstacles and trends in implementation of electronic healthcare records in Europe. *Internernational Journal of Medical Informatics, 52*(123), 105-117.

Iakovidis, I. (2000). Towards a health telematics infrastructure in the European Union. In *Information technology strategies from US and the European Union: Transferring research to practice for healthcare improvement*. Amsterdam: IOS Press.

Inamdar, N., Kaplan R.S. & Bower, M. (2002). Applying the balanced scorecard in healthcare provider organizations. *Journal of Healthcare Management, 47*(3), 179-195.

Information Society SA (2003, May 15th). Healthcare information system for the 2nd Regional Healthcare Authority of Central Macedonia, *Request For Proposal co-funded by the 3rd CSF under the EU decision C(2001)551/14-3-2001*, Greece.

Kaplan, R.S. (2001). Strategic performance measurement and management in non-profit organizations. *Non-Profit Management and Leadership, 11*(3), 353-371.

Kaplan, R.S. & Norton, D.P. (1996). *The balanced scorecard: Translating strategy into action*. Harvard Business School Press.

McKee, M. & Healy, J. (2002). *Hospital in a changing Europe*. European Observatory on Health Care Systems Series, Open University Press.

Norris, T.E., Fuller, S.S., Goldberg, H.I. & Tarczy-Hornoch P. (2002). *Informatics in primary care, strategies in information management for the healthcare provider*. Springer Editions.

Object Management Group (2001). *The Common Object Request Broker Architecture (CORBA) specification*. OMG Document Number 2001-09-01. Information retrieved on December 15, 2003 from *http://www.omg.org/technology/documents/formal/corba_iiop.htm*

Oliveira, J. (2001). The balanced scorecard: An integrative approach to performance evaluation. *Healthcare Financial Management Journal, 55*(5), 42-46.

Oxley, H. & MacFarlan, M. (1994). Healthcare reform controlling spending and increasing efficiency. *Economics Department Working Papers No. 149*, OCDE/GD(94)101, OECD.

Pink, G.H., Mc Killop, I., Schraa, E.G., Preyra, C., Montgomery, C. & Baker, G.R. (2001). Creating a balanced scorecard for a hospital system. *Journal of Health Care Finance, 27*(3), 1-20.

Polyzos, N. (1999). *Hospital efficiency based upon the use of DRGs*. TYPET Edition [in Greek].

Schneiderman, A.M. (1999). Why balanced scorecard fail. *Journal of Strategic Performance Measurement*, January(4), 6-11.

Shortliffe, E.H., Perreault L.E., Wiederhold, G. & Fagan, L.M. (2001). *Medical informatics, computer applications, healthcare and biomedicine* (2nd edition). Springer Editions.

Spyrou, S., Berler, A. & Bamidis, P. (2003). Information system interoperability in a regional healthcare system infrastructure: A pilot study using healthcare information standards. *Proceedings of MIE 2003,* Saint Malo, France (pp. 364-369). IOS Press.

Stegwee, R. & Spil, T. (2001). *Strategies for healthcare information systems.* Hershey, PA: Idea Group Publishing.

Strawser, C.L. (2000). Building effective knowledge management solutions. *Journal of Healthcare Information Management, 14*(1), 73-80.

Tarantino, D.P. (2003). Using the balanced scorecard as a performance management tool. *Physician Exec, 29*(5), 69-72.

The TopicMap.Org (2001). Retrieved August 6, 2001, from *http://www.topicmaps.org/xtm/1.0/*

Vagelatos, A. & Sarivougioukas, J. (2003). Regional healthcare authorities delivering application services to primary health care units. *MIE 2003,* Saint Malo, France.

Wall, D.K. & Proyect, M.M. (1998). *Critical pathways development guide.* Chicago: Precept Press Editions.

Weber, D. (1999). Performance management – The balanced scorecard: A framework for managing complex and rapid changes. *Strategies for Healthcare Excellence, 12*(11), 1-7.

The World Wide Web Consortium – W3C Web site (2003). Information retrieved March 23, 2003, from *http://www.w3.org*

Chapter VIII

Multimedia Capture, Collaboration and Knowledge Management

Subramanyam Vdaygiri, Siemens Corporate Research Inc., USA

Stuart Goose, Siemens Corporate Research Inc., USA

Abstract

This chapter presents methods and technologies from Siemens Corporate Research that can assist in the process of creating multimedia collaborative knowledge bases: capture, querying, visualization, archiving, and reusability of multimedia knowledge bases. A selection of Siemens products in the healthcare and communication domains are introduced, above which novel multimedia collaboration and knowledge management technologies have been developed by the authors. With examples, it is explained how in concert these technologies can contribute to streamlining the processes within healthcare enterprises, telemedicine environments and home healthcare practices.

Copyright © 2005, Idea Group Inc. Copying or distributing in print or electronic forms without written permission of Idea Group Inc. is prohibited.

Introduction and Motivation

The networked healthcare enterprise is providing unprecedented opportunities for healthcare workers to collaborate and make clinical decisions in an efficient manner. Significant progress has been made to enable healthcare personnel to obtain answers to simple clinical questions by using medical databases of evidence-based answers. This approach allows reuse and sharing of knowledge to help healthcare professionals to save time and effort and help patients in an efficient manner. For situations where healthcare workers have simple questions with simple answers, this approach is perhaps overkill. However, when the questions are of a more complex nature, by capturing and archiving complex answers in a rich multimedia form they can be exploited multiple in the future to explain solutions in a manner that can be easily digested by healthcare workers.

A contemporary healthcare enterprise involves complex media elements (images, videos, documents, etc.) and volumes of documentation both digital and on paper. The healthcare knowledge base should incorporate these media elements and easily allow users to search, extract, and reuse. Some of the modern knowledge management systems allow building of communities of practice around documents. But there is a need to move beyond regular office documents to address rich media and encompass specialized medical and clinical data.

The networked enterprise is also enabling a plethora of ways for healthcare personnel to communicate and collaborate. The next generation of communication technologies will bring converged voice and data solutions on a single network. This is helping integrate the healthcare IT (Information Technology) systems with Web-based communications. In recent years we have witnessed a proliferation of communication and data devices like GPRS cellular phones and PDAs, thus providing an opportunity for accessing clinical information anywhere/anytime and allowing users to collaborate over clinical information to reach decisions quickly.

The concept of presence and availability offered by various instant messaging tools is changing the manner in which people are communicating with each other. *Presence* enables a user to know who in their contact or buddy list is available or not at any given point in time. *Availability* options allow a user to signal whether they are available to be contacted and which form of communication they favor. Presence and availability information allow users to interact in various ways in offline, real-time or in near-real time modes. Mobile communication technologies are being developed that enables mobile location and presence. The integration of the healthcare enterprise content repository with a Web-based infrastructure and presence and availability represent the three pillars of modern unified, or converged, communication.

Although the potential for a rich communications and IT infrastructure is high, there remains a need to streamline the communications and collaborations between healthcare personnel to ensure that valuable knowledge gained from daily interactions between healthcare personnel is not lost. This chapter presents methods and technologies from Siemens Corporate Research that can assist in the process of creating multimedia collaborative knowledge bases: capture, querying, visualization, archiving, and reusability of multimedia knowledge bases. Throughout the chapter, a number of Web-based technologies are introduced that enable healthcare personnel to interact in a variety of modes regardless of whether they are mobile or sedentary.

Background: Streamlining Healthcare and Telehealth

Since the advent of Web-based workflows, there has been a growing emphasis in healthcare enterprises on methods to increase organizational efficiencies, reduce errors and focus on patient care. One such platform is Soarian (Siemens Soarian) from Siemens Medical (Siemens Medical) that offers an integrated workflow technology that can streamline the operational processes of healthcare. Soarian's infrastructure has been engineered based on clinical processes that enable physicians to focus less on administrative duties and more care by providing them with access to all clinical data in a single view. The goals being to offer actionable guidelines to clinicians based on best practices and to support them in reaching accurate, evidence-based decisions promptly.

A nascent area for which technology can be a significant enabler is that of telemedicine (Hibbert, Mair, May, Boland, O'Connor, Capewell, & Angus, 2004), where clinical needs are extended beyond the boundary of the hospital. A key requirement to facilitate a telehealth consultation is to have medical personnel in remote locations to communicate and collaborate with each other in a quick, efficient, and effective manner.

Some of the issues that an integrated communications and healthcare medical IT infrastructure can help address are:

- Reducing time and effort wasted in daily communications (paging, phone calls) between clinicians, nurses, and patients
- Reducing the communication pathways by a combined infrastructure can streamline how users can be reached at a given point in time
- Complex solutions to questions can be assembled in multimedia fashion to succinctly convey the message
- Knowledge exchanged and gained from daily interactions between people can captured and archived with high granularity
- Reuse, by querying and retrieving relevant segments from past collaborative activities, can avoid recreating the wheel when a recurrent problem reappears
- Reduce paper work by having one consolidated IT-communications infrastructure
- Reduce cost by managing a single converged voice and data network

Remote communications in contemporary telehealth systems allows users to collaborate in two different modes: store & forward and in real time. The asynchronous mode, or store

& forward method, allows for the sharing of medical information in an offline mode, but the absence of human interactivity prevents healthcare personnel from augmenting this with personal comments, insight and knowledge. The real time collaboration mode requires the session to be scheduled in advance, and all participants must be dedicated to the collaboration session for the duration.

The current mode of document/data sharing involves all users within telehealth WAN (Wide Area Network) cluster to upload documents, data and code to a central repository from where other users can extract materials of interest. This mode of collaboration among users is not sufficient for many users to exchange complex ideas and viewpoints regarding various images and other documents in intricate detail. For example, medical researchers frequently need to locate and reference information that is not only physically distributed across the sites of their collaborators, but also in an array of formats (images, reports generated based on experiments, or code execution).

Often users collaborate by exchanging e-mail messages along with data/documents downloaded from a central Web server. Although this allows individuals to participate at their own pace, users invest much time typing descriptive text to discuss a particular topic, especially when expressing complex thoughts within context of a document or image. The problem becomes particularly acute when users are collaborating using document attachments. This mode leads to large documents being exchanged as attachments back and forth between users, with the potential for inconsistencies to arise between successive versions. Moreover, with clinicians collaborating across time zones on documents or images, a need has been observed for conveying complex information via rich multimedia.

To anticipate and streamline the future communication and IT needs of healthcare enterprises and burgeoning telehealth market, the various requirements were distilled:

- **Communication**: It is clear that there exists a need to combine communications within the healthcare processes to optimize the clinicians' time to manage better the inexorable stream of phone calls, paging, exchange of paper and digital documents. Presence-based communications enables the healthcare personnel to communicate in a very effective manner.

- **Collaboration:** The tools should provide a consistent manner to collaborate irrespective of the mode, all online in a real time conference, or a single user making her analysis and comments on an image in offline mode. By allowing users to collaborate in different modes within a browser-based environment on diverse medical data can facilitate a seamless collaboration workflow. These technologies can enhance typical telemedicine scenarios and also extend conventional doctor-patient interactions to include web-based interactions such as chronic care or follow-ups.

- **Knowledge Management**: There should also be a consistent way to archive the multimedia annotations made on such documents. This would not only allow researchers to collaborate based on their time and schedule, but also to search, retrieve and filter all comments and analysis made previously by collaborators on any multimedia document.

The remaining sections describe ongoing efforts at Siemens Corporate Research to develop technological solutions driven by these anticipated future needs.

Technological Building Blocks

OpenScape (Siemens OpenScape) from Siemens Communications Networks (Siemens Information and Communications Networks) is a suite of communication applications designed to increase the productivity of information workers. It aims to control the panoply of communication applications and devices, on both fixed and wireless networks, connected via local and wide area networks. OpenScape addresses the fragmented nature of communication modalities and their separation and provides a unifying framework for integrating, managing and streamlining communication in the enterprise.

It obviates redundant communication sessions, where a person calls multiple phone numbers and/or leaves duplicate voice, email, and instant messages in an effort to communicate urgent issues. Hence, it is possible to avoid:

- Unnecessary cell phone intrusions into client meetings, work sessions or personal time;
- Wasted time setting up conference calls, communicating call-in information, sending and synchronizing documents, and establishing separate sessions for voice, Web and video collaboration; and
- The difficulty of mobilizing all key colleagues that may be equipped with different applications, or because setting up a collaborative session is too complex and time consuming.

It provides healthcare personnel the experience of a single, synchronized set of communication resources, sharing common controls and shared communication rules and intelligence. These capabilities can be accessed via a wide range of devices and interfaces to serve the constantly changing needs of mobile employees. OpenScape provides personal and workgroup communications portals for multiple healthcare domains. OpenScape makes extensive use of presence-based communication. There are different ways of showing presence and the user's status:

Device Presence: indicates the presence of an application or communications device. The user could be online or off-line. For example:

- Off hook/on hook status of an IP desk phone or mobile phone
- Instant Messaging (IM) application presence
- Collaboration application presence indicating whether an individual has signed on to the application or not
- On which terminal can the user be reached

User Presence: associates presence with an individual rather than the user's device

- The willingness of the user to be reached
- The activity that the user is currently engaged
- The location of the user—working remotely, out of office, and so on.
- Mood—happy, good or bad tempered or even annoyed

Device presence or terminal status will become less important as terminals become more advanced and intelligent, with the ability to handle a multitude of multimedia content adaptively. A mobile phone or PDA, for instance will be able to negotiate with the remote communication party the type of information it can support, for instance video, audio, and so on.

IMS (Siemens IMS) from Siemens Communications Mobile (http://www.siemens-mobile.com) provides new presence and communications platform for mobile devices and networks. The IMS platform was standardized for new multimedia applications and services that could be rapidly deployed by mobile network operators, such as audio/ video conferencing, chat, and presence services over new mobile devices. Moreover, IMS is positioned to voice and multimedia communication. Different from the legacy circuit-switched voice/data communication in regular phone calls, IMS is based on IP technology that can control real-time and non-real-time services on the same IP network.

Together OpenScape, IMS and Soarian provide the core technological building blocks that enable the integration of the healthcare enterprise content repository with a Web-based infrastructure and presence and availability. Our research work is layered above these three pillars and the following sections illustrate the benefits of a converged communication healthcare enterprise.

Figure 1. OpenScape: A way of communication

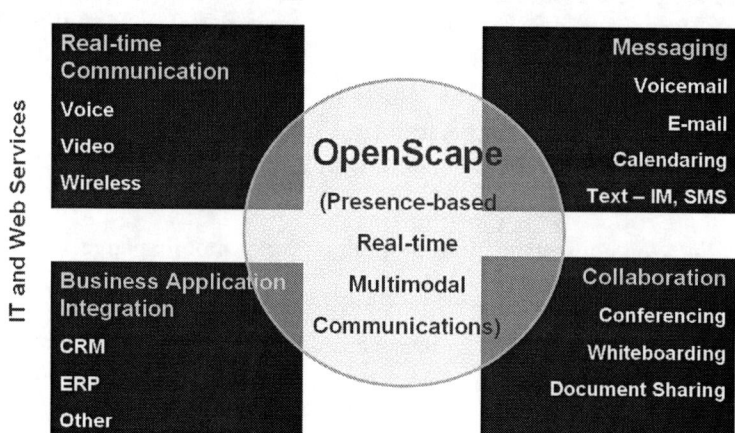

Enriching Collaborative Consultations

During a telehealth consultation between patient and clinician various medical documents might be used including laboratory reports, photos of injury/skin lesions, x-rays, pathology slides, EKGs, MR/CT images, medical claim forms, prescriptions, clinical results, case reports, and other documents. Some of the documents like photos and images might have been captured using camera during patient visits, while other clinical (lab reports) or financial information (insurance claim) were gleaned from other information systems. Face-to-face video conferencing is commonly used in telemedicine not only for personalized remote communications, but also for regulatory reasons as evidence of a consultation the costs for which can subsequently be reimbursed.

It would be convenient if a clinician could combine, interface, convert, and extract disparate medical information such as those listed above from peripheral devices like photo cameras, video conferencing session, content management systems, PACS, and regular office documents into a Web-based *composite* document. Thereafter, it would be convenient if this composite document could be used as the basis for browser-based collaboration (whether it is offline or real time) between various participants. Such a composite document generated should combine all relevant information needed for a particular collaboration session into one seamless document so that effective offline or real time collaboration can occur.

Let us now look at some of the details involved in realizing our solution. A user could choose specific pages from document(s) stored locally and automatically have the selected pages converted to HTML format and hyperlinked with each other to form a composite document. Then, he or she could highlight important parts of the document and add personal comments with the help of voice and graphic annotations using our multimedia presentation software, called *ShowMe* (Sastry, Lewis, & Pizano, 1999). The multimedia annotation technology developed is unique as it not only captures the spatial nature but also the temporal aspects. For instance, the multimedia annotation on a document would capture a synchronized temporal voice, graphic and mouse pointer annotations. Finally, the user could save the composite document along with the annotations on the local web server, and this document is referred to as the collaboration document. Associated with this collaboration document is some metadata in the form of an XML schema that describes this document. This metadata is finally uploaded to the central server and its URL on this central server can then be sent to other participants.

Participants can view the collaboration document via the URL of the metadata stored on the central Web server using a lightweight Web-browser player. As the collaboration documents are stored locally, they are amenable to document management tasks, including deleting, moving, and so on. However, any such document management task must be accompanied by making an appropriate change in the metadata located at the central Web server. The above process allows various users to collaborate over documents quickly and easily by only sharing information relevant to the topic in question. In addition, there is increase in productivity as users can quickly exchange information without having to exchange e-mail to explain problem/solution.

Figures 2, 3, and 4 show a particular workflow implemented to demonstrate the spectrum of modes of browser-based collaboration using office documents and Web content. The components can be re-used in several other collaboration workflows and processes. As only specific parts of different documents might be needed for a particular collaborative session, participants are able to combine, on the fly, specific pages from several documents with different formats into one seamless Web-based composite document. Using synchronous voice and graphic annotations along with mouse as a pointer, a clinician can continuously narrate a patient's case history across the entire composite

Figure 2. Multimedia Enhanced Store & Forward Collaboration: A clinician can quickly create one seamless web-based composite multimedia document by combining various medical information segments like images, photos, EKGs together. Synchronous voice, graphic and mouse annotations can then be easily added on top of the composite document before sending it via regular e-mail. The recipient needs only a web browser to view the composite document along with the voice and graphic annotations.

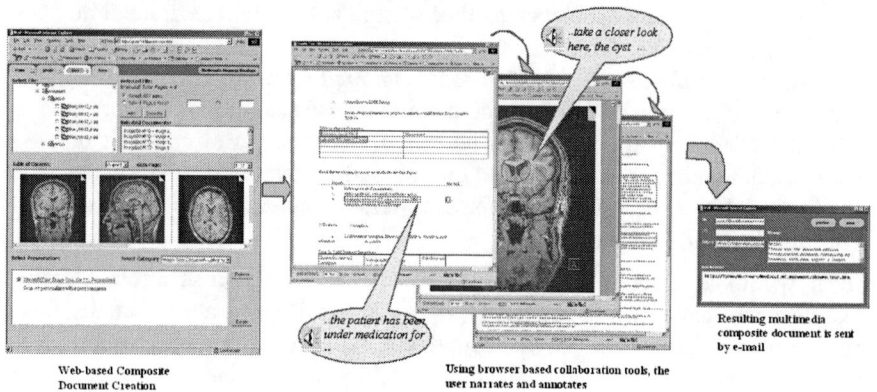

Figure 3. Instant Messaging Based Document Collaboration: This uses presence and availability of different participants to setup collaboration. Also, this allows asynchronous messaging of voice and graphic annotations within a real-time collaboration session. This enables users to exchange information at one's own pace and yet participate in a near real-time collaboration session.

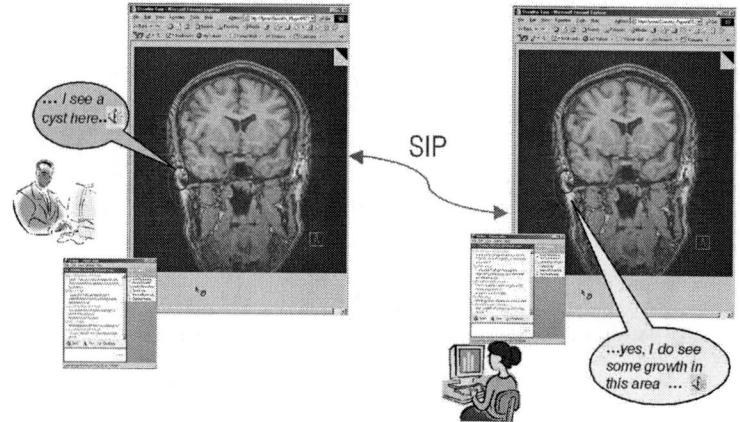

Copyright © 2005, Idea Group Inc. Copying or distributing in print or electronic forms without written permission of Idea Group Inc. is prohibited.

Figure 4. Multi-participant Real-Time Collaboration: Several remote participants using their web browser can collaborate on the generated composite medical document. The collaboration is accompanied by a voice conference.

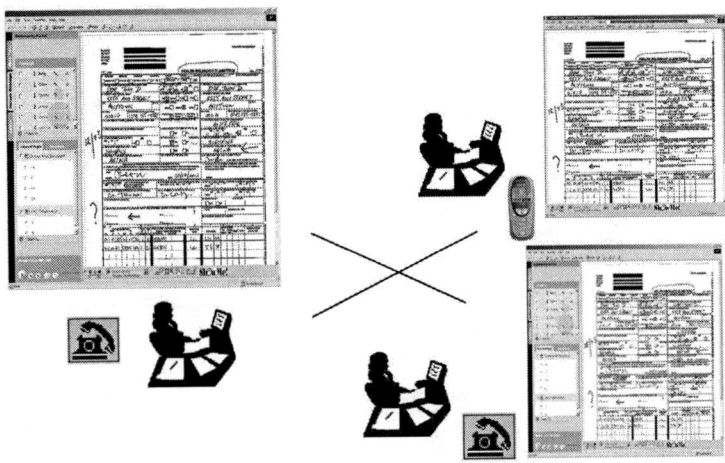

document while using the mouse to gesture or annotate on different parts of the aggregated document.

These modes of collaboration—whether off-line, IM-based, or real-time—take place using a regular web browser with the requisite collaboration plug-ins. This allows the participants to collaborate within the familiar environment of a Web browser. The annotations by various participants during a collaboration session on composite medical documents can be archived into a database in a very lightweight manner. The voice and graphic annotations along with the composite document can be archived with high granularity along with meta-data describing the various annotations. This would allow for searching and retrieval of not only medical information/documents exchanged during collaboration sessions, but also to obtain information regarding comments, interactions and dialogue between various participants. For instance, one can query and retrieve specific comments/annotations made by particular participant from a past collaboration session on a specific medical document. In addition, one can easily follow the changes in the medical condition of the patient by comparing the medical documents and associated annotations made during successive patient visits.

Collaboration Knowledge Access and Management

Access to and the management of the knowledge contained within the collaboration documents introduced above is reviewed in this section.

Filtering Collaboration Annotations

The ability to save annotations from a real time, offline or IM based document collaboration enables participants to filter and view a conference in several different ways. The figure below illustrates how an archived conference may be viewed in different ways by applying different filters. For example, the filters can be to show the conference view by participants, by documents collaborated on, by threaded view, by type of annotations, by session, and so on. New filters can be easily created based on the use case scenario.

Navigation, Access and Reusability of Knowledge Documents

The ability to archive annotations (both temporal and spatial) allows the users to retrieve and reuse segments of existing annotated documents from any mode of collaboration. For example, during a real time conference, several participants collaborated on different composite pages (where the composite pages might have come from different original documents). A user can select a particular page from an existing collaboration document, filter annotations based on any criteria (as described above), add any more composite pages, add more temporal/spatial annotations and collaborate in off-line, IM, or real time mode. This allows the enterprise to preserve the knowledge gathered during several collaboration sessions performed in any mode, while still enabling users to select parts of annotated documents from previous sessions for further reuse in future collaboration sessions.

Figure 5. The clinician can filter and quickly access specific segments from past collaborative sessions and reuse to address the current problem.

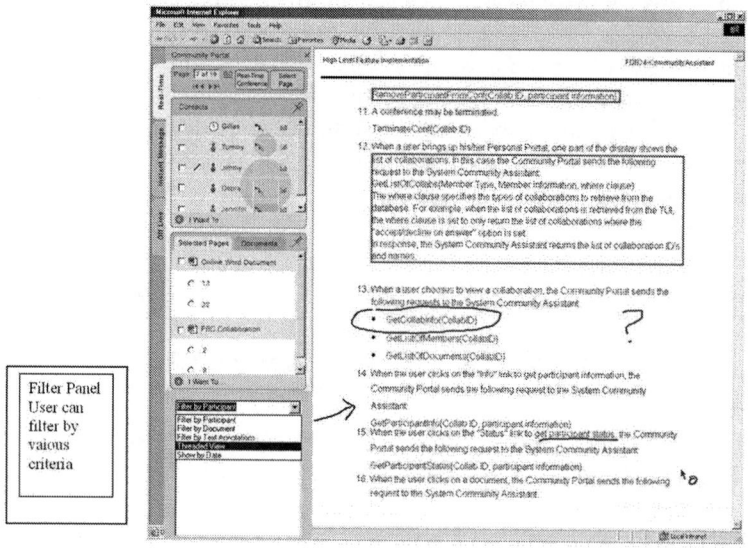

Multimedia Response and Discussion Boards

The structure of the collaboration archive allows one user to respond to any particular comment or annotation made by another user. The response could be in the form of a temporal or spatial annotation, or both. One scenario is that a sender can compose a composite document, make his or her annotations and send to the recipient(s). The annotated document could be sent as an attachment, or accessed from a central server. In the latter case, the annotated document is sent as a URL link. The recipient(s) can overlay their replies (annotations) and send back the annotated document, or the annotated document can be further annotated by other users (forwarding via e-mail.) By quickly overlaying temporal or spatial annotations it allows users to imbue information and temporal context to their responses.

Figure 6 shows document(s) being collaborated upon by several users in different collaboration modes using a threaded approach. These annotations could have been made in offline mode by different users or some of the annotations could have happened in a real time or IM conference and then imported into the discussion thread. For example, a user creates a composite document, adds various annotations in offline mode and sends it to her workgroup. The workgroup meets later and discusses in real time conference mode where annotations are captured as temporally with high granularity. During the real-time conference, they add more composite pages and perform collaborations above them. When finished, one user adds the resulting composite document with the temporal and spatial annotations to the discussion group. Subsequently, participants can add/edit/delete more comments (or annotations), or respond to any particular comment—hence a rich threaded discussion is facilitated. This scenario illustrates how group of users can work in a newsgroup discussion manner and leverage rich multimedia to express their thoughts.

Figure 6. A multimedia threaded discussion board

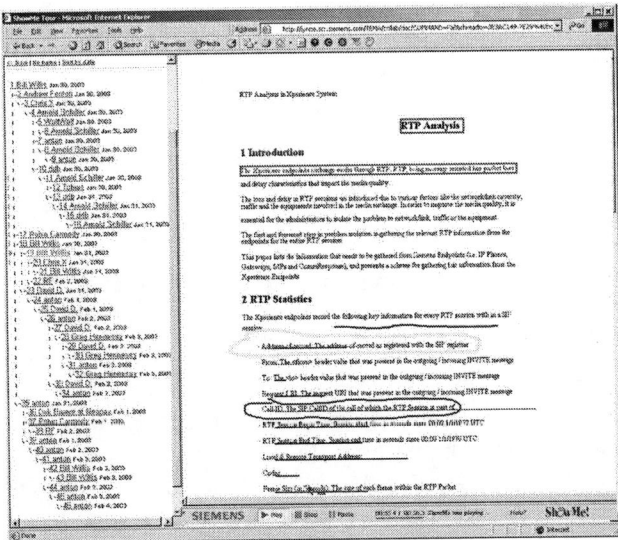

Query and Retrieval of Multimedia Knowledge Base

As reported in the previous sections, the collaboration archive of interactions—voice, graphic, text, or mouse pointer annotations—between participants over several documents in different collaboration modes being captured and archived in a structured manner over time. As knowledge in the form of annotations is captured and reused in high granularity, there is a need for a simple search mechanism to locate appropriate segments of annotated documents from a large archive of annotated composite documents.

Queries to the knowledge base can be based on various criteria, for example, using metadata captured during collaboration session, the annotations using voice, text and the document content, and also by using feature relative annotations one can get the textual content of the document on which annotations were overlaid.

Representation, Visualization and Browsing of Collaborations

To leverage the knowledge contained within archived collaboration sessions, we have also explored various approaches for representing, visualizing and browsing archived collaboration sessions. As collaborations often involve two dimensional documents and are by nature temporal, we are investigating representing the structure and content of collaborations in a three dimensional space. Arguably the most prevalent standard in the 3D graphics arena is VRML: a high-level textual language for describing the geometry and behavior of 3D scenes. Fortunately, a plethora of browser plug-ins are readily available. Following an iterative design/evaluation cycle, we converged on the design of a single VRML model with an interface that embodied and combined many of the characteristics that we sought:

- A viewpoint providing a visual overview/summary of the session
- A viewpoint providing a detail view of the session
- Interactivity and browsing capabilities at both overview and detail levels

Figures 7 and 8 show two viewpoints on the same VRML model, with each viewpoint having a distinct purpose.

The viewpoint shown in Figure 8 conveys the structure of the session (three topics were discussed), the documents used in the course of the discussion (indicated by the thumbnail icons on the right of the timeline), and that two action items were assigned (the red columns on the right of the timeline).

The viewpoint shown in Figure 8 exposes more detail regarding the content of the session. An audio waveform running along the timeline illustrates the presence of noise and is also color-indexed to the participants. An exploded view of the current document under discussion is show in the middle of the visualization. A small portrait image of the

Figure 7. A visual overview of a session that conveys structure, documents used, and action items assigned

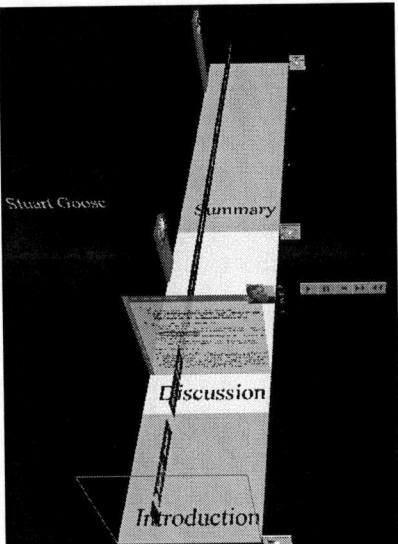

Figure 8. The detail view of a session

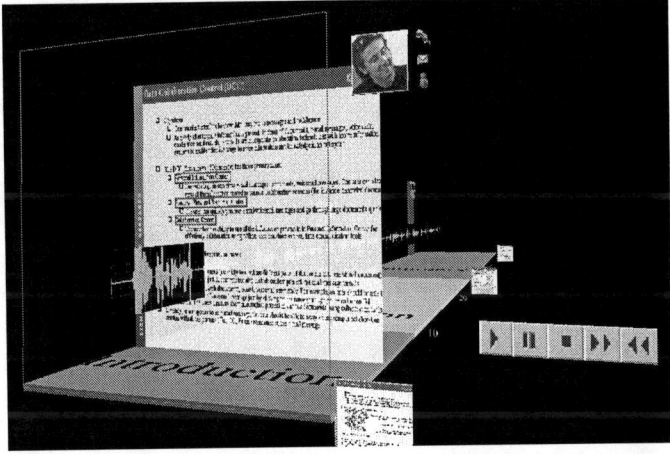

current speaker is shown in the corner of the current document, along with an icon that can be clicked on to initiate a phone call, e-mail, or instant message to that person. In addition, red columns along the left side of the timeline indicate assigned action items with a portrait image of the person responsible.

A VCR metaphor is provided to assist with navigation, and a VCR control panel can be seen at the right of the interface. When the play button is activated, the exploded view of the current document begins to glide backwards through time in synchrony with the audio stream played at normal speed. The forward and rewind buttons move the current position to the beginning of the respective session topic.

Copyright © 2005, Idea Group Inc. Copying or distributing in print or electronic forms without written permission of Idea Group Inc. is prohibited.

In Support of Home Health: Nomadic Nurse

Home healthcare nursing is an increasingly widespread type of healthcare that typically involves a nurse driving to a patient's place of residence to provide the necessary monitoring and treatment. There are a wide variety of cases in which home healthcare is recommended by physicians, and recent years have seen a steady increase in the need for this type of care (Giles, 1996). While many industry observers believe home healthcare will increase in the future, many current home healthcare providers are either non-profit agencies or operate with conservative profit margins. In addition, as nurses' time is a scarce resource and IT spending limited, any new technologies that are not economically priced or that require nurses to increase the time spent with patients have a high probability of being passed over or abandoned.

Home healthcare nursing is by nature a peripatetic profession, and as such nurses need to transport their equipment from one home to the next. From our studies described previously, it was observed that the preponderance of nurses used paper and pencil notes augmented by their memory to record the interactions with their patients throughout their working day. The written and memorized notes were entered into the computer-based system upon their return home or to the office. There were a plethora of reasons as to why some nurses chose to operate in this mode. For example, some asserted that anything that breaks the contact between the clinician and the patient breaks the treatment involved in the visit. "There is a healing involved in the physical touch between the nurse and the patient" was the expression used. To avoid breaking this important connection, the application would need to use less obtrusive techniques that still provide for sterile-hands operation. In addition, the application would need to integrate well with the nurses' workflow and their frequent need for non-linear information access.

Hardware and Software Selection

The hardware selection was influenced by the following factors:

- Laptops were identified as the device of choice for the majority of home healthcare clinicians interviewed during the study. Lowering costs of laptops make them relatively affordable even for low-budget agencies; lightweight of the devices is appropriate in context of patient visits.

- The study revealed that clinicians were open to the idea of using a PDA for reasons of size, weight, and no boot-up latency. However, concern was expressed as to whether the screen size would be adequate for their needs.

- New government regulations (HIPPA) place significant emphasis on ensuring secure access to sensitive patient information in order to preserve patient privacy. Towards satisfying the HIPPA regulations, a Siemens biometric mouse was incorporated (Siemens Biometric Mouse).

- To reduce the impact of the interaction upon the nursing tasks, a discreet wireless Bluetooth earpiece was incorporated. This enables the nurse to issue spoken commands, but also to receive spoken feedback without the patient hearing. The absence of wires was crucial so as not to inhibit the nursing duties.

The software technology selection was influenced by the following factors:

- Clinicians may access the application from a variety of different locations, including the office, their homes, patient homes, from the car while driving, etc.
- Administering clinical care requires sterile conditions, which makes traditional input devices unsuitable. This requirement indicates a multimodal interface (Multimodal Interaction Working Group), which, if necessary, can be controlled entirely by voice.
- Home healthcare practitioners would prefer continuous access to the patient database, however the existing wireless coverage of rural areas remains too fragmented to be reliable. The application needs to be able to work in either offline or online modes of connectivity.

Collectively, these requirements led us to conclude that a WWW-based solution was feasible, and that recent advances in multimodal technologies provide support on various devices. As such, the user interface was developed using a combination of HTML, Java and Javascript.

Notebook Design: Capturing Patient Vital Signs

The goal of our prototype is to show how cost-effective technology can be integrated into the workflow of a nurse to prove that patient vital signs can be captured unobtrusively. If successful, this could obviate the need for the nurse to enter this data into a computer system upon returning to the office.

In order to address the majority of agencies that expressed interest in notebook and PDA platforms, our design sought to offer a similar user experience while attempting to leverage the respective advantages of each. To provide hands-free operation, the nurse's notebook computer is equipped with a Bluetooth (Bluetooth) capability. Bluetooth wireless headset (Bluetooth headset from Siemens) supports mobile speech interaction with the application within an adequate radius. The initial implementation was developed for a notebook, as seen in Figure 9.

As can be seen from Figures 10 and 11, a multimodal interface allows the nurse to use the keyboard/mouse and/or speech to navigate and enter values into the visualization. For the laptop, SALT (SALT Forum) was selected as the technology used to develop the multimodal interface. In the classical SALT paradigm, speech recognition is initiated using either a keyboard or a mouse, but to offer truly hands-free operations we introduced some novel approaches to support continuous recognition and pause/resume function-

Figure 9. Nurse's laptop, Bluetooth PCMCIA card and Bluetooth enabled headset

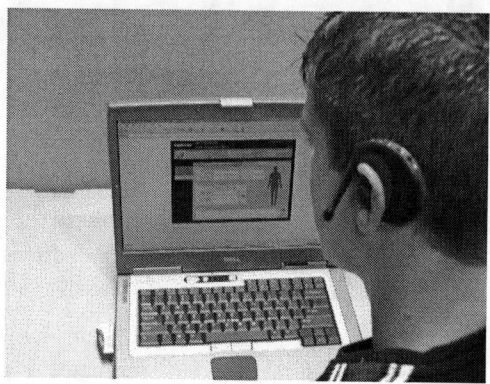

Figure 10. While taking measurements from the patient, the nurse can speak the clinical measurements via the Bluetooth enabled headset directly into the HTML form

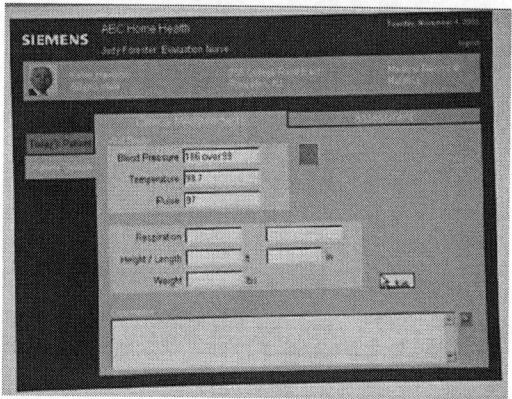

Figure 11. While interviewing the patient, the nurse can speak the diagnosis and functional assessment data via the Bluetooth enabled headset directly into the HTML form

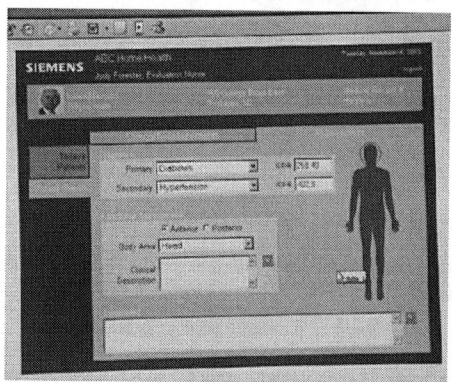

ality. This enables nurses to use verbal commands to indicate to the application that they are about to start dictating commands, or whether they are engaged in the conversation with a patient. Additionally, after temporarily deactivating the speech recognition, the nurse can resume working by simply issuing a voice command.

PDA Design: Less is More

The notebook implementation was heavily leveraged for the subsequent implementation for the Pocket PC PDA. While the functionality was preserved, the HTML was simplified and modified for appropriate consumption and interaction on the PDA form factor, as can be seen in Figure 12.

Although SALT technology can be demonstrated using a Pocket PC PDA, the speech processing is not performed locally on the mobile device but redirected to a server machine on the LAN. While this is not practical for deployment, it enabled us to experiment with the approach. It is anticipated that speech processing on the PDA will become available in the near future.

Speech is not the most appropriate input mechanism for every occasion, but one tool in the palette of a multimodal interface designer. As many of the nurses that we studied rely on paper and pencil to record patient notes, we sought to exploit handwriting recognition technology as a means to harness this activity and increase nurses' productivity by capturing this text and entering in directly into HTML form fields. This approach can be seen in Figure 12.

In our interviews we found that the old adage of a picture capturing a thousand words is not lost on nurses. Hence, we sought to offer in our prototype the seamless support for capturing images and integrating them directly into the patient information. A small and inexpensive digital camera peripheral connected via the SDIO port can be used for this purpose. The process of capture, abstraction through a thumbnail representation, and the viewing of the image can be seen in Figures 13 and 14.

Figure 12. Leveraging handwriting recognition to capture vital signs into an HTML form

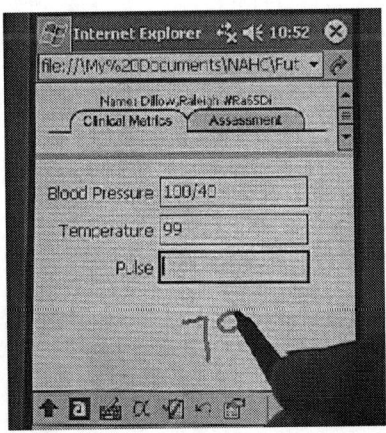 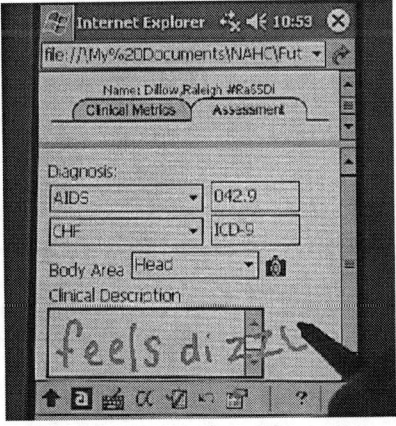

Figure 13. Convenient image capture is seamlessly integrated into the patient record

Figure 14. A thumbnail icon of the captured image can simply be clicked-on for closer inspection

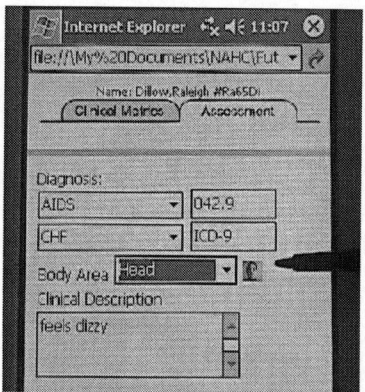
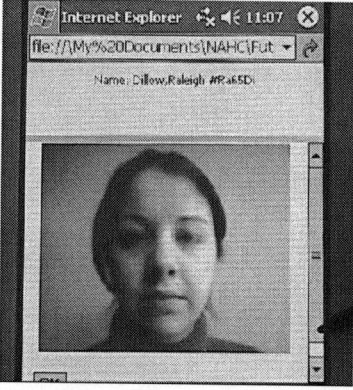

In addition to capturing an image, it is often desirable to be able to augment an image with a synchronized audio commentary and associated pen markings to describe specific aspects of the patient's condition. Hence, a multimedia annotation capability is offered that can support this requirement and be stored along with the patient record.

Current and future work includes conducting formal user studies with nursing practitioners from two home healthcare agencies to evaluate the usability and efficacy of the devices and approaches described above to inform the subsequent iteration of the prototype design.

Conclusions

The manner in which a networked enterprise can facilitate a plethora of ways for healthcare personnel to communicate and collaborate was explored within this chapter. The next generation of communication technologies augurs well for converged voice and data solutions on a single network. We anticipate a closer union between healthcare IT systems and web-based communications. Technology innovation has spawned a proliferation of communication and data devices, such as GPRS cellular phones and PDAs, and with it a significant opportunity for accessing clinical information anywhere, anytime, allowing healthcare practitioners to collaborate upon clinical information and reach conclusions more effectively.

Throughout the chapter a selection of technologies has been presented that can enable healthcare personnel to interact in a variety of styles. These approaches demonstrated how multimedia technologies can be harnessed to capture information and enable users to collaborate in range of modes in an effective manner. It was also described how the knowledge captured during collaboration interactions can be saved and archived for future search, reference and reusability. In addition, technology support for home healthcare nurses is presented that shows potential for streamlining the capture and entry of information into the patient record.

Acknowledgments

The authors would like to thank our team members at Siemens Corporate Research for their contributions to the technologies described.

References

Bluetooth (n.d.). Online *http://www.bluetooth.com*

Bluetooth headset from Siemens. Online *http://www.siemens-mobile.com/cds/frontdoor/ 0,2241,hq_en_0_49735_rArNrNrNrN,00.html*

Giles, T. (1996). The cost-effective way forward for the management of the patient with heart failure. *Cardiology, 87*(1), 33-39.

Hibbert, D., Mair, F.S., May, C.R., Boland, A., O'Connor, J., Capewell, S., & Angus, R.M. (2004). Health professionals responses to the introduction of a home telehealth service. *Journal of Telemedicine and Telecare 10*(4), 226-230.

HIPPA (n.d.). Online *http://www.hhs.gov/ocr/hipaa/*

Multimodal Interaction Working Group (n.d.). Online *http://www.w3.org/2002/mmi/*

SALT Forum (n.d.). Online *http://www.saltforum.org*

Sastry, C., Lewis, D. & Pizano, A. (1999). Webtour: A system to record and playback dynamic multimedia annotations on Web document content. *Proceedings of the ACM International Conference on Multimedia,* Orlando, October (pp. 175-178).

Siemens Biometric Mouse (n.d.). Online *http://www.siemensidmouse.com/*

Siemens IMS (n.d.). Online *http://www.siemens-mobile.com/cds/frontdoor/ 0%2C2241%2Chq_en_0_860_rArNrNrNrN%2C00.html*

Siemens OpenScape (n.d.). Online *http://www.siemensenterprise.com/prod_sol_serv/products/openscape/*

Siemens Soarian (n.d.). Online *http://www.medical.siemens.com/webapp/wcs/stores/servlet/CategoryDisplay?storeId=10001&langId=-1&catalogId=-1&catTree=100001,19051,19027&categoryId=19027*

Siemens Information and Communications Mobile (n.d.). Online *http://www.siemens-mobile.com*

Siemens Information and Communications Networks (n.d.). Online *http://www.icn.siemens.com*

Siemens Medical (n.d.). Online *http://www.medical.siemens.com*

Chapter IX

Biomedical Image Registration for Diagnostic Decision Making and Treatment Monitoring

Xiu Ying Wang, The University of Sydney, Australia and
Heilongjiang University, China

David Dagan Feng, The University of Sydney, Australia and
Hong Kong Polytechnic University, Hong Kong, China

Abstract

The chapter introduces biomedical image registration as a means of integrating and providing complementary and additional information from multiple medical images simultaneously to facilitate diagnostic decision-making and treatment monitoring. It focuses on the fundamental theories of biomedical image registration, major methodologies and contributions of this area, and the main applications of biomedical image registration in clinical contexts. Furthermore, discussions on the future challenges and possible research trends of this field are presented. The chapter aims to assist in a quick understanding of main methods and technologies, current issues, and major applications of biomedical image registration, to provide the connection between biomedical image registration and the related research areas, and finally to evoke novel and practical registration methods to improve the quality and safety of healthcare.

Copyright © 2005, Idea Group Inc. Copying or distributing in print or electronic forms without written permission of Idea Group Inc. is prohibited.

Introduction

Clinical knowledge management is a challenging and broad discipline related to the collection, processing, visualization, storage, preservation, and retrieval of health-related data and information to form useful knowledge for making critical clinical decisions. As an important part of clinical knowledge, medical images facilitate the understanding of anatomy and function, and are critical to research and healthcare. Medical imaging modalities can be divided into two major categories: anatomical modalities and functional modalities.

Anatomical modalities, mainly depicting morphology, include X-ray, computed tomography (CT), magnetic resonance imaging (MRI), ultrasound (US). Functional modalities, primarily describing information on the biochemistry of the underlying anatomy, include single photon emission computed tomography (SPECT) and positron emission tomography (PET). With the advances in medical imaging technologies, these imaging modalities are playing a more and more important role in improving the quality and efficiency of healthcare. For example, the functional imaging techniques can be used to image physiological and biochemical processes in different organs, such as brain, lung, liver, bone, thyroid, heart, and kidney (Figure 1). In such clinical settings, PET aids clinicians in choosing the most appropriate treatment and monitoring the patients' response to these therapies. Since information from multiple medical imaging modalities is usually of a complementary nature, proper extraction registration of the embedded information and knowledge is important in the healthcare decision making process and in clinical practice.

The combination of more advanced and user-friendly medical image databases is making medical imaging results more accessible to clinical professionals. Starting in the early 1990s, the Visible Human Project and Human Brain Project at the US National Library of Medicine have produced a widely available reference of multimodal images of the human body. These projects provide users with labeled data and the connection of structural-anatomical knowledge with functional-physiological knowledge (Ackerman, 2001; Riva, 2003), and assist in making image data more usable for clinical training and surgery simulation and planning. A significant step in these virtual reality projects is the collection and registration of medical images from multiple imaging modalities.

Clinical practice often involves collecting and integrating considerable amounts of multimodality medical imaging data over time intervals to improve the optimization and precision of clinical decision making and to achieve better, faster, and more cost-effective healthcare. For example, in neurosurgical planning, the proper registration of the functional information with the detailed anatomical background enables the surgeon to optimize the operation with minimal damage to the healthy organs. The accurate and efficient registration of the complementary information available from different imaging modalities provides a basis for diagnostic and medical decision-making, treatment monitoring, and healthcare support.

A key issue in clinical knowledge management is biomedical image registration, which provides an effective mechanism to integrate the relevant information and knowledge in clinical and medical decision-making, operation planning, and image guided surgery. Registration algorithms also offer new possibilities to analyze and visualize multimodal image datasets simultaneously.

Figure 1. Positron Emission Tomography (PET) (Courtesy of Hong Kong Sanatorium & Hospital)

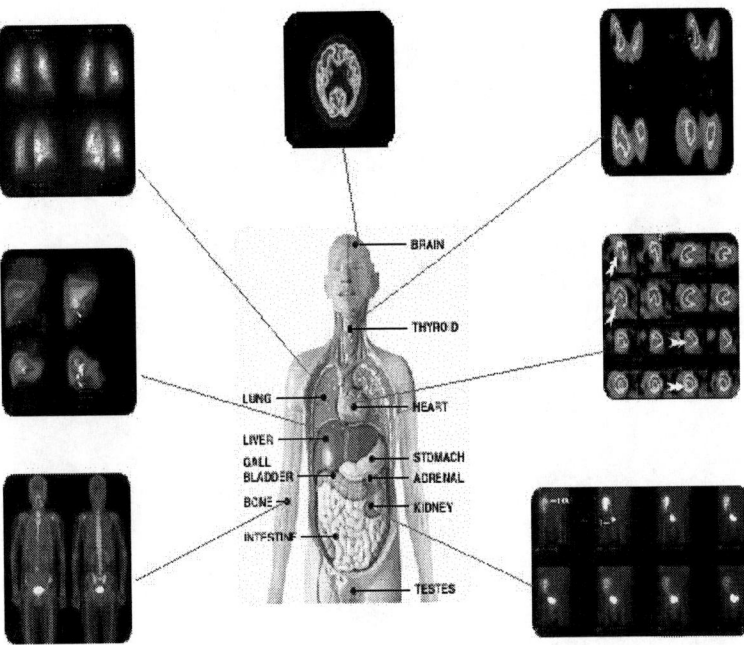

Using these algorithms, image data from multiple imaging modalities can be matched and presented in a common coordinate system, therefore, anatomical and functional image information can be visualized simultaneously. Hence, biomedical image registration enables clinical professionals a complete insight into the patient data and can help to improve medical diagnosis and patient treatment (Handels, 2003).

Applications of biomedical image registration include radiation therapy, interventional radiology, diagnostic and clinical decision-making, image-guided surgery, procedure planning, and simulation, dynamic structural and functional change measurement, treatment, and disease progression monitoring, minimally invasive procedures, and the correlations between the function and morphology of human body. Furthermore, image registration is widely used in biomedical imaging, which includes methods developed for automated image labeling and pathology detection in individuals and groups. Moreover, registration algorithms can encode patterns of anatomic variability in large human populations, and can be used to create disease-specific, population-based atlases (Bankman, 2000).

Although biomedical image registration has been intensively investigated and enormous advances in imaging techniques have been achieved, the ever-increasing growth of imaging data and their applications in medical and clinical environments ensures the existence of future challenges in more precise and efficient biomedical image registration.

Figure 2. Registration of PET image scan with anatomic maps (Courtesy of Hong Kong Sanatorium & Hospital)

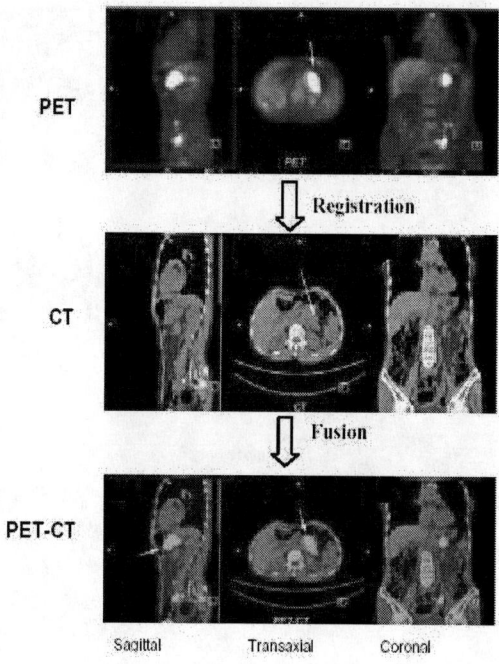

Background

Accurate and efficient biomedical image registration can lead to additional clinical information not apparent in the isolated images and provide clinical professionals with sufficient information for diagnostic and medical decision-making. For example, functional imaging such as PET cannot provide very high-resolution image data, but by the registration of these functional images with anatomical images, for example, CT scanning, physiological and functional regions can be located more precisely (Figure 2). After automatic image registration to localize and identify anatomy and lesions, accurate diagnostic and clinical decision-making can be achieved. Such functional-to-anatomical data registration is very useful for clinical diagnosis and surgical operation, especially for telesurgery. By presenting relevant clinical information to clinicians at the point of care, biomedical image registration can improve the quality of care, patient safety, and healthcare benefits.

As a fundamental task in image processing, the process of registration aims to match two data sets that may differ in time of acquisition, imaging sensors, or viewpoints. Because of its crucial role in improving healthcare quality, medical image registration has been studied extensively for decades, which has resulted in a bulk of reviews, surveys, and books, for example, Bankman (2000); Brown (1992); Fitzpatrick, Hill, and Maurer (2000); Lester and Arridge (1999); Maintz and Viergever (1998); Mäkelä (2002); Maurer and Fitzpatrick (1993); Rohr (2000); and van den Elsen, Pol, and Viergever (1993), . According to the registration feature space, principally, medical image registration can be distinguished into intensity-based registration and feature-based registration (Brown, 1992).

As one principal medical image registration methodology, intensity-based registration has attracted significant attention in the research community. As a result, numerous registration approaches have been proposed and used, for example, correlation based methods, Fourier-based approaches, the moment and principal axes methods [Alpert, Bradshaw, Kennedy, and Correia,(1990)], minimizing variance of intensity ratios (Woods, Mazziotta, and Cherry , 1993; Hill, Hawkes, Harrison, and Ruff, 1993), and mutual information methods (Collignon, Vandermeulen, Suetens, and Marchal 1995; Viola and Wells, 1995). Directly exploiting the image intensities, the intensity-based registration algorithms have the advantages of no segmentation required and few user interactions involved, and most importantly, these methods have potential to achieve fully automated registration. However, this category of schemes does not make use of a priori knowledge of the organ structure and the registration computation is not efficient. In order to improve the registration performance, speed, accuracy, and at same time to avoid the local minima, hierarchical medical image registration has been proposed, for example, Thevenaz and Unser (2000), and Pluim, Mainze, and Viergever (2001).

The other principal medical image registration category is based on corresponding features that can be extracted manually or automatically. The feature-based medical image registration methods can be classified into point-based approaches, for example, Besl and MaKey (1992); Bookstein (1992); and Fitzpatrick, West, and Maurer (1998); curve-based algorithms, for example, Maintz, van den Elsen, and Viergever (1996) and Subsol (1999); and surface-based methods, for example, Audette, Ferrie and Peters (2000) and Thompson and Toga (1996). One main advantage of feature-based registration is that the transformation can be stated in analytic form, which leads to efficient computational schemes. However, in the feature-based registration methodologies, a preprocessing step of detecting the features is needed and the registration results are highly dependent on the result of this preprocessing.

In medical image registration, a transformation which maps datasets obtained from different times, different viewpoints, and different sensors, must be determined. Depending on the characteristics of the differences between the medical images to be registered, generally, the registration transformations can be divided into rigid and non-rigid transformations. The rigid transformations can be used to cope with rotation and translation differences between the images. But usually, patient postures, tissue structures, and the shapes of the organs cannot always remain the same when they are imaged with different imaging devices or at different times, therefore, elastic or non-rigid registrations are required to cope with these differences between the images (Rohr, 2000). The elastic medical image registration was first introduced by Bajcsy (1989). As a challenging and active research topic, elastic medical image registration has attracted extensive attentions of researchers and a number of novel methods have been proposed; for example, a block matching strategy was used by Lin et al. (1994), and a flexible fluid model was proposed by Christensen, Kane, Marsh, and Vannier (1996). Elastic biomedical image registration is still an ongoing and challenging research topic and a lot of efforts are needed in this area.

Copyright © 2005, Idea Group Inc. Copying or distributing in print or electronic forms without written permission of Idea Group Inc. is prohibited.

Technological Fundamentals of Biomedical Image Registration

Biomedical Image Registration Definition

The problem of registration arises whenever images acquired from different sensors, at different times, or from different subjects need to be combined or compared for analysis or visualization. Biomedical image registration is the primary tool for comparing two or more medical images to discover the differences in the images or to combine information from multimodality medical images to reveal knowledge not accessible from individual images. Its main task is to determine a mapping to relate the pixels of one image to the corresponding pixels of a second image with respect to both space and intensity.

$$I_2 = g(I_1(f(x,y,z))) \tag{1}$$

I_2 and I_1 are 3-D images, indexed by (x,y,z);

$f: (x,y,z) \circledR (x',y',z')$, spatial transformation;

g: One-dimensional intensity transformation.

Figure 3 illustrates the basic registration steps and the corresponding functions of each step.

Figure 3. Biomedical image registration procedure

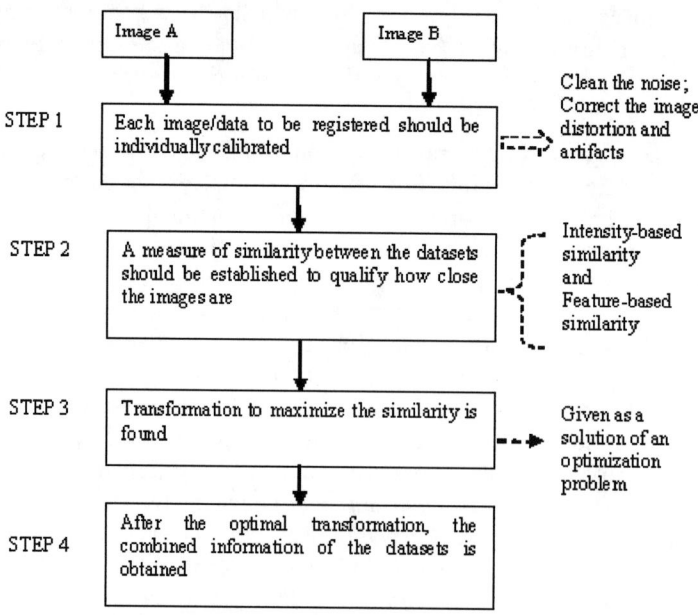

Biomedical Registration Transformations

Biomedical image distortions must be taken into consideration when two sets of medical images are to be registered. There are many factors that can result in medical image distortions, for example, different underlying physics of imaging sensors, inter-subject differences, voluntary and involuntary movements of the subject during imaging. These distortions impose many challenges for biomedical image registration because image characteristics and distortions determine the registration transformations. For more details on medical imaging deformation characteristics and registration transformations, please refer to Bankman (2000); Fitzpatrick, Hill, and Maurer (2000); Turner and Ordidge (2000). According to its transformation type, biomedical image registration can be divided into rigid registration and non-rigid registration.

The rigid registration is used to correct the simplest distortions caused by rotation and translation. Because of the rigid structure of the skull, the distortions of the brain images are often assumed as rigid distortions. When the brain image registration is carried out, the rigid transformation, which preserves the lengths and angle measures, is often used to correct these translation and rotation displacements.

Affine transformation, which maps parallel lines into parallel lines, is used for the correction of translation, rotation, scaling, and skewing of the coordinate space. For example, affine transformation is useful for the correction of skewing distortion in CT caused by tilted gantry.

Non-rigid medical image deformations can be caused by the dramatic changes of the subject positions, tissue structures, and the shapes of the organs when the subject is imaged with different imaging devices or at different times. Usually, different imaging devices require the subject to pose differently to get optimal imaging results, therefore, rigid and affine transformations are not sufficient for correcting these non-rigid deformations. The involuntary motions of the lung and the heart lead to elastic deformations which cannot be registered using rigid and affine transformations as well. Even brain structure cannot always be considered as the same over time because of the differences between pre-operation and post-operation. Non-rigid image registration is an active research area, which is important for correcting the soft tissue deformations, temporal displacements due to disease progression and surgical intervention, and individual variations. In elastic transformations, the straightness of lines cannot be preserved and the transformations can be arbitrarily complex.

Implementation Issues

Interpolation is required when an image needs to undergo transformations. Usually, the intra-slice resolution is higher than the inter-slice resolution, hence, the interpolation operation should be carried out to compensate for this difference. Images from different imaging modalities have different resolution, therefore in a multimodal image registration, lower resolution images are often interpolated to the sample space of the higher resolution images. Lehmann, Gönner, and Spitzer (1999) presented a survey of interpolation methods in medical image processing.

In biomedical image registration, the most frequently used interpolation methods include nearest neighbor, linear, bilinear, trilinear, cubic, bicubic, tricubic, quadrilinear, and cubic convolution interpolation. The more complex the interpolation methods, the more surrounding points concerned, and the slower the registration speed. In order to speed up the registration procedure, low cost interpolation techniques are often preferred. Because of its good trade-off between accuracy and computational complexity, the bilinear interpolation is the most commonly used method (Zitová and Flusser, 2003). According to the research of Thévenaz, Blu, and Unser (2000), in cardiac and thorax image registration, trilinear interpolation can help to achieve good registration performance.

The optimization algorithm is required by almost every registration procedure, which serves as a searching strategy. There are several optimization algorithms often used in the biomedical image registration. The exhaustive searching method has been selected as optimization strategy by researchers. However, because of its high computational complexity, the exhaustive method for searching for the global optimization is not an efficient choice. The Powell algorithm by Powell (1964) and Simplex method by Nelder and Mead (1965) are more efficient than the exhaustive searching strategy in finding an optimum solution.

The Powell algorithm has been used frequently as an optimization strategy for biomedical image registration, for example, Collignon et al. (1995), Maes, Collignon, Vandermeulen, Marchal, and Suetens (1997), Wang and Feng (2005). The Powell algorithm performs a succession of one-dimensional optimizations, finding in turn the best solution along each freedom degree, and then returning to the first degree of freedom. The algorithm stops when it is unable to find a new solution with a significant improvement to the current solution.

The Downhill-Simplex algorithm has been used by, for example, Hill et al. (1993) and van Herk and Kooy (1994). Rohlfing and Maurer (2003) adopted a variant of the Downhill-Simplex algorithm restricted to the direction of the steepest ascent.

In order to search a vast number of parameters, which represent the complex deformation fields, multi-resolution optimization algorithms have been adopted by researchers in the biomedical image registration community, for example, Penny (1998). Initially, the registration is performed at coarse spatial scales, then to the finer ones. These multi-resolution or coarse-to-fine optimization algorithms can accelerate computation and help to escape from the local minima.

Performance Validation of Biomedical Image Registration

For all types of registration, assessment of the registration accuracy is very important. A medical image registration method cannot be accepted as a clinical tool to make decisions about patient management until it has been proved to be accurate enough. Important criteria for assessing the performance of registration schemes are accuracy, robustness, usability, and computational complexity. The often used validation methods include Fiducial landmarks, Phantom studies, and Visual inspection.

Fiducial landmarks, which can predict the expected error distribution, have been devised to assess the registration accuracy. To a certain extent, assessment using fiducial landmarks provides a "gold standard" for medical image registration. However, the fiducial landmarks can either suffer from the movement of skin mobility or are highly invasive. Also, these validation measures cannot be applied retrospectively.

Phantom studies are important for the estimation of the registration accuracy because the data and displacement information is fully known beforehand. Phantom-based validations provide measures for mean transformation errors and computational complexity for different methods, and they are especially useful for estimating the accuracy of intra-modality registration methods.

Visual inspection, which provides a qualitative assessment, is the most intuitive method for evaluation of the registration accuracy. This assessment method may involve the inspection of subtraction images, contour overlays, or viewing anatomical landmarks. It has been used widely in both rigid and non-rigid registration assessment, but it may be considered as an informal and insufficient approach.

Researchers have been developing novel and practical validation techniques, for example, Hellier, Barillot, Memin and Perez (2001) proposed a hierarchical estimation method for 3-D registration. The measurement of the consistency of transformations has been proposed to serve as accuracy qualification method by, for example, Holden et al. (2000). Wang, Feng, Yeh, and Huang (2001) proposed a novel automatic method to estimate confidence intervals of the resulting registration parameters and allow the precision of registration results to be objectively assessed for 2-D and 3-D medical images. Fitzpatrick, Hill, and Maurer (2000); Mäkelä, Clarysse, Sipilä, Pauna, Pham, Katila, and Magnin (2002), and Zitová and Flusser (2003) presented good discussions and summaries about performance validation methods for medical image registration.

However, validation of registration accuracy is a difficult task because of the lack of the ground truth. Objective performance validation still remains a challenge in the field of biomedical image registration.

Major Biomedical Image Registration Methodologies

As described previously, the biomedical image registration methods can be divided into intensity-based category and feature-based category.

Intensity-Based Medical Image Registration

Intensity-based medical image registration fully and directly exploits the image raw intensities, and an explicit segmentation of the images is not required. Therefore, this category of registration provides high registration accuracy. Because of little or no user interaction involved in this kind of registration, fully automated registration and quantitative assessment become possible. However, this category of schemes does not make use of knowledge of the organ structure and also the registration computation is not very efficient. The well-established intensity-based similarity measures used in the biomedical image registration area include minimizing the intensity differences, correlation-based techniques, and entropy-based techniques.

Similarity measures by minimizing the intensity differences include the Sum of Squared Differences (SSD) and the Sum of Absolute Differences (SAD), which exhibit a minimum in the case of perfect matching. Although they are efficient to calculate, these methods are sensitive to intensity changes.

$$SSD = \sum_{i}^{N} |R(i) - T(S(i))|^2 \qquad (2)$$

$$SAD = \frac{1}{N} \sum_{i}^{N} |R(i) - T(S(i))| \qquad (3)$$

Where $R(i)$ is the intensity value at position i of reference image R and $S(i)$ is the corresponding intensity value in study image S; T is geometric transformation.

Correlation techniques were proposed to aim at multimodal biomedical image registration, for example, Maintz, van den Elsen, and Viergeve (1996). The cross-correlation technique has also been used for rigid motion correction of SPECT cardiac images, for example, Mäkelä et al. (2002). However, because usually the geometric deformations of the image modalities are not likely to be linear, these correlation methods, which require a linear dependence between the intensity of the images, cannot always achieve reliable registration results. The normalized cross correlation is defined as:

$$CR = \frac{\sum_{i}(I_R(i) - \bar{I}_R)(I_S(i) - \bar{I}_S)}{\sqrt{\sum_{i}(I_R(i) - \bar{I}_R)^2} \sqrt{\sum_{i}(I_S(i) - \bar{I}_S)^2}} \qquad (4)$$

Where $I_R(i)$ is the intensity value at position i of reference image R and $I_S(i)$ is the corresponding intensity value in study image S; $\bar{I}_R$ and $\bar{I}_S$ are the mean intensity value of reference and study image respectively.

Information theoretic techniques play an essential role in multimodality medical image registration.

The Shannon entropy is widely used as a measure of information in many branches of engineering. It was originally developed as a part of information theory in the 1940s and describes the average information supplied by a set of symbols $\{x\}$ whose probabilities are given by $\{p(x)\}$.

$$H = -\sum_{x} p(x)\log p(x) \tag{5}$$

In image registration area, when the images are correctly aligned, the joint histograms have tight clusters and the joint entropy is minimized. These clusters disperse as the images become less well registered, and correspondingly, the joint entropy is increased. Because minimizing the entropy does not require that the histograms are unimodal, the joint entropy is generally applicable to multimodality registration and obviates the need of segmentation of images.

Mutual information (MI) was first proposed by Collignon et al. (1995) and Viola et al. (1995), and it is a promising and powerful criterion for multimodality medical image registration, for example, Maes, et al. (1997); Roche, Malandain, and Ayache (2000); Thevenaz and Unser (2000); and Likar and Pernus (2001).

Let R be the reference data presented by m samples $\{r_0, r_1, ..., r_{m-1}\}$ with a marginal probability distribution $P_R(r)$. Analogously, the study data S consists of n samples $\{s_0, s_1, ..., s_{n-1}\}$ with a marginal probability distribution $P_S(s)$. The mutual information I of the reference image R and study image S measures the degree of dependence of R and S by measuring distance between the joint distribution $P_{RS}(r,s)$ and the distribution associated to $P_R(r)$ and $P_S(s)$. MI can be defined as:

$$I_{R,S} = \sum_{(r,s)} P_{RS}(r,s) \log\left(\frac{P_{RS}(r,s)}{P_R(r)P_S(s)}\right) \tag{6}$$

With $H(R)$ and $H(s)$ being the entropy of R and S, respectively, $H(R,S)$ is their joint entropy.

$$H(R) = -\sum_{r} P_R(r) \log P_R(r) \tag{7}$$

$$H(S) = -\sum_{s} P_S(s) \log P_S(s) \tag{8}$$

$$H(R,S) = -\sum_{r,s} P_{RS}(r,s) \log P_{RS}(r,s) \tag{9}$$

MI is related to entropy by the equation:

$$I_{R,s} = H(R) + H(S) - H(R,S) \tag{10}$$

Under the assumption that the mutual information of the two images is maximum when the images are in registration, registration can be performed by maximizing the mutual information as a function of a geometric transformation T of the study image S:

$$S_T = T(S) \tag{11}$$

$$I_{R,S_T} = \sum_{(r,T(s))} P_{R,S_T}(r,T(s)) \log(\frac{P_{RS_T}(r,T(s))}{P_R(r)P_{S_T}(T(s))}) \tag{12}$$

$$T_{reg} = \arg\max_T (I_{R,S_T}) \tag{13}$$

T_{reg} is the transformation that will bring the images into registration. Mutual information registration does not assume a linear relationship among intensity values of the images to be registered and is one of the few intensity-based measures that are well suited to the multimodality image registration.

Feature-Based Medical Image Registration

In feature-based registration approaches, transformations often can be stated in analytic form, hence efficient computational schemes can be achieved. However, in most of these methods, the preprocess step is needed and the registration results are highly dependent on the result of this preprocess. Because registration algorithms using landmarks often require users to specify corresponding landmarks from the two images manually or semi-automatically, such methods cannot always provide very accurate registration.

The feature-based medical image registration methods can be classified into point-based approaches, for example, Fitzpatrick, West, and Maurer (1998), curve-based algorithms, for example, Maintz et al. (1996), and Subsol (1999), and surface-based methods, for example, Chen, Pellizari, Chen, Cooper, and Levin (1987); Borgefors (1988), and Pellizari, Chen, Spelbring, Weichselbaum, and Chen (1989). Figure 4 illustrates the feature-based registration procedure. Point-based registration involves identifying the corresponding points, matching the points, and inferring the image transformation.

The corresponding points are also called homologous landmarks to emphasize that they should present the same feature in the different images. These points can either be anatomical features or markers attached to the patient, which can be identified in both images modalities. Anatomical landmark based registration methods have the drawback of user interaction being required. Registration algorithms based on extrinsic landmarks which maybe invasive or non-invasive, are comparatively easy to implement, fast, and can be automated, but they may have drawbacks of invasiveness and less accurate results. As a successful example, iterative closest points (ICP) method proposed by Besl and McKay (1992), maybe the most widely used medical image registration approach in medical imaging applications, for example, Fitzpatrick, West, and Maurer (1998).

When points are available, Thin-Plate Splines (TPS) which produce a smoothly interpolated spatial mapping, are often used to determine the transformation for 2-D medical image registration, for example, Bookstein (1989).

Boundaries or surfaces are distinct features in medical image registration due to various segmentation algorithms which can successfully locate such features. Surface-based registration methods can be rigid or deformable.

Figure 4. Feature-based registration procedure

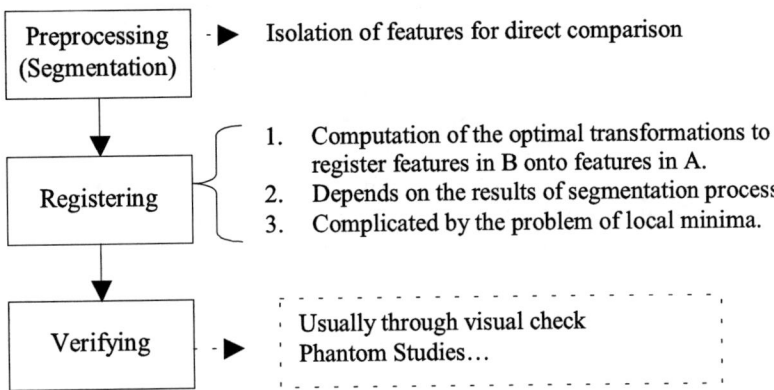

In rigid surface-based registration methods, the same anatomical structure surfaces are extracted from the images and used as input for the registration procedure. The Head-and-Hat algorithm, proposed by Chen et al. (1987) and Pellizari, Chen, Spelbring, Weichselbaum, and Chen (1989), is one successful surface fitting technique for multimodal image registration. In this method, two equivalent surfaces are identified in the images. The first surface extracted from the higher-resolution images, is represented as a stack of discs, and is referred to as "head". The second surface, referred to as "hat", is represented as a list of unconnected 3D points. The registration is determined by iteratively transforming the hat surface with respect to the head surface, until the closest fit of the hat onto the head is found. Because the segmentation task is comparatively easy, and the computational cost is relatively low, this method remains popular. However, this method is prone to error for convoluted surfaces.

In deformable surface-based registration methods, the extracted surfaces or curves from one image is elastically deformed to fit the second image. The deformable curves are known as snakes or active contours which help to fit contours or surfaces to image data. Snakes operate by simulating a controllable elastic material, much like a thin, flexible sheet. We can initially position the model by using information from anatomical atlases; then, the model is allowed to relax to a stationary position. This minimum energy position seeks to find the best position to trade off internal and external forces. The internal forces are due to the elastic nature of the material and the external forces stem from sharp boundaries in image intensity. Important deformable surface-based registration approaches include, for example, the elastic matching approach proposed by Bajcsy and Kovacic (1989), and the finite-element model technique proposed by Terzopoulos and Metaxas (1991), and so on. Deformable surface-based registration is suited for intersubject and atlas registration. A drawback of these methods is that a good initial pre-registration is required to achieve a proper convergence.

Applications of Biomedical Image Registration

Clinical and Surgical Applications

Biomedical image registration algorithms can combine information from multiple imaging modalities, allowing the evaluation of the progress of disease or treatment from time series of data, and the measurement of dynamic patterns of structural changes, tumor growth, or degenerative disease processes. Biomedical image registration can be used in medical and surgical areas, including diagnostic planning and simulation, treatment monitoring, image-guided surgery, pathology detection, and radiotherapy treatment.

Multimodal image registration methods play a central role in therapeutic systems, for example, Hill et al. (2000), and Lee, Nagano, Duerk, Sodee, and Wilson (2003). By registering the information from different imaging modalities, better and more accurate information can be obtained to aid the therapy planning. For example, the registration of functional images with anatomical images helps early diagnosis and better localization of pathological areas. Also, by quantitative comparison of images taken at different times, the information about evolution over time can be referred, for example, the monitoring of tumor growth in image sequences.

Many surgical procedures require highly precise 3D localization to extract deeply buried targeted tissue while minimizing collateral damage to adjacent structures (Grimson, 1995). Image-guided surgery (Figure 5) emerged to meet this end. In surgical practice, surgeons usually examine 2-D anatomical images (MRI or CT) and then mentally transfer the information to the patient. Thus, there is a clear need for registered visualization technique which allows the surgeons to directly visualize important structures to guide the surgical procedure.

Biomedical image registration is important for telemedicine, which is the integration of telecommunication technologies, information technologies, human-machine interface technologies, and medical care technologies, when distance separates the participants. In the case of healthcare, a telemedicine system should be able to register multiple sources of patient data, diagnostic images, and integrate other information to enhance healthcare delivery across space and time. For example, biomedical image registration is an important component in teleradiology, which is a primary image-related application.

Biomedical Image Registration for Different Organs

Medical image registration has been applied to the diagnosis of breast cancer, cardiac studies, wrist and other injuries, and different neurological disorders including brain tumors, and so on.

Brain image registration is the most intensively studied subject in the biomedical image registration field. Numerous rigid and non-rigid registration algorithms for monomodal

Figure 5. Image-guided surgical system

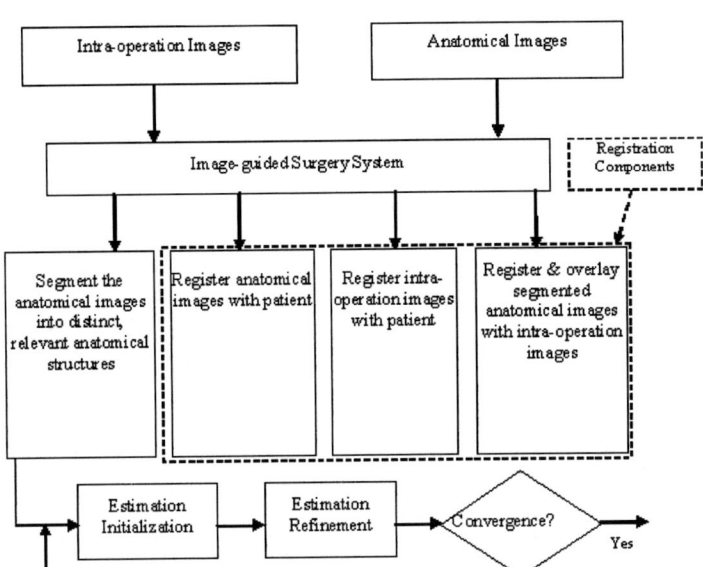

and multimodal images have been proposed. Feature-based biomedical image registration includes, for examples, crest-line-based registration (Guéziec and Ayache 1992, etc.), chamfer-matching-based registration (van Herk and Kooy 1994; Xiao and Jackson 1995), head-hat surface matching technique (Pelizzari et al. 1989), and ICP algorithm (Besl and McKay 1992). Intensity-based head image registration algorithms include: minimization of variation ratios (Woods 1993; Hill 1993), correlation-based registration (Collins, Neelin, Peters, and Evans 1994), and mutual information registration (Collignon et al. 1995), and so on. Figure 6 illustrates the main methodologies of brain medical image registration.

The registration of cardiac images from multiple imaging modalities is a preliminary step to combine anatomic and functional information. The integration of the complementary data provides a more comprehensive analysis of the cardiac functions and pathologies, and additional and useful information for physiologic understanding and diagnosis.

Because of the lack of anatomical landmarks and the low image resolution, the cardiac image registration is more complex than brain image registration. The non-rigid and mixed motion of the heart and the thorax structures makes the task even more difficult. Researchers have proposed numerous registration approaches for cardiac images, for example, Mäkelä et al. (2002); McLeish, Hill, Atkinson, Blackall, and Razavi; Pallotta et al. (1995); Thirion (1998) published a good review of cardiac image registration methods. However, cardiac image registration remains a challenge because of a number of problems related to the existing registration methods. For example, point-based registration approaches for the heart are not always accurate because of the lack of accurate anatomical landmark points in the cardiac; using heart surfaces can result in better registration of the region of interest, but the registration result is highly dependent on

Figure 6. Brain image registration

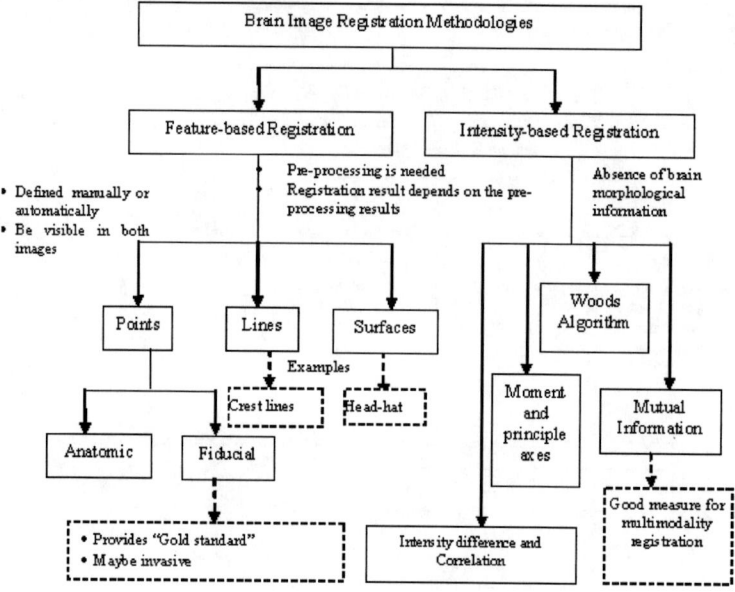

the surfaces selected and the imaging modalities. In intensity-based cardiac image registration, the use of image intensity difference and correlation methods relies on the assumption that intensity values in the registered images are strongly correlated. However, this assumption, especially in multimodal registration, is frequently violated, which would lead to unsatisfactory results.

Challenges and Future Trends

Although it has attracted considerable researchers, biomedical image registration is not widely applied in routine clinical practice. With automatic continuous developments of medical imaging techniques and their applications in clinical areas, biomedical image registration will remain a challenge in the future. Duncan and Ayache (2000) presented an excellent prospective of challenges ahead in medical image analysis area. In this section, we summarize a few of the many possible and potential research trends in the biomedical image registration area.

Active Research Areas and Open Issues

Although an enormous number of biomedical image registration methods have been proposed, researchers are still facing challenges of producing registration approaches with high precision, efficiency, and validity, which can be used in clinical practice. Hierarchical biomedical image registration and hybrid biomedical image registration are two registration schemes that have advantages of both increased computation efficiency and the ability to find better solutions.

In the hierarchical registration methods, the images are first registered at coarse, lower-resolution scales, and then the transformation solution obtained at this resolution scale is used as the initial estimation for the registration at a higher-resolution scale. The advantages of the hierarchical biomedical image registration approaches include accelerating computation efficiency and avoiding local minima, and therefore, improving the registration performance (Lester & Addridge, 1999).

The challenges created by inter-subject variations in the organ structures promote researchers to explore the hybrid approaches for biomedical image registration. Hybrid registration approaches, combining the intensity-based algorithms with landmark-based methods and making use of the merits of both these methods, have potential to achieve automatic and high performance biomedical registration results. Hence, objective criteria can be defined to identify how organ structure is altered by aging, gender, disease, and genetic factors. Deformable organ registration remains a challenge because of the differences in organ shape and volume, complex motion sources, and specific characteristics of different imaging modalities.

Mental integration of image information from different modalities is subjective, less accurate, and time-consuming. Therefore, in order to benefit clinical safety and facilitate clinical decision making, automatic registration, especially for the deformable organs such as heart, lung, and liver, is highly desired. Elastic registration approaches are particularly promising for the integration of deformable organ information from multiple imaging modalities. Currently, there is still no general automatic approach for the registration of heart images, lung images, and liver images. Hybrid methods, combining similarity measures with morphological information may provide possibilities for elastic registration.

The validation of the registration performance is particularly important. Although a wide variety of registration approaches have been proposed, objective validation of these methods is not well established. Image databases may in the future provide a source for the objective comparison of different registration methods.

Future Trends

Precise and efficient biomedical image registration is not only a big challenge, but also provides exciting opportunities to improve the quality and safety of diagnostic and medical decision making, treatment monitoring, and healthcare support. Although the more advanced imaging system, the PET scanner containing a CT scanner, has been developed, there is still a need for multi-dimensional, multimodality image registration techniques to assist the analysis of temporal changes and the integration of necessary information from different imaging modalities. With ever-increasing growth of medical datasets with higher resolution, higher dimensionality, and wider range of scanned areas, the demand for more efficient biomedical image registration will increase.

The applications of multimedia techniques, for example, electronic patient records and medical images, greatly push the advance of telemedicine and e-health. As an important component and technique of telemedicine and e-health, accurate and efficient biomedical image registration will play a more and more important role in remote diagnosis, patient monitoring, teleradiology, and overcoming the barriers of distance in healthcare service.

Although the existing image-based virtual human can provide the healthcare professionals with a quality of anatomical information and knowledge, there is a need to produce virtual humans with both anatomical and functional information and knowledge. Hence, whole-body multimodality image registration needs further efforts to support the virtual human projects which are essential in surgery simulation and virtual and augmented reality in medicine.

As an active research area, biomedical image registration will continue attracting researchers to develop automatic, non-rigid registration methods, which will facilitate clinical decision making, treatment monitoring, and surgical planning. The research of biomedical image registration will greatly promote the development and advance of medical imaging techniques, patient care service, and medical education. The multimodality biomedical image registration will be more and more important in medical diagnosis, surgery planning as well as intraoperative navigation, and in the future, biomedical image registration will play a more essential role in helping people to discover the mysteries of the human body and its complicated functions.

Conclusions

Biomedical image registration of different medical images, which aims to extract and combine the complementary and useful information and knowledge provided by the individual images, is an important step to a more comprehensive and accurate analysis of the organ functions and pathologies. Biomedical image registration has been extensively investigated and numerous methodologies have been proposed in this area.

Monomodality image registration is essential for follow-up treatment and disease monitoring, while multimodality image registration provides complementary and additional information for diagnosis, surgical planning, and treatment assessment. Rigid registration is often used to correct translation and rotation displacement, but it is not sufficient for correcting the complex, non-linear distortions, which may result from factors such as differences between the imaging modalities, the temporal displacements due to disease progression and surgical intervention, and individual variations. Non-rigid biomedical image registration remains an active and challenging area for both brain and other deformable organs.

As an important issue of clinical knowledge management, biomedical image registration is helpful in preparing medical data to be more useful for diagnostic and clinical decision making, treatment monitoring, and healthcare supporting, and it provides critical solutions in some cases for achieving better and more efficient healthcare.

Acknowledgments

This work is supported by the ARC, UGC, HLJNSF and HLJE grants.

References

Ackerman, M.J., Yoo, T., & Jenkins, D. (2001). From data to knowledge: The visible human project continues. *Medinfo, 10*(2), 887-890.

Alpert, N.M., Bradshaw, J.F., Kennedy, D. & Correia, J.A. (1990). The principal axes transformation: A method for image registration. *Journal of Nuclear Medicine, 31,* 1717-1722.

Audette, M., Ferrie, F. & Peters, T. (2000). An algorithm overview of surface registration techniques for medical imaging. *Medical Image Analysis, 4*(4), 201-217.

Bajcsy, R. & Kovacic, S. (1989). Multiresolution elastic matching. *Computer Vision, Graphics, and Image Processing, 46,* 1-21.

Bankman, I. (2000). *Handbook of medical imaging: Processing and analysis.* Academic Press.

Besl, P.J. & MaKey, N.D. (1992). A method for registration of 3-D shapes. *IEEE Transactions on Pattern Analysis and Machine Intelligence, 14*(2), 239-256.

Bookstein, F. L. (1989). Principal warps: Thin-plate splines and the decomposition of deformations. *IEEE Transactions on Pattern Analysis and Machine Intelligence, 11*(6), 567-585.

Borgefors, G. (1988). Hierarchical chamfer matching: A parametric edge matching algorithm. *IEEE Transactions on Pattern Analysis and Machine Intelligence, 10,* 849-865.

Brown, L.G. (1992). A survey of image registration techniques. *ACM Computing Surveys 24*(4), 325-376.

Chen, C., Pellizari, C.A., Chen, G.T.Y., Cooper, M.D. & Levin, D.N. (1987). Image analysis of PET data with the aid of CT and MR images. *Information Processing in Medical Imaging,* 601-611.

Christensen, G.E., Kane, A.A., Marsh, J.L. & Vannier, M.W. (1996). Synthesis of an individual cranial atlas with dysmorphic shape. *Mathematical methods in biomedical image analysis* (pp. 309-318). Los Alamitos, CA: IEEE Computer Society Press.

Collignon, A., Vandermeulen, D., Suetens, P. & Marchal, G. (1995). 3D multi-modality image registration using feature space clustering. In N. Ayche (Ed.), *Computer Vision, Virtual Reality, and Robotic Mechine, 905, of Lecture Notes in Computer Science* (pp. 195-204). Berlin: Springer-Verlag.

Collins, D.L., Neelin, P., Peters, T.M. & Evans, A.C. (1994). Automatic 3D intersubject registration of MR volumetric data in standardized Talairach space. *Journal of Computer Assisted Tomography, 18*(2), 192-205.

Davatzikos, C.A., Prince, J.L. & Bryan, R.N. (1996). Image registration based on boundary mapping. *IEEE Transactions on Medical Imaging, 15*, 112-115.

Dey, D., Slomka, P.J., Hahn, L.J., & Kloiber, R. (1999). Automatic three-dimensional multimodality registration using radionuclide transmission CT attenuation maps: A phantom study. *Journal of Nuclear Medicine, 40*(3), 448-455.

Duncan, J.S. & Ayache, N. (2000). Medical image analysis: Progress over two decades and the challenges ahead. *IEEE Transactions on Pattern and Machine Intelligence, 22*(1), 58-106.

Fitzpatrick, J.M. (2001). Detecting failure, assessing success. In J.V. Hajnal, D.L.G. Hill & D.J.E. Hawkes (Eds.), *Medical image registration* (pp. 117-139). CRC Press.

Fitzpatrick, J.M., Hill, D.L.G. & Maurer, C.R. (2000). *Handbook of medical imaging.* Bellingham, WA: SPIE Press.

Fitzpatrick, J.M., West, J.B. & Maurer, C.R. (1998). Predicting error in rigid-body point-based registration. *IEEE Transactions on Medical Imaging, 17*, 694-702.

Grimson, W.E.L. (1995). Medical applications of image understanding. *IEEE Expert, 10*(5), 18-28.

Guéziec, A. & Ayache, N. (1992). Smoothing and matching of 3-D space curves. In R.A. Robb (Ed.), *Visualization in biomedical computing,* 1808, Proc. SPIE (pp. 259-273). Bellingham, WA: SPIE PRESS.

Hajnal, J., Saeed, N., Soar, E., Oatridge, A., Young, I. & Bydder, G. (1995). Detection of subtle brain changes using subvoxel registration and subtraction of serial MR images. *Journal of Computer Assisted Tomography, 19*(5), 677-691.

Handels, H. (2003). Medical image processing: New perspectives in computer supported diagnostics, computer aided surgery and medical education and training. *Year Book of Medical Informatics*, 503-505.

Hellier, P., Barillot, C., Memin, E. & Perez, P. (2001). Hierarchical estimation of a dense deformation field for 3-D robust registration. *IEEE Transactions on Medical Imaging, 20,* 388-402.

Hill, D. L. G., Batchelor, P. G., Holden, M. H. & Hawkes, D. J. (2001). Medical image registration. *Physics in Medicine and Biology, 46*(1), 1-45.

Hill, D. L. G., Hawkes, D. J., Harrison, N. A. & Ruff, C.F. (1993). A strategy for automated multimodality image registration incorporating anatomical knowledge and imager characteristics. In H.H. Barrett & A.F. Gmitro (Eds.), *Lecture Notes in Computer Science. Proceedings of the 13th International Conference on Information Processing in Medical Imaging* (pp. 182-196). New York: Springer-Verlag.

Hill, D. L. G., Smith, A.D., Mauerer, C.R., Cox, T.C.S., Elwes, R., Brammer, M.J., Hawkes, D.J. & Polkey, C.E. (2000). Sources of error in comparing functional magnetic resonance imaging and invasive electrophysiology recordings. *Journal of Neurosurgery, 93,* 214-223.

Holden, M., Hill, D.L.G., Denton, E.R.E., Jarosz, J.M., Cox, T.C.S., Rohlfing, T., Goodey, J. & Hawkes, D.J. (2000). Voxel similarity measures for 3D serial MR image registration. *IEEE Transactions on Medical Imaging, 19*, 94-102.

Lee, Z., Nagano, K.K., Duerk, J.L., Sodee, D.B. & Wilson, D.L. (2003). Automatic registration of MR and SPECT images for treatment planning in prostate cancer. *Academic Radiology, 10*, 673-684.

Lehmann, T.M., Gönner, C. & Spitzer, K. (1999). Survey: Interpolation methods in medical image processing. *IEEE Transactions on Medical Imaging, 18*(11), 1049-1075.

Lester, H. & Arridge, S.R. (1999). A survey of hierarchical non-linear medical image registration. *Patter Recognition, 32*, 129-149.

Likar, B. & Pernus, F. (2001). A hierarchical approach to elastic registration based on mutual information. *Image and Vision Computing, 19*, 33-44.

Lin, K. P., Huang, S. C., Baxter, L. & Phelp, M.E. (1994). A general technique for interstudy registration of multifunction and multimodality images. *IEEE Transactions on Nuclear Science, 41*, 2850-2855.

Maes, F., Collignon, A., Vandermeulen, D., Marchal, G. & Suetens, P. (1997). Multimodality image registration by maximisation of mutual information. *IEEE Transaction on Medical Imaging, 6*(2), 187-198.

Maes, F., Vandermeulen, D. & Suetens, P. (1999). Comparative evaluation of multiresolution optimisation strategies for multimodality image registration by maximisation of mutual information. *Medical Image Analysis, 3*(4), 373-386.

Mäkelä, T., Clarysse, P., Sipilä, O., Pauna, N., Pham, Q.C., Katila, T. & Magnin, I.E. (2002). A review of cardiac image registration methods. *IEEE Transaction on Medical Imaging, 21*(9), 1011-1021.

Maintz, J.B.A. & Viergever, M.A. (1998). A survey of medical image registration. *Medical Image Analysis, 2*(1), 1-36.

Maintz, J.B.A., van den Elsen, P.A. & Viergever, M.A. (1996). Evaluation of ridge seeking operators for multimodality medical image registration. *IEEE Transactions on Pattern Analysis and Machine Intelligence, 18*(4), 353-365.

Maurer, C. R. & Fitzpatrick, J. M. (1993). A review of medical image registration. In R.J. Maciunas (Ed.), *Interactive image guided neurosurgery* (pp. 17-44). Parkridge, IL: American Association of Neurological Surgeons.

McLeish, K., Hill, D.J.G., Atkinson, D., Blackall, J.M. & Razavi, Reza. (2002). A study of motion and deformation of the heart due to respiration. *IEEE Transactions on Medical Imaging, 21*(9), 1142-1150.

Nelder, J. & Mead, R.A. (1965). A simplex method for function minimization. *Computer Journal, 17*, 308-313.

Pallotta, S., Gilardi, M.C., Bettinardi, V., Rizzo, G., Landoni, C., Striano, G., Masi, R., & Fazio, F. (1995). Application of a surface matching image registration technique to the correlation of cardiac studies in positron emission tomography by transmission images. *Physics in Medicine and Biology, 40*, 1695-1708.

Pellizari, C.A., Chen, G.T.Y., Spelbring, D.R., Weichselbaum, R.R. & Chen, C.T. (1989). Accurate three-dimensional registration of CT, PET, and/or MR images of the brain. *Computer Assisted Tomography, 13*(1), 20-26.

Penny, G.P., Weese, J., Little, J.A., Desmedt, P., Hill, D.L.G. & Hawkes, D.J. (1998). A comparison of similarity measures for use in 2D-3D medical image registration. *IEEE Transactions on Medical Imaging, 17,* 586-595.

Pluim, J.P.W., Maintz, J.B.A. & Viergever, M.A. (2001). Mutual information matching in multiresolution contexts. *Image and Vision Computing, 19*(1-2), 45-52.

Powell, M.J.D. (1964). An efficient method for finding the minimum of a function of several variables without calculating derivatives. *Computuer Journal, 7,* 155-163.

Riva, G. (2003). Review paper: Medical applications of virtual environments. *Year Book of Medical Informatics,* 159-169.

Roche, A., Malandain, G. & Ayache, N. (2000). Unifying maximum likelihood approaches in medical image registration. *International Journal of Imaging Systems and Technology, 11,* 71-80.

Rohlfing, T. & Maurer, C.C. (2003). Nonrigid image registration in shared-memory multiprocessor environments with application to brains, breasts, and bees. *IEEE Transactions on Information Technology in Biomedicine, 7*(1), 16-25.

Rohr, K. (2000). Elastic registration of multimodal medical images: A survey. *Auszug aus: Kunstliche Intelligenz, Heft.*

Subsol, G. (1999). Crest lines for curve-based warping. *Brain Warping* (pp. 241-262). San Diego: Academic.

Subsol, G., Thirion, J.-P. & Ayache, N. (1998). A scheme for automatically building three-dimensional morphometric anatomical atlases: application to a skull atlas. *Medical Image Analysis, 2*(1), 37-60.

Terzopoulos, D. & Metaxas, D. (1991). Dynamic 3D models with local and global deformations: Deformable superquadrics. *IEEE Trans. PAMI, 13*(7), 703-714.

Thévenaz P., Blu, T. & Unser, M. (2000). Image interpolation and resampling. In I. Bankman (Ed.), *Handbook of medical imaging: Processing and analysis* (pp. 393-418). Academic Press.

Thévenaz, P. & Unser, M. (2000). Optimization of mutual information for multiresolution registration. *IEEE Transaction on Image Processing, 9*(12), 2083-2099.

Thirion, J.P. (1998). Image matching as a diffusion process: An analogy with Maxwell's demons. *Med. Image Anal., 2,* 243-260.

Thompson, P. & Toga, A.W. (1996). A surface-based technique for warping three-dimensional images of the brain. *IEEE Transaction on Medical Imaging, 15*(4), 402-417.

Turner, R. & Ordidge, R.J. (2000). Technical challenges of functional magnetic resonance imaging: The biophysics and technology behind a reliable neuroscientific tool for mapping the human brain. *IEEE Engineering in Medicine and Biology,* 42-54.

van den Elsen, P.A., Pol, E.J.D. & Viergever, M.A. (1993). Medical image matching—A review with classification. *IEEE Engineering in Medicine and Biology, 12,* 26-39.

van Herk, M. & Kooy, H.M. (1994). Automatic three-dimensional correlation of CT-CT, CT-MRI, and CT-SPECT using chamfer matching. *Medical Physics, 21*(7), 1163-1177.

Viola, P. & Wells, W.M. (1995). Alignment by maximization of mutual information. *The Fifth International Conference on Computer Vision* (pp. 16-23).

Wang, H.S., Feng, D., Yeh, E. & Huang, S. C. (2001). Objective assessment of image registration results using statistical confidence intervals. *IEEE Transactions on Nuclear Science, 48*, 106-110.

Wang, X. & Feng, D. (to be published in Vol.3, 2005). Automatic elastic medical image registration based on image intensity. *International Journal of Image and Graphic (IJIG)*, World Scientific.

Xiao, H. & Jackson, I.T. (1995). Surface matching: Application in poat-surgical/post-treatment evaluation. In H.U. Lemke, K. Inamura, C.C. Jaffe & M.W. Vannier (Eds.), *Computer assisted radiology* (pp. 804-811). Berlin: Springer-Verlag.

Zitová, B. & Flusser, J. (2003). Image registration methods: A survey. *Image and Vision Computing, 21*, 977-1000.

Chapter X

Clinical Knowledge Management:
The Role of an Integrated Drug Delivery System

Sheila Price, Loughborough University, UK

Ron Summers, Loughborough University, UK

Abstract

Issues and complexities that arise from the adoption of clinical knowledge management are explored within the context of delivering drugs to the lungs. The move towards electronic data capture and information retrieval is documented together with cross-organisational working and sharing of clinical records. Key drivers for change are identified and their effects on, for example, the patient-clinician relationship are investigated. Conclusions drawn indicate the crucial role that all stakeholders play to bring about effective and efficacious patient care.

Introduction

Clinical knowledge management is the discipline concerned with the collection, processing, visualisation, storage, preservation, and retrieval of health related data and information, whether on an individual patient or a clinical specialty. Its successful adoption into daily clinical practice requires the use of new technologies such as: electronic health records; standardised medical terminologies, tools and methods to support speedier retrieval and dissemination of clinical information; and reliable networks to facilitate electronic communication in "real time." Recent advances in Information and Communication Technologies (ICTs) have facilitated the development of mobile communications, offering the opportunity for integration into existing and planned clinical information systems, and offer solutions for data capture and information retrieval at the point of care. Hidden organisational and cultural complexities arise from the anticipated use of ICTs, such as training needs, clinical acceptance, and a shift in empowerment within the patient-clinician relationship towards the patient. Addressing these issues requires commitment, application and tenacity from all stakeholders involved in the healthcare process.

The management of any clinical condition has traditionally been viewed as a clinician's domain. Data collection and information dissemination is still seen in some places as the preserve of an individual clinician with little evidence of sharing clinical knowledge. However, this established model of care in the UK is now being replaced by patient-centred care in which the patient is viewed as an equal partner in the care process. The Department of Health (DH) is committed to bringing together patients, carers and clinicians to create informal and formal relationships that support the self-management of chronic conditions. The DH also recognises this shift in responsibility for healthcare provision in its publication of documents such as "The Expert Patient" (Department of Health, 2001a). The "NHS Plan" (Department of Health, 2001b) developed from consultations with patients, is committed to the development of 'high quality patient-centred care' along with the development of modern ICT systems in primary and secondary care. It acknowledges the effective role that patients can play in the effective management of chronic diseases. Key elements of patient-centred care include, a patient's right to have access to their own medical records, access to identified and appropriate clinicians for treatment, access to information relating to waiting times, information on adverse drug effects and care planning, and the expectation that clinicians will keep accurate and accessible healthcare records. Patient knowledge is recognised as a beneficial asset to the healthcare process and as such is becoming a valuable tool in terms of the management of chronic conditions. In the specific area of management of chronic diseases, the opportunity to capture home-based data offers patients, carers and clinicians the possibility to add value in terms of increasing knowledge to manage the care process more effectively and as a consequence, empower key stakeholders in the healthcare experience.

Clinical Records Management

A medical record relating to an individual patient is a collection of information that relates to distinct episodes of care. In the past, medical records were associated with one particular GP Practice or secondary healthcare organisation. In the early days of the NHS, patient notes were exclusively hand-written. Recipients of the information complained that they could not read handwriting, the information contained in the document was not always appropriate to the clinical need at that moment, and that too many documents did not reach the intended recipient in time to be of use for planning of ongoing treatment and care. Medical records such as the "Lloyd George" paper folder can still be found in daily use in some primary healthcare organisations. The transition to an electronic healthcare record is not quite complete, as issues such as scanning of paper notes and training staff to use computers is claimed to be an impediment to progress. Both issues require resource allocation that is not always seen as a priority in organisations with limited resources that exist in an environment of constant change.

As far as clinical records management is concerned, prior to the introduction of an electronic healthcare record, all documents relating to an individual patient remained with the originating organisation. Clinical information was collected and recorded by both nurses and doctors for use within one clinical domain. The main problem with paper note keeping in secondary care was ensuring that the patient notes were in the right place at the right time and in a format in which people could easily retrieve information required.

Electronic Care Records

Roll out of electronic healthcare records in the UK is currently underway. In 1998 the NHS launched an IT strategy, "Information for Health" (1998), that aimed to link all GPs with the NHS's own intranet, NHSnet, and to procure basic electronic records for all hospitals and healthcare communities. Financial planning, interest in clinical outcomes data, and computer support for clinical decision making at the point of care are examples of some of the original key drivers that led to this development. To expedite progress of this IT initiative, a national programme for NHS IT (NPfIT) has been created (National Programme for Information Technology). The programme has developed a new vision of an integrated care records service (ICRS) which will be designed around patients rather than institutions, and will be accessible 24 hours a day, seven days a week. An electronic clinical noting system reflects a degree of flexibility in order to provide clinicians with the ability to retrieve clinical information in a format that they require. Certain items of information that relate to the patient (such as name and address) that lie outside the scope of clinical information should always appear in the same place in all clinical documents. Electronic healthcare records that can be shared between staff seem like an obvious solution to support clinical knowledge management; although their potential goes far beyond that of recording clinical information.

NPfIT is setting national standards to ensure that clinicians are provided with real-time access to information, which allows them to share patient and relevant documents with other clinicians and allied health professionals. A key concept in this change is that patient information will cross organisational boundaries allowing documents relating to clinical care, care in the community and social care to be integrated. This concept relies on the establishment of best practice in knowledge management. By 2010, it is envisaged that every NHS patient in England will have an electronic healthcare record that they will be able to access via the Internet regardless of location. To support the key drivers and enable successful integration into the healthcare process, electronic healthcare records must be web-based, and be web-enabled allowing interaction with Internet technologies such as web services.

The current situation regarding clinical note keeping in the NHS shows that communication between organisations remains sketchy and fragmented. The technology is available to support communication across organisational boundaries, however it is fair to say that not all secondary healthcare organisations are as advanced as primary care in their adoption of ICT to support clinical note keeping. Integration of primary, secondary and tertiary healthcare electronic healthcare records is a prerequisite of the practice of clinical knowledge management. Failure to link systems to enable seamless transfer of clinical information between relevant organisations will create a situation little different to that brought about by existing paper based systems.

Opportunities for Supporting Clinical Care: Integrated Drug Delivery

The development of home-based drug delivery systems offers the opportunity to capture data relating to the use of a device by a patient, collect data to support the assessment of clinical conditions, as well as empowering patients to become more involved in the management of chronic conditions. The data captured by a drug delivery device can be used to create useful information for both clinicians and patients/carers. Care should be taken to ensure that the information generated is about the right person, stored in a medium that is accessible to all who are nominated to use the system, and delivered at the right time to any specified delivery device, such as mobile phones, palmtops as well as desktop PC's. A critical element to the successful integration of informatics into the device must be commitment by all stakeholders, at all stages from development, implementation through to adoption into daily use.

Data integration from electronic healthcare record systems is essential in order to achieve state of the art clinical communication between stakeholders. The data collected must be able to be represented in a ubiquitous format that is acceptable to all those who take part in the care process. Critical elements in successful adoption of a home based drug delivery system into daily clinical use include the identification of the right data, the creation of the right structure, and the integration into the right clinical processes to add value. The care of asthmatic patients will be used to exemplify issues associated with home-based delivery of clinical treatment. If the above elements are addressed then the information architecture developed can be transferred to any clinical condition that requires delivery of drugs to the lungs.

Technology can support the diffusion of healthcare to the home through the development of home-based drug delivery systems. However, patient use of these systems may reveal issues that alter the traditional patient/clinician relationship. For example, clinicians who are not specialist in asthma may be confronted with patients who are considerably more knowledgeable about their condition. Acceptance and adoption into daily care will only be achieved through detailed and thorough integration into clinical practice, along with rigorous evaluation of the impact ICT has upon the care process. The effect of cultural issues will require assessment and evaluation and will influence adoption of novel approaches to patient centred care.

Home-based drug delivery systems are one way of managing clinical decision making at the point of care. They offer the opportunity for healthcare professionals to have instant access to a wide range of clinical information that may refer to individual patients, but the ability to access information such as clinical guidelines and protocols will only enhance and support the care process.

Clinical Exemplar

Asthma is a condition that affects the airways (bronchi) of the lungs. From time to time the airways constrict in people who have asthma. The condition can commence at any age, but it most commonly starts in childhood. At least 1 in 10 children, and 1 in 20 adults, have asthma in the UK. Around 1500 people die from asthma in the UK each year (National Asthma Campaign, 2001).

Asthma in the UK places a high economic burden on both primary and secondary healthcare systems, at an estimated annual cost of over £850 million (Office for National Statistics). One in ten asthmatics that experience severe or moderately severe symptoms fail to control their condition adequately, even with the best clinical and preventative management available. Integrated drug delivery systems have the potential for easing the economic burden on the NHS with improved patient care. Improved delivery of drugs to the lungs in terms of both the ability to deliver larger quantities of drugs to the correct region of the lung, combined with improved patient compliance to prescribed treatment regimens, will result in better disease management with less need to resort to expensive "relief care" strategies and will slow down progression to severe disease status.

Also, very few asthmatics have a documented self-management plan that explains when to take their medication, the use of peak flow meters to measure the speed of air blown out of the lungs, and what to do if their asthma condition deteriorates (Lorig, K.R., Sobel, D.S., Stewart, A.L., Brown, B.W., Bandara, A., Ritter P., et al., 1999). Traditionally patients are asked to keep a manual diary over a defined time period such as two weeks, charting peak flow readings as a method of assessing a patient's current clinical state. Regular peak flow readings are a recognised method of assessing how well prescribed treatments are working. As a chronic disease, it is essential to involve patients in decisions relating to the development of a tailored care plan for the treatment of asthma, this will result in the more effective delivery of healthcare to the individual patient and to asthmatics as a whole. Patients (and clinicians) must learn how to use their inhalers correctly. The practice nurse or doctors have pivotal roles to play in educating patients how to use an inhaler properly. It is easy to demonstrate in a healthcare environment but ongoing home-based monitoring may ensure better device compliance for specific groups such as the newly diagnosed or teenagers. Empirical evidence shows that these groups may have compliance issues relating to the use of inhaled therapies. Patients must agree on an "asthma action plan" with a nominated clinician. This means that adjustments to the dose of the prescribed inhalers, depending on the symptoms and/or peak flow readings can be made at a relevant point in time.

Stakeholders

Patients are now increasingly recognised for the central role they play in the healthcare process. They must be involved and consulted at every stage of the process. For a home-based drug delivery device to provide effective treatment and collect relevant information, the patient and/or carer must understand why the device is being used, agree to its use and understand and accept a degree of responsibility for their role in the changing relationship between clinician and patient. The partnership that develops between patient and clinician will benefit patients' individual knowledge and will complement clinicians' general knowledge about a specific chronic disease. The pooling of this knowledge will in effect support the effective management of chronic diseases.

Clinicians should have skills that allow them to embrace novel innovations in drug delivery systems. Many current clinical practitioners are still resistant to the roll out of ICT within the NHS. This resistance may be explained by a fear of change, the inability to embrace changes in clinical practice, and the empowered role of the patient in the care process.

Home-based drug delivery devices are one of the tools that can be utilised to support clinical knowledge management in the modern healthcare setting by delivering concise, appropriate and timely clinical information relating to an individual patient's clinical state in "real time" to key stakeholders involved in the care and management of chronic clinical conditions. Collection and storage of clinical data electronically at the point of care offers healthcare professionals instant access to clinical data such as medication history and test results from remote locations. Mobile access also allows healthcare professionals to access and update care plans, clinical guidelines and protocols, thus tailoring care by having access to detailed information relating to an individual patient.

However, not all clinicians are comfortable or ready to use mobile devices in clinical practice. The benefits of using mobile health technology at the point of care will only be demonstrated by widespread adoption into clinical practice. For example, the use of handheld computers to access clinical guidelines and protocols etc, especially by junior doctors in secondary care is becoming more widespread. Use by all clinical stakeholders at the point of care should be encouraged, in order to benefit patient care. As adoption of these devices becomes more commonplace, the opportunity arises for innovative uses in the clinical setting to be explored.

The management of chronic clinical conditions often involves community healthcare practitioners. Access to timely information relating to a patients condition will support home-based care. In the past, relatives and neighbours often provided clinical care at home, clinicians and allied health professionals played a small role in the care of the average patient due to the related cost of healthcare. When the NHS was established, the focus of care for chronic conditions transferred to the "free" primary and secondary healthcare systems. Today, home- and self-care are re-emerging in response to cost pressures, the emergence of the Internet as a conduit of health information to patients, and by the diffusion of inexpensive computer technology as an aid to medical decision-making at the point of care. The benefits of home based drug delivery systems will be judged on patient outcomes as well as the reduction in costs relating to the provision of healthcare to support the ongoing treatment of chronic conditions.

Copyright © 2005, Idea Group Inc. Copying or distributing in print or electronic forms without written permission of Idea Group Inc. is prohibited.

Standards

Issues surrounding the security of electronic patient records may impede development and implementation of innovative home-based drug delivery systems. Time spent debating this issue without the underpinning knowledge of international standards can often lead to an insular focus on ICT development and its ongoing benefits to the provision of patient centred care. Security and privacy of clinical data are at the forefront of any debate concerning mobile health. It is essential that uniform data standards for patient information and the electronic exchange of that information be adopted on a worldwide basis and are embedded in system development.

Recently, the US government has recommended that Health Level Seven (HL7) be recognised as the core-messaging standard (Health Level 7). Standards development must address issues such as security and the protection of data privacy, while facilitating communication between individuals. In developing a system for home-based drug delivery systems, patient–clinician communication must be structured and secure to ensure reliability and relationships underpinned by trust to be developed. Essential data must be clearly represented within a document structure (Nygren et al., 1998) Adherence to standards will allow analysis of the data collected and enable clinical outcomes to be communicated to the wider clinical community without an individual patient's privacy being compromised.

Communication of electronic clinical data and information across platforms should not be device specific. The electronic provision of clinical information, using an extensible mark up language (XML) format for interoperability, is a stated UK Government target that is endorsed by the National Health Service Information Authority (NHSIA). Regardless of which device is chosen, the use of XML facilitates electronic document management and workflow, and allows tasks to be distributed amongst multi-professional healthcare personnel that will result in the optimisation of patient care.

Current Prescribing Practice

There is a wide range of inhalers available for the delivery of drugs to the lung. Studies have shown that asthmatics do not use prescribed inhalers correctly (Thompson, Irvine, Grathwohl, & Roth, 1994). The most effective inhaler is one that will be used by the patient on a regular basis and in an effective manner. As "poorly" compliant patients are often at risk of frequent attacks, it may be desirable to develop drug delivery devices that can capture data relating to use and feed this information into an electronic healthcare record system, accessible by all stakeholders in the care process, in order to support home based patient care and facilitate clinical knowledge management.

Mobile Devices

Mobile health solutions allow for a point of care interaction regardless of whether the patient is in hospital, at home or elsewhere in the community (Price & Summers, 2004). Mobile devices can assist health professionals in their day-to-day clinical environment by facilitating the provision of timely information. Mobile devices can also be of use in medical research, used for clinical trials to record and transmit data that are costly to capture by traditional methods. It is essential that the clinical information is delivered in "real time" and can be readily accessed and understood by all stakeholders, including patients, who are involved in the care process. The development of an integrated drug delivery system enables all stakeholders, including patients and clinicians, to add value to the care process. The smart interface and the use of a patient diary provide a solution to the management of complex clinical communication issues by allowing documents to be stored electronically and retrieved at will. It is crucial that these documents are available in various formats and have the capability to be delivered to a wide range of platforms, depending on preference and circumstances of use. The device holds information that can be read in context of the recipient—this means that careful choice terms have to be used so as to avoid misinterpretation. It is clear that the integration of mobile devices into the healthcare process will demonstrate considerable benefit to UK plc by reducing the economic burden on primary, secondary and tertiary care.

Clinical Knowledge Management

Data captured by the drug delivery device can be used to create useful information for both clinicians and patients. The information generated must be about the right person, stored in a medium that is accessible and delivered at the right time by any specified delivery device, such as mobile phones, palmtops as well as desktop PC's. A critical element to the successful integration of mobile-devices must be commitment by all stakeholders, at all stages from development, implementation through to adoption into daily use. Home-based drug delivery systems must provide efficient and cost effective care for patients, focusing on preventing complications through the use of alerts generated by the system when recorded results fall outside the 'agreed' patient profile, negotiated by individual patients and their nominated clinician.

Issues relating to the adoption of home based drug delivery devices include awareness of users previous experience relating to ICT in general, the device used to transmit and receive data, the identification of data that has specific relevance to a nominated clinician and the acceptability of filtering information in order to generate appropriate clinical information to the right person at the right point in time. A rigorous evaluation process will identify further issues generated when the device is integrated into the care process.

In order to take advantage of the developments surrounding electronic clinical note keeping, a device has been developed to deliver drugs to the lungs, initially to treat the symptoms of chronic asthma, but with planned exploitation routes to include diabetes

Figure 1. E-medic platform

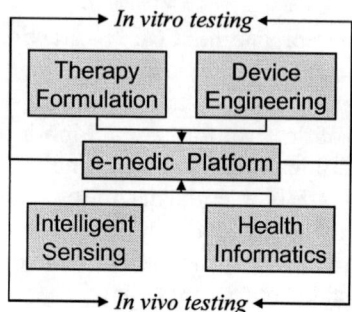

and pain management. An integrated drug delivery system including a smart interface and patient diary capability allows documents to be stored electronically and retrieved at will. It is crucial that these documents are available in various formats and have the capability to be delivered to a wide range of platforms, depending on preference and circumstances of use. The device will also hold information that crosses various organisational boundaries, to be read in context of the recipient—this means that careful choice of terms has been used so as to avoid misinterpretation. It is clear that the device possesses considerable benefit to UK plc by reducing the economic burden on primary and secondary care.

In order to investigate the elements to be utilised for data capture and dissemination of clinical information a set of typical Use Cases was generated, describing patient and clinical aspects, the device to systems interfaces, and administrative functions. Investigation and mapping of e-health capabilities, such as generic search, retrieval and visualisation took place. The first draft of the functional specification required for the integration of data generated and captured from the drug delivery device, is described below.

The system is configured to accept sign-on from patients, carers and clinical staff. Demographic information will initially be populated locally, but later will be drawn from the local Community Index (Exeter system). Patient ID, initially populated locally using NHS numbers where possible, will later be drawn from the Community Index. Patient registration can take place by registering the device against a patient—which would need to be managed carefully. If the device is issued to a new patient without the system being updated, then the data will be associated with the wrong patient. Alternatively, all data are supplied with a patient ID (e.g., NHS number). This would mean that the device is either capable of holding an ID for the user or that the patient submits data in a two-phase process, by signing onto the system and then providing a password. Patient folders are automatically created when the patient is positively identified. In order to achieve maximum coverage for "alerts" to nominated health professionals, it is essential that clinicians are able to select a delivery device that they feel most comfortable with. The options available include: mobile phones, PDAs, land line phones, and fax.

Figure 2. Informatics metrics

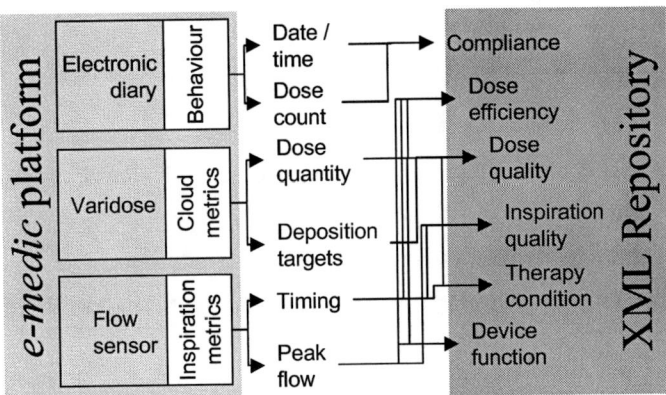

Factors to take into consideration when deciding upon appropriate channels of communication include ability to access "real time" information, the need to communicate quickly with the nominated clinician and to agree the method of communication with the patient. The device supports the delivery of clinical information delivered in "real time", which can be readily accessed and understood by all stakeholders, including patients and their carers who are involved in the management of chronic conditions.

The information held within the system covers:

- Schedules for prescription against which alerts/actions may be triggered. These alerts will be negotiated at an individual patient level and after consultation with the nominated clinician. It is envisaged that these alerts will eventually be sent to a nominated pharmacy to ensure and support efficient and timely prescription of the required drugs to patients.
- Drug dosing information (prescription) received from the devices, these data will be tested against the schedule
  - Drug name—the patient may be taking more that one type of drug therapy and it is therefore important to distinguish which drug has been taken.
  - Dose prescribed
  - Frequency—will enable accurate data to be collected to map actual use against planned use
  - Alerting rules
  - Remaining inhalations
  - Date next prescription is due

- Drug taken—information from the inhaler device
  - Patient ID—NHS number, or device number that matches to the patient
  - Drug name
  - Breathing profile indicator
  - Device malfunction status flag
  - When a drug event is received, it is checked against the most recent schedule.
  - Check for remaining inhalations; if down to last inhalation send "in-tray" alert to nominated clinician
  - Check against Alerting rules; if fails check create "in-tray" item to nominated clinician
  - Check value of Breathing Profile indicator; if outside normal profile create "in-tray" item for nominated clinician
  - Create drug event and store in electronic healthcare record

Structured Evaluation

Integration of home based drug delivery devices to capture and disseminate clinical data and information into daily clinical practice must be subjected to thorough and rigorous evaluation at every stage of the procurement and implementation process. A key element of evaluation is the need to identify "best practices" relating to device use and patient feedback, ensuring that is rolled out to the wider clinical community and embedded in the development of future asthma management plans, along with clinical protocols and guidelines. Unless it can be demonstrated that mobile devices have improved organisational effectiveness as well as contributing to an improvement in patient care, and have subscribed to the development of clinical care in general, their contribution to the clinical knowledge management process will be devalued.

Academic clinical studies will be required to evaluate the benefits of home based drug delivery systems focusing on the possible clinical benefit to the patient and additional benefit to clinical care, looking at the use of the device by patients with little or no ICT skills, an increase in patient involvement in the care management process and the level of patient compliance compared with those patients not using the device. Randomised trials indicate improvement in patient health and reduction in healthcare costs (Simon, Von Korff, Rutter, & Wagner, 2000). The return on investment should be measured in terms of any increase in the level of patient care and any advancement in the knowledge relating to chronic clinical conditions.

Future Uses

As adoption of integrated devices into daily clinical practice becomes more commonplace, the opportunity arises for innovative uses in the clinical setting to be explored. The device can be used for clinical trials to record and transmit data that are costly to capture by traditional methods. The ability to record the patients breathing profile could enable the relationship between a successful dose deposition of the drug being delivered to the lungs and the patient ability to match their pre-determined "perfect profile". This information would be of use in the future development of novel drug therapies and would also provide the clinical community with invaluable information relating to patterns of inhaler use and effectiveness.

Outcomes

Successful implementation of the device into daily clinical use will require commitment from patients/carers, individual healthcare professionals as well as healthcare organisations. A tailored training infrastructure for all is essential, not only to demonstrate correct use, but also demonstrate benefits that can be achieved from the device on an individual basis. The integration of the device into daily clinical practice will only be achieved if issues associated with it are incorporated into clinical teaching practice and continuous professional development programmes. Although training can be expensive and time consuming, it is an investment that is critical to the acceptance of all home based clinical systems. Home-based drug delivery devices are likely to generate data to populate mature information databases to inform clinical practice and may facilitate the growing trend to decentralisation of care.

Conclusions

Uptake of novel drug delivery devices will depend on many factors, including organisational support, effective training infrastructure, reliable systems, integration with legacy systems, data standards, workflow patterns, privacy and security, and healthcare standards. In the US, drivers to ensure successful implementation will be improved clinical outcomes and the ability to bill patients for each element of a clinical episode at the point of care. In the UK, more emphasis will be placed on the ability to input and retrieve data and information using an "integrated electronic healthcare record" accessible by clinicians and allied health professionals at the point of care, regardless of location.

It is essential to involve all stakeholders in the debate relating to the goals of clinical knowledge management and the strategies required to achieve them. Presently, clinical information is not currently disseminated in a way that facilitates understanding by patients. Patient awareness of their rights to access clinical data is patchy and channels of access to information are not well developed. The use of abbreviations and codes acts as a barrier to understanding by non-clinicians. Patients with specific knowledge relating to chronic conditions require tailored communication infrastructures to be developed before they can take their place as equal partners in the care process. It is essential that all stakeholders are comfortable with clinicians adopting a supportive role to enable patients to recognise deteriorating asthma and for patients be relaxed about taking a pivotal role in the self-management of their own medication to prevent deterioration in their condition. Discontinuity and lack of communication between primary care, emergency departments, and even within secondary care may hinder the development of an integrated care programme for asthma.

The way to address this is through greater involvement of patients in their own management. The use of a drug delivery device that includes the use of a patient diary will go some way to support an integrated care programme for the management of the asthmatic patient. Thus the drug delivery device is the integrator that underpins the integrated care process. The electronic healthcare record is the repository that is pivotal to the knowledge sharing activity that is required to empower all stakeholders in the care process and capture "real time" clinical information to add to knowledge already held about a chronic clinical condition. Many clinicians complain that they suffer from "information overload"; mobile communication devices are one of the tools that can be utilised to support clinical knowledge management in the modern healthcare setting.

References

Department of Health (DOH). (2001a). *The expert patient: A new approach to chronic disease management for the 21st century.* London: DOH.

Department of Health (DOH) (2001b). *Building the information core: Implementing the NHS Plan.* London: DOH.

Health Level 7. Retrieved March 24, 2004 from *http://www.hl7.org*

Information for Health. (1998). London: NHS Executive.

Lorig, K.R., Sobel, D.S., Stewart, A.L., Brown, B.W., Ritter P., Gonzalez, V.M., Laurent, D.D. & Holman, H. (1999). Evidence suggesting that a chronic disease self-management program can improve health status while reducing hospitalisation – A randomised trial. *MedCare, 37*(1), 5-14.

National Asthma Campaign (2001). Asthma audit. Retrieved March 24, 2004, from *www.asthma.org.uk*

National Programme for Information Technology (2004). Retrieved March 24, 2004, from *http://www.dh.gov.uk/PolicyAndGuidance/InformationTechnology/NationalITProgramme/fs/en*

Nygren, E., Wyatt, J. & Wright, P. (1998). Helping clinicians to find data and avoid delays. *The Lancet, 352,* 1462-6.

Office for National Statistics (n.d.). Retrieved from *http://www.statistics.gov.uk*

Price, S. & Summers, R. (2002). Clinical knowledge management and m-health. *Proceedings of the 24th IEEE Engineering in Medical and Biological Sciences Annual International Conference*, Houston, Texas (pp. 1865-1866) [CD-ROM].

Simon, G.E., Von Korff, M., Rutter, C. & Wagner, E. (2000). Randomised trial of monitoring, feedback and management of care by telephone to improve treatment of depression in primary care. *BMJ, 320*, 550-554.

Thompson, J., Irvine, T., Grathwohl, K. & Roth, B. (1994). Misuse of metered dose inhalers in hospitalised patients. *Chest, 105*(3), 715-717.

Chapter XI

Medical Decision Support Systems and Knowledge Sharing Standards

Srinivasa Raghavan, Krea Corporation, USA

Abstract

This chapter discusses about the concept of medical decision support system and the knowledge sharing standards among medical decision support systems. The author discusses the evolution of decision support in the healthcare arena, the characteristics and components of a medical decision support system, the medical decision support problem domains, and the popular medical decision support systems. Furthermore, a unique challenge in the healthcare arena—sharing of knowledge among medical decision support systems is discussed. The author discusses about the need for knowledge sharing among medical decision support systems, the evolution of various knowledge sharing standards, and the application of the knowledge sharing standards by the medical decision support systems. Finally, interesting aspects about the future trends in the medical decision support systems, its awareness, its usage and its reach to various stakeholders are discussed.

Introduction

The healthcare industry has been a pioneer in the application of decision support or expert systems capabilities. Even though the area of medical informatics and decision support has been around for more than four decades, there is no formal definition for a medical decision support system. Wyatt & Spiegelhalter (1991) describes a medical decision support system as a computer-based system using gathered explicit knowledge to generate patient specific advice or interpretation. The healthcare industry has witnessed a phenomenal growth and advances both in the areas of practice and research. This rapid growth of medical science has made the practice of medicine both challenging and complex. To address these challenges, the recognized medical standards organizations developed medical practice guidelines to simplify the research findings to practical applications in order to improve the overall healthcare quality and delivery. Despite such initiatives, it has been difficult for the physicians to keep up with the guidelines and to tune it to their practice settings. A clear gap began to develop between developing of practice guidelines and the implementation of the same. Grimshaw and Russell (1993) identified that the knowledge dissemination and implementation strategies are critical to the impact the developed guidelines will have on the physician behavior.

The natural evolution was to develop paper-based decision support workflow or protocols. Paper-based decision support models were developed and are widely used at most physician practices to apply the guidelines into practice. Paper-based decision support models, while effective, are also error prone. For example, dose recommendation models could require tedious calculations and can be error prone when performed manually. Such errors could defeat the purpose of using a decision support model. Also, paper-based models require education and training to the personnel performing those calculation and require an additional layer to review the calculations. Most paper-based models are used by the medical personnel as a batch work flow process.

Thus computer-based decision support systems were developed to provide accurate guideline compliance and to enhance physician performance (Hunt, Haynes, Hanna, & Smith, 1998). Computerized decision support system can be extremely valuable for treatment or diagnosis support and compliance accuracy when used at the point of care (Lobach & Hammond, 1997). This feature of computerized medical decision support system is a key differentiator that makes the paper-based decision support models inferior. A well designed computerized medical decision support system can be used to provide patient specific support at the desired time and location with the adequate content and pace. When decision support systems are blended into the day-to-day practice workflow, these systems have the potential to function as a valuable assistant and also as an educational tool (Thomas, Dayton, & Peterson, 1999).

The computerized decision support systems make decisions based on the clinical practice guidelines. Clinical practice guidelines are the rule-based knowledge that guides the decision makers in a medical setting. These guidelines have been developed over the years to reduce the variations among medical practices with a common goal to provide cost effective and high quality healthcare services (Field & Lohr, 1992). The availability of several decision support systems and the use of common knowledge and rules triggered the need for a common method of sharing the knowledge. This technique can save the cost of developing the same medical practice guidelines for multiple decision support systems.

While it is common among organizations to share the data and information between computer systems, sharing the knowledge was a challenge. The medical arena rose to the occasion and various associations were involved in development of a standard way of sharing the medical knowledge and clinical practice guidelines among systems. After reading this chapter, the reader should be able to:

- Understand the complexity involved in medical decision making process
- Identify the commonly used decision support techniques in medical decision support systems
- Identify the characteristics of medical decision support systems
- Identify the various phases of the decision making process
- Identify the components involved in the development of a medical decision support system
- Analyze the various decision support functions addressed by medical decision support systems
- Know about the popular medical decision support systems in use around the world
- Understand the importance of knowledge sharing standards in the medical decision support arena
- Know about the popular medical knowledge sharing standards
- Understand the future trends in decision support in the field of medicine

Background

The invention of computers resulted from the dream of creating an electronic brain. This quest to artificial intelligence has driven several technological advances in various industries. The potential of intelligence and decision support in the area of medicine was set forth more than four decades ago by Lusted and Ledley (1959). While several computer systems have been developed to improve the administrative efficiency and medical records access, the significant challenges have always been in the development of medical decision support systems. Knowledge-based decision support systems acquire, formalize, and store the expert knowledge in a computer system and infer the represented knowledge in a problem area by modeling the decision process of experts.

One of the first expert systems developed is in the field of medicine. This expert system called MYCIN, provides decision support to physicians on treatment related advices for bacterial infections of the blood and meningitis. The basic structure of the MYCIN expert system is to present a question and answer dialog to the physician. After collecting the basic information about the patient, it asks about the suspected bacterial organisms, suspected sites, presence of relevant symptoms, laboratory results and recommends the course of antibiotics (Buchanan & Shortliffe, 1984). Even though this was a primitive expert system, it brought out the power of such decision support tool in the area of medicine and the potential for more growth in the area of artificial intelligence in medicine. After the introduction of MYCIN in the field of medicine, various decision support systems were introduced and have been successfully used. The commonly used decision support techniques in medical decision support systems are intelligent agents, rule-based engines, heuristics, and decision algorithms.

The use of an intelligent agent or softbot (intelligent software robot) is a popular decision support technique that has the potential to become one of the most important in the next decade (Turban & Aronson, 2000). There are several definitions of an intelligent agent, but the general concept is that the agent carries out a set of operations with some degree of autonomy (Murch & Johnson, 1999). To perform these tasks, an intelligent agent should contain some knowledge. The key characteristics of an intelligent agent are that it should be autonomous, goal-oriented, collaborative, and flexible (Brenner, Zarnekow, & Wittig, 1998). For example, an appropriate task for an intelligent agent would be to use a patient's history and problem list to assist the physician with diagnostic coding during the order entry process.

A heuristic is a rule of thumb that is used primarily to arrive at a "good enough" solution to a complex problem. This technique is used primarily when input data are limited and incomplete. When the problem or the reality is extremely complex and optimization techniques are not available, a heuristic may be used. Heuristics do not yield definitive solutions, but assist decision makers in arriving at provisional solutions (Camms & Evans, 1996). For example, the determination of the choice of initial doses of both medication and dialysis treatment for a renal patient relies on population-based pharmacokinetic and solute kinetic heuristics, respectively. Subsequent dosing does not require heuristic reasoning, but may be calculated on the basis of measurements of the individual patient's response to the initial dose (Raghavan, Ladik, & Meyer, 2005).

Rule-based decision support systems primarily store knowledge in the form of rules and problem solving procedures. In expert systems, the knowledge base is clearly separated from processing. Inputs may come from the user, or may be collected from various other programs. Rules are applied on the data collected using an inference engine, decisions are suggested, and explanations supplied (Turban & Aronson, 2000). For example, DARWIN, a renal decision support system, uses rule-based reasoning to advise dialysis technical staff regarding the response to bacteriologic monitoring of fluids used during hemodialysis. Another example of a rule-based or expert system is the use of clinical protocols to generate suggestions regarding anemia treatment using erythropoietin and iron protocols. On the basis of knowledge of new and historical laboratory results and medications, the expert system can suggest the modification of drug doses or the discontinuation of a medication. The inference engine can also provide an explanation of the basis of the recommendation in the protocol.

A decision algorithm is a set of instructions that is repeated to solve a problem. Highly intelligent algorithms include the capability to learn and to perform several iterations of the algorithm or of parts of the algorithm until an optimal solution is reached (Turban &

Aronson, 2000). A very simple decision algorithm may alert physicians on receipt of out-of-range laboratory results. If the decision algorithm verifies the patient's history and decides if the standard abnormal range is indeed abnormal for the patient, the algorithm can be classified as an intelligent decision algorithm.

In the next sections, we will discuss about the characteristics of a medical decision support systems, generic components, problem areas in which decision support systems are used and knowledge sharing standards.

Characteristics of Medical Decision Support Systems

To understand the characteristics of a medical decision support system, it is important to analyze the medical decision making process. A clear understanding of the medical decision making process is essential to appreciate the value and the characteristics of a medical decision support system. Decision making in medicine is not a single point event. It is an extremely complex sequence of inter related and differentiated activities that occur over a period of time. The vastness of the knowledge area presents itself infinite paths to reach a decision. Thus, the importance of finding an optimal path is extremely important in the decision making process. Any assistance in this critical step can be of immense value to the medical decision makers. Hastie (2001) identifies three components to a decision making process. The first component is the choice of options and courses of actions. The second is the beliefs and opinions about the objective states, and the processes including input and outcomes states. The third is the values and consequences attached to each outcome of the event-choice-action combination.

The decision support literature classifies problems into three major categories: structured, semi-structured, and unstructured (Gorry & Morton, 1971). Structured problems are routine and repetitive: solutions exist, are standard, and pre-defined. Unstructured problems are complex and fuzzy; they lack clear and straight-forward solutions. Semi-structured problems combine the features of the two previous categories; their solution requires human judgment as well as the application of standard procedures (Simon, 1971).

Thus the complexity involved in a decision making process can be grouped as a knowledge gathering process, knowledge storage, knowledge retrieval, and information processing. The literature categorizes the decision making process into four distinct phases: intelligence, design, choice, and review (Stohr & Konsynski, 1992).

During the intelligence phase of the decision making process, the need for a decision making (or the trigger) event is recognized and the clinical problem or opportunity is properly identified and defined. This is done by eliciting two different types of knowledge domains: public and private. Private knowledge generally is heuristic and experience-based knowledge that usually comes from the clinical practitioners. Public knowledge can be gathered from standards, guidelines, text books and journals. The office of technology assessment (1995) suggests usage of the following questions in identifying the problems and opportunities for medical decision support.

1. Can the solution to the problem/opportunity assist in diagnosing a patient's condition?
2. Can the solution to the problem/opportunity assist in determining what the proper drug dosage level should be?
3. Can the solution to the problem/opportunity remind the appropriate care giver about the preventative services to be administered to a patient or to patient care related function?
4. Can the solution to the problem/opportunity assist in carrying out diagnostic procedure by recommending specific treatments or tests?
5. Can the solution to the problem/opportunity assist in carrying out medical procedures by alerts regarding potential adverse events?
6. Can the solution to the problem/opportunity assist in providing cost effective medical care by reminding previous orders, results, frequency rule checks, and schedule of treatment or procedure?

During the design phase, the problems or opportunities identified during the intelligence phase are further analyzed to develop possible courses of actions to construct decision models. A thorough search for ready made solutions, customized off the shelf solution, or a decision to custom solution development is made. Modeling of a decision problem involves identification of variables, identification of the relationship between those variables, and developing an abstraction into quantitative or qualitative forms.

During the choice phase, the actual decision to follow a certain course of action is made. If a single option results from the design phase, the choice phase is a simple acceptance or rejection of the option. If multiple courses of actions are identified in the design phase, then the decision makers are challenged in choosing from multiple conflicting objectives. Some decision rules for multivariable problems have applied strategies like holistic evaluation methods, heuristic elimination, and holistic judgment.

During the review phase, the outcomes of the decision making process are reviewed for validity and applicability for the case in hand. Any deviation and changes to these outcomes could very well become an expert knowledge that can be fed back into the knowledgebase to help future decision making cases.

Several studies were performed to identify the characteristics and functions of a medical decision support system. Turban and Aronson (2000) and Beatty (1999) have performed extensive research in this area of medical decision support system. Turban and Aronson (2000) summarizes the characteristics into few categories as listed below.

1. Support decision makers in semi structured and unstructured situations by bringing medical expert judgment and computer knowledgebase together.
2. Support decision making when the problem area consists of various inter-dependent and/or sequential paths.
3. Should be adaptive and flexible in nature. This allows new learning, adjustments to current knowledge and knowledge pattern recognition.

4. User friendliness is a key for success. Even a great decision support system can be useless if the users are challenged by its interface. This is particularly true in the case of healthcare arena. The system should be easy, fast, and require few mouse clicks by physicians.
5. Allow decision makers to construct simple decision constructs.
6. Should provide both automated and interactive decision support depending on the nature of the problem/opportunity.
7. Should provide complete control to the medical decision maker and should not attempt to replace the clinician's judgment.

Beatty (1999) conducted an empirical study about the characteristics and functions of a medical decision support system. This study provided several interesting findings on what support do medical decision makers expect during a decision making process. Similar to Turban and Aronson (2000) study, Beatty's (1999) finding also indicates that the decision makers favored characteristics in a decision support system that advised or guided rather than controlled. Thus, medical decision makers prefer to be in total control during the decision making process and would use the computerized system as a support or assistance during the process.

Beatty's (1999) finding also differentiated between the expectations of a medical user from a non medical user in medical decision making. This is an important variation in a medical setting as several non medical people perform various routine functions in a medical setting and medical decision support systems could assist their functions as well. The interesting finding is that the non medical users prefer to give more control and trust to the decision support system while the medical users prefer to keep control.

Medical decision support systems should be capable to operate in both an offline mode and a real time mode. This allows the system to be used as an educational tool during offline mode and as an advisor during the real time mode. Beatty (1999) indicates that the users prefer a detailed advice about a medical problem, treatment option, and so on, preferably through flow diagrams or case studies but a real time support should be clear, precise, alarms, and very sophisticated. The knowledge base for the decision support systems should be gathered from literature, case studies, and wide group of experts.

Turban & Aronson (2000) used a schematic view to describe the components of a decision support system. This view was used as a guideline to develop a schematic view to identify the components of a medical decision support system. As depicted in Figure 1, a typical medical decision support system consists of a medical decision support model database, a medical data management component, a medical decision support engine, third party objects including medical databases and rules, decision support sub systems, and a decision support user interface. The decision support model database contains all the tables and data needed to support the decision support models. The data needed for development of rules engine, prediction models, and protocols are stored in the model database.

The medical data management component is the critical part of the medical decision support system. This interfaces with the databases external to the decision support

Figure 1. Components of a medical decision support system

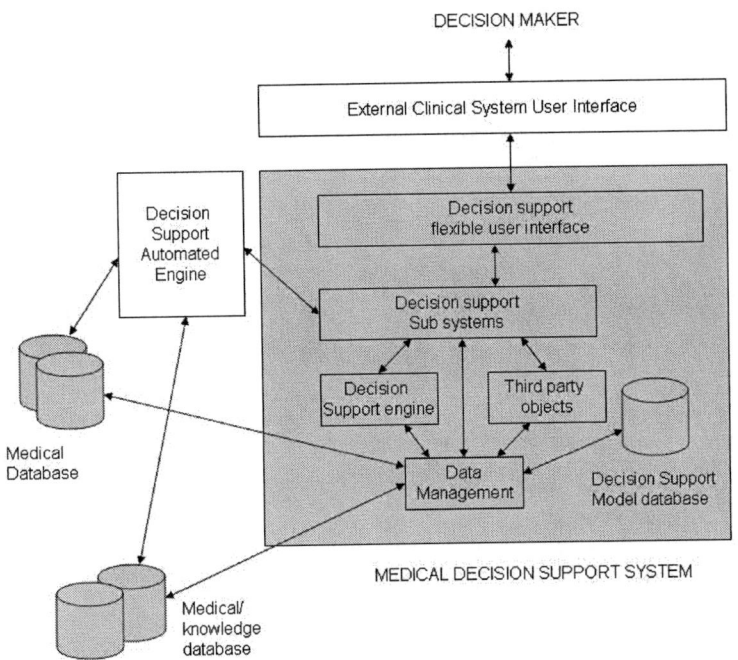

system to allow the decision support engine and third party interfaces to apply model database rule on the medical data to arrive at intelligent decisions.

The medical decision support engine is a set of programs that constitute the model for decision support. It was important to develop a flexible engine that could be data-driven. Several day-to-day decisions in patient care are based on changing regulations and practice guidelines. These guidelines are periodically reviewed and changed. In order to expedite such decision rule changes, it is important to maintain the rules themselves in databases, thus making the model database a meta-database.

There are a few decision support processes that are available commercially. These third party objects may or may not include the model database bundled within the object. If they are bundled, it would become a direct interface with the decision support subsystem and the data management system. If they are not bundled, then the data is carefully placed either in the model database or in the medical database as appropriate.

Decision support subsystems are independent decision support systems that can interface with any external subsystem directly or through the decision support user interface. The medical decision support system generally consists of both automated and interactive subsystems. The automated decision support subsystems may use the automated engine to integrate the automated decision making capabilities with existing automated functions. For decision support functions that require user interaction, the decision support system flexible interface is called from the user interface layer and presented to the decision maker.

Table 1. Decision support functions and clinical problems

Function	Sample Clinical Problems/Opportunities
Alert	Based on laboratory results with the range customizable at various levels
Diagnosis	Identify the possible diagnosis based on the history, physical, results, and evaluation inputs
Reminder	Reminding practitioners in order approvals, and schedules
Notification	Non conformance, risks, abnormal events, and episodes of care
Suggestion	Drug adjustments based on the recent lab values, trends, and current drug levels
Interpretation	Guidelines as applicable to the current situation – lab test schedule, protocol development
Prediction	Predicting results based on some independent variables
Assistance	Providing drug to drug interaction and drug-allergy interaction
Critique	The usage of diagnostic codes for medical procedure justification based on the applicable guidelines and the patient's medical history

In the past four decades, numerous medical decision support systems were developed using a wide range of techniques such as automated reasoning, natural language processing, game playing, automatic programming, robotics, machine learning, and expert systems. These decision support systems generally address the medical decision support functions listed in Table 1 (Raghavan, Ladik, & Meyer, 2005). These functions are generic in nature, while the knowledge base of the decision support systems is generally specific to a medical domain area like oncology, nephrology, pediatrics, and so on.

The most common decision support function found in medical decision support systems is alerts and reminders. In a real time environment, these decision support functions are attached to the monitoring devices to provide immediate alerts as and when the trigger condition occurs. For example, the oxygen and blood pressure monitors in an acute setting can alert the nurses if the patient's condition goes beyond the set threshold value. In a chronic setting, a simple scan of laboratory results and an email or pager alert to the corresponding decision maker are valuable decision support functions. A few medical decision support systems can provide image recognition and interpretation functions. These are extremely helpful in large hospitals where various radiology reports can be interpreted and alerts can be generated to gain the attention of the experts.

Diagnostic support is a key function that several medical decision support systems attempt to offer to help the physicians in detecting the problem based on symptoms and etiology. Such systems are commonly used to detect rare diseases and also as an aid for inexperienced practitioners. A few medical decision support systems offer care plan critiques by looking at data inconsistencies, errors, and omissions against practice guidelines.

In the next section, the problem domain, and the decision support functions provided by various popular medical decision support systems will be discussed. While there are close to hundred known medical decision support systems in use, the most popular medical decision support systems based on the existing literature and usage will be discussed in this chapter.

Medical Decision Support Systems

HELP (Health Evaluation through Logical Processes)

This is the most successful and popular medical decision support system in the United States. The HELP system of the Latter Day Saints Hospital in Salt Lake City is a hospital-based medical information system that gives practitioners a comprehensive patient record with decision support capabilities. HELP is a comprehensive medical information system with full fledged medical records, physician order entry, charges, radiology, pharmacy, ICU monitoring, laboratory, and robust decision support functions (MedExpert-HELP, 2004).

The success of HELP information system was due to the fact that the decision support functions were integrated with the hospital information system. The availability of required data and the knowledge base makes this a powerful system. The decision support capabilities made available to the decision makers at the point of care made this an effective product. This system has also been implemented at more than twenty hospitals.

The HELP information system supports the following decision support functions: alerts and reminders, decision critiquing, patient diagnosis, care suggestions, and protocols. The integrated laboratory information system allows automatic monitoring and alerts for abnormalities or out of range values. The critiquing feature is integrated within the transfusion ordering module to critique the reason provided against the strict guidelines. The system also includes modules such as automated surveillance system that uses various data elements to diagnose nosocomial infections. As an extension of this module, the antibiotic assistant recommends the antibiotics that can produce optimal benefit to the patient.

DXplain

Dxplain is a diagnostic decision support system that is owned by Massachusetts General Hospital and can be licensed and accessed over the internet as well. DXplain is a powerful diagnosis decision support system used in general medicine. The power of DXplain is its knowledge base that can diagnose close to 2000 diseases emphasizing the signs and symptoms, etiology, pathology and prognosis. This system was developed during the early 1990s by Massachusetts General Hospital and has been used by thousands of physicians and medical students.

DXplain uses a set of signs, symptoms, and laboratory results to produce a ranked list of diagnoses which might explain the clinical manifestations. DXplain uses Bayesian logic to derive the clinical diagnosis interpretation (MedExpert-DxPlain, 2004). It is important to note that the system only provides suggestion and not definitive conclusions. Since this system is in production and licensed as a product, the knowledge base should be continually updated.

DXplain is routinely used at medical schools for clinical education. DXplain diagnosis decision system has characteristics of an electronic medical textbook and a reference system. After receiving the inputs, the system develops a list of ranked diagnosis based on the input and provides the justification. This is one of the main reasons why this system is used widely for education purposes.

DXplain system provides about 10 appropriate references for each suggested diagnosis. The user interface for the DXplain system is very intuitive with graphical user interface requiring only clicks from the lists to drill down to the recommendations. This is one of the main reasons suggested for the success of this product.

RMRS

Regenstrief Medical Records System (RMRS) is one of the oldest, popular, and commonly cited decision support system in the United States. RMRS, developed at Indiana University, is a hospital-based medical information system with decision support capabilities like preventive care reminders and advice regarding cost-effectiveness details during physician order entry.

As early as 1974, RMRS began to deliver paper generated automatic reminders to the physicians the night before a patient's visit. This intelligent system reviewed the patient's medical record against the set of pre-defined protocols to generate reminders about the patient's condition and to suggest corrective actions. In 1984, this report-based decision support system was programmed into an interactive Gopher-based system to break the long list of reminders into logical groups, allowing physicians to step through the reminders during the physician order entry process. The success of RMRS system was due to the fact that the decision support functions were integrated with the hospital information system. The availability of required data and the knowledge base makes this a powerful system. The decision support capabilities made available to the decision makers at the point of care made this an effective product.

In the past decade, RMRS went through a lot of developments and has included various decision support techniques such as pattern recognition and intelligent agents to help physicians assess the patterns of care and identify patients with particular risk factors and adverse outcomes. This system is used at more than 40 inpatient and outpatient facilities and the commercial version is marketed internationally.

PRODIGY

PRODIGY (PROject prescribing rationally with Decision support In General practice study) is a prescribing decision support system that offers evidence-based, cost effective general practice prescribing. The Sowerby Center of Health Informatics at Newcastle University was approached by the English National Health Service Executive during 1995 to develop a medical decision support system to guide UK general practitioners on therapeutic actions covering prescribing. While developing PRODIGY, the research team realized the importance of integration with already existing general practice software used by the general practitioners. Hence integration with all the five major general practice software in the UK was performed to make PRODIGY a powerful medical decision support system.

While several medical decision support systems assist in diagnosing the problem, PRODIGY starts after the diagnosis is made by providing medical advice and therapeutic recommendations (MedExpert-Prodigy, 2004). The system also stores patient information leaflets containing information about the disease in simple terms. The trigger for decision support within PRODIGY is the input of diagnosis code. Once the physician enters the diagnosis code in their general practice system, it activates PRODIGY decision support to check if suitable medical recommendations are available. The decision maker at that point has the option to explore the recommendations or ignore it. When explored, the decision maker will be presented with therapy scenarios, evidence-based prescribing, and availability of patient information leaflets along with justification and references for the recommendations. PRODIGY medical decision support system has been successfully implemented nation wide at all general practitioners in the UK.

CADIAG – II (Computer Assisted DIAGnosis)

CADIAG – II is a hospital-based medical decision support system developed by the Department of Medical Computer Sciences at University of Vienna. CADIAG decision support system has been integrated with the medical information system of the Vienna General Hospital.

CADIAG decision support system focuses on the colon diseases, rheumatic diseases, gall bladder, pancreatic diseases, and bile duct diseases (MedExpert-Cadiag, 2004). This medical decision support system provides online consultation support for the physician to assist in diagnosis of the diseases given the signs, symptoms, etiology, and laboratory results. This system contains a very strong knowledge database in the specialty areas of colon diseases and rheumatic diseases.

NeoGanesh

NeoGanesh is a widely cited knowledge-based decision support system that is used primarily to manage the mechanical ventilation in Intensive Care Units. This decision support system is one of the very few automated and controlled system with very limited human intervention. This real time medical decision support system checks for the real time data and controls the mechanical assistance provided to the patients suffering from lung diseases in a pressure supported ventilation mode (Dojat, Pachet, Guessoum, Touchard, Harf, & Brochard, 1997). The other interesting feature of the system is to develop a therapeutic strategy to respond to the patients and evaluate the capacity to breathe. This system is used at Henri Mondor Hospital in France. This system is being planned to be released for commercial purpose as well.

Medical Decision Support Knowledge Sharing

Overview

The field of medicine has been a pioneer in the introduction of decision support systems. One of the first expert systems to be developed was in the field of medicine. The complexity of decision making in the field of medicine and the challenge of keeping up to date with the new findings and research has been a key motivator for the usage of medical decision support systems. As discussed in the previous sections, numerous medical decision support systems exist in the market. The knowledge for most of these medical decision support systems are acquired from the clinical practice guidelines issued by various government and medical associations. Clinical practice guidelines are the rule-based knowledge that guides the decision makers in a medical setting. These guidelines have been developed over the years to reduce the variations among medical practices with a common goal to provide cost effective and high quality healthcare services (Field & Lohr, 1992).

Healthcare organizations historically have focused more on the development of the clinical practice guidelines compared to the implementation of the guidelines (Grimshaw & Russell, 1993). Past research studies have proved that the computerized medical decision support systems when integrated with the clinical workflow can improve the practitioners' compliance with clinical practice guidelines and outcomes (Johnsons, Langton, Haynes & Mathieu, 1994). Thus, it was clear that the medical decision support systems are best vehicles to promote compliance. The availability of several decision support systems and the use of common knowledge and rules triggered the need for a common method of sharing the knowledge. This technique can save the cost of developing the same medical practice guidelines for multiple decision support systems.

While it is common among organizations to share the data and information between computer systems, sharing the knowledge was a challenge. While knowledge sharing could be considered as losing competitive advantage, the healthcare industry was one of the few industries where several guidelines are common as they are generally developed by government agencies or medical associations. The medical arena rose to the occasion and various associations were involved in development of a standard way of sharing the medical knowledge and clinical practice guidelines among systems. In the following sections, the various prominent clinical practice guideline models and knowledge sharing standards are discussed.

ARDEN Syntax

ARDEN Syntax is an ASTM (American Society for Testing and Materials) and ANSI (American National Standard Code for Information Interchange) standard a hybrid knowledge representation format that was created with the main aim to address the ability to share, reuse, and understand the medical knowledge base. It is first introduced in 1989 at the Columbia University's Arden homestead conference. The initial idea for the conference was to create a knowledge representation format that can facilitate the definition of knowledge bases and thus could result in an effective implementation of new knowledge consistently (Hripcsak, Pryor & Wigertz, 1995). The development of knowledge bases is the biggest challenge faced by the ever changing medical field and introducing a knowledge representation format was thought to be of significant help to allow various institutions to share their knowledge among medical decision support systems.

The ARDEN syntax was largely derived from the logical modules used by the HELP and RMRS medical decision support systems. The concept of the ARDEN syntax is to develop rule-based modular logical rules called medical logical modules (MLM). The medical decision support system that is based on Arden Syntax polls for the occurring events and executes the MLM's out of its knowledge base that has defined the occurring event as the triggering condition.

In ARDEN syntax, the MLM's are composed of slots that are grouped into three categories called maintenance, library, and knowledge. Each category has a set of predefined slots. The slots are broadly classified as textual slots, textual list slots, coded slots, and structured slots. The maintenance category contains the slots that specify general information about the MLM like title, mlmname, arden syntax version, mlm version, institution, author, specialist, date, and validation. The library category contains slots about the knowledge base maintenance that is related to the module's knowledge. This category contains the relevant literature, explanation, and links that were used in defining the MLM. The library category slots are purpose, explanation, keywords, citations and links. The knowledge category contains the slots that specify the real knowledge of the MLM. The knowledge category dictates the triggering event of the MLM and the logic of the MLM. The knowledge category slots are type, data, priority, evoke, logic, action, and urgency.

Example

Let us look at a simple anemia management protocol for a renal patient as an example. On receipt of new hemoglobin laboratory result, if the value is greater than 13.3, and if the patient is on Erythropoietin dose, then recommend discontinuation of the dosage. The MLM for this simple rule is given below.

maintenance:

    title: Screen for anemia management (triggered by hemoglobin storage);;

    filename: renal_anemia_Hemo_Epo;;
    version: 1.00;;
    institution: Sample Institute; Sample Medical Center;;
    author: Sri Raghavan, PhD.;;
    specialist: ;;
    date: 2004-03-01;;
    validation: testing;;

library:

    purpose:
      Warn the patient care personnel of the level of hemoglobin and EPO dosage.;;

    explanation:
      Whenever a hemoglobin blood result is stored, it is checked for anemia management. This simple protocol checks if the value is greater than 13.3, and if EPO is administered to the patient, then an alert is generated to recommend the practitioner to discontinue the dosage.;;

    keywords: anemia; hemoglobin; renal; Erythropoietin;;

    citations: ;;

knowledge:

    type: data_driven;;

    data:

      /* evoke on storage of a hemoglobin result*/
      storage_of_hgb := EVENT {storage of hemoglobin};

      /* read the potassium that evoked the MLM */
      hgb:= READ LAST {hemoglobin level};

      /* get the last active Erythropoietin order */
      epo_order := read last epo order};

      ;;

    evoke:
      /* evoke on storage of a hemoglobin */

```
    storage_of_hemoglobin;;

logic:

    /* exit if the hemoglobin value is invalid */
    if hemoglobin is not number then
        conclude false;
    endif;

    /* exit if there hemoglobin is <=13.3 */
    if hemoglobin <= 13.3 then
        conclude false;
    endif;

    /* exit if Erythropoietin order cannot be found */
    if (epo_order is null) then
        conclude false;
    endif;

    /* send an alert */
    conclude true;

    ;;

action:

    write "The patient's hemoglobin level on (" ||time of potassium|| ") is" ||hemoglobin|| ". The patient is currently on Erythropoietin dosage. As per clinical guideline, it is recommended to discontinue the Erythropoietin dosage.";

    ;;

    urgency: 90;;

end:
```

ARDEN syntax-based decision support systems are widely in use. Since this syntax became an ANSI standard, it was embraced by several medical information system vendors to provide decision support capabilities. Vendors who have developed ARDEN compliant applications include: Cerner Corporation, Healthvision, McKesson, SMS, and Micromedex.

ASBRU

Asbru is a knowledge representation language to capture the clinical guideline and protocols as time oriented skeletal plans. The development of Asbru was part of the Asgaard project developed by the Technical University of Vienna, Stanford Medical Informatics, University of Newcastle, and University of Vienna. The goal of the Asgaard project is to develop task specific problem solving methods that perform medical decision support and critiquing tasks (Seyfang, Miksch, & Marcos, 2002).

Asbru is a difficult language to be understood by a physician and hence a graphical knowledge acquisition tool is used to gather the knowledge or guideline and then output that into Asbru syntax behind the scenes for computer interpretation. The graphical tool used is called Protégé. The Asbru output created by Protégé can be shared with other medical decision support systems that are compliant with Asbru format.

EON

EON is a clinical guideline modeling system developed by Stanford University for creating guideline-based decision modules. The EON guideline model system was developed to address the problem of modeling clinical guidelines and protocols to provide patient specific decision support. This modeling system allows creating a knowledge engineering environment for easy encoding of clinical guidelines and protocols. EON guideline model allows association of conditional goals with guidelines. It provides three criteria languages to allow the medical practitioners to use and express their decision making complexities: a simple object oriented language that medical practitioners can use to encode the decision criteria, a temporal query type language, and a predicate logic.

EON also uses Protégé as the medical knowledge acquisition tool to help medical practitioners provide medical decision criteria. The EON language is generated at the back end by the Protégé tool. This code is platform independent and can be shared between decision support systems that are EON compliant.

EON is widely used in European countries and one of the major decision support system that uses EON is PRODIGY. The decision support system ATHENA, which provides hypertension related advices, also uses the EON guideline model as its decision support architecture.

GEM (Guideline Elements Model)

GEM, an XML- (Extensible Markup Language) based guideline document model was introduced in 2000 by Yale center of medical informatics. This initiative is intended to facilitate the translation of medical guideline documents to a standard computer interpretable format. The key difference between the other techniques and GEM is that it uses the industry standard XML for knowledge transfer purposes. GEM's XML hierarchy is made up of 100 discrete tags and 9 major branches. These branches are identity, developer, intended audience, target population, method of development, testing, review plan, and knowledge components (Shiffman, Karras, Agrawal, Chen, Marenco, & Nath, 2000). Several of these branches appear identical to the slots of ARDEN syntax. While the goal of all of these knowledge transfer techniques is the same, they differ in their implementation strategies.

The strength of this model is that it encodes both the recommendations and adequate information about the guideline recommendations, including its reason and quality of evidence (Shiffman et al., 2000). GEM was accepted as an ASTM standard in 2002. Knowledge extraction in GEM is a simple document markup rather than a programming task. GEM uses a knowledge acquisition tool called GEM Cutter to interactively gather he knowledge data and store it in a XML format.

GLARE (Guide Line Acquisition Representation and Execution)

GLARE is a medical domain independent guideline model system for acquiring, representing, and executing clinical guidelines. GLARE was introduced in 1997 by the Dipartimento di Informatica, Universita del Piemonto Orientale, Alexandria, Italy. This system boasts a modular architecture with acquisition and execution modules. The GLARE Knowledge representation language is designed to satisfy both the complexity and expressive rules. The format is limited but focused set of primitives. It consists of plans and atomic actions that can be queries, decisions, work actions, and conclusions (Open Clinical-Glare, 2004).

GLARE knowledge acquisition system is an easy to use graphical user interface with syntactic and semantic verifications to check the formulated guidelines. Artificial intelligence temporal reasoning techniques are applied to weed out inconsistencies. The GLARE system goes one step ahead and also incorporates a decision support system for knowledge execution purposes. GLARE system has been tested for applicability in medical domains like bladder cancer, reflux esophagitis, and heart failure at the Laboratorio di Informatica Clinica, Torino, Italy. The commercial version of this product is still under development.

Future Trends

While medical decision support systems have been around for the past four decades, its usage among clinicians is still questionable. Several variables play a role in a success of a medical decision support system and it is important to address these to move ahead from where we stand now. Issues like physician reluctance, intimidating systems, proprietary interests, local practice, technological factors, and outdated knowledgebase are all limiting factors for the growth and acceptance of computer-based decision support systems. In addition to these limitations, the decision making functions are no longer limited to the physicians. The patients, healthcare administrators, insurance carriers, healthcare professionals, policy makers, risk analysts, and the government at large are all stakeholders in this medical decision making process. It will be extremely limiting to tie down these requirements to local proprietary computer systems.

While the field of medicine pioneered the knowledge sharing techniques among decision support systems, the introduction of several sharing standards segmented market as a net result defeated the purpose of such standard. The growth, availability, and accessibility of the Internet opened the door for development of a new model for sharing medical knowledge and decision support systems across networks. Internet-based decision support systems with focus on a particular medical domain will be prevalent in the future.

The various knowledge sharing standards are naturally meeting a common merger. For example, several medical guideline modeling languages are using a common interface for knowledge acquisition purposes. Protégé is the most commonly used tool to acquire the knowledge and then formulate the output in various modeling languages. This converging trend will begin to happen at the knowledge representation and knowledge execution modules as well. This will give raise to a true acceptance of a common knowledge sharing technique.

The next key area of medical decision support systems is the unique characteristics of its users. Great medical decision support systems when seldom used solves no purpose. Most users get intimidated by the complexities and challenges posed by the user interface. While medical decision making is not a simple process, the decision support systems should help reducing the complexity and not increasing them. Past studies indicate that the complexity of customizing the knowledge or a simple rule forces the physicians to refrain form using the system. Training clinicians in usage of complex decision support process is a worthy investment. The benefits of the usage of medical decision support systems outweigh the costs incurred in development, implementation, and maintenance of such products.

Chapter Summary

Medical decision support systems, when used along with the clinical information systems at the point of care can improve the quality of care provided. Despite evidence of the benefits of decision support, and despite endorsement by a national patient safety panel (Kohn, Corrigan & Donaldson, 2000), computerized clinical decision support has not achieved wide diffusion. A recent survey found that such systems were used in less than five percent of all healthcare facilities (Wong, Legnini, Whitmore, & Taylor, 2000). When the user interface, availability, and accessibility issues of the medical decision support systems are properly addressed, the usage of the systems will grow multi fold! When such medical decision support systems are blended into the day-to-day practice workflow, then these systems are destined to succeed!

References

Beatty, P.C.W. (1999). User attitudes to computer-based decision support in anesthesia and critical care: A preliminary survey. *The Internet Journal of Anesthesiology, 3*(1).

Brenner, W., Zarnekow, R. & Wittig, H. (1998). *Intelligent software agents: Foundations and applications.* New York: Springer-Verlag.

Buchanan, B.G. & Shortliffe, E.H. (1984). *Rule-based expert systems: The MYCIN experiments of the Stanford heuristic programming project.* Reading, MA: Addison-Wesley.

Camms, J.D. & Evans, J.R. (1996). *Management science: Modeling, analysis and interpretation,* Cincinnati, OH: South-Western.

Dojat, M., Pachet, F., Guessoum, Z., Touchard, D., Harf, A., & Brochard, L. (1997). NeoGanesh: A working system for the automated control of assisted ventilation in ICUs. *Artificial Intelligence in Medicine, 11,* 97-117.

Field, M.J. & Lohr, K.B. (1992). *Guidelines for clinical practice: From development to use.* Washington, DC: National Academy Press.

Frenster, J.H. (1989). *Expert systems and open systems in medical artificial intelligence.* Presented at the Congress on Medical Informatics. May 11-13.

Gorry, G.A. & Morton, S. (1971). A framework for management information systems, *Sloan Management Review, 13*(1).

Grimshaw, J.M. & Russell, I.T. (1993). Effect of clinical guidelines on a medical practice: A systematic review of rigourous evaluations. *Lancet,* 342(8883).

Hastie, R. (2001). Problems for judgment and decision making. *Ann Rev Psychol, 52,* 653-683.

Hripcsak, G., Pryor, A.T. & Wigertz, O.B. (1995). Transferring medical knowledge bases between different HIS environments. In H.U. Prokosch & J. Dudeck (Eds.), *Hospital information systems: Design and development characteristics, impact and future architectures.* Amsterdam: Elsevier Science B.V.

Hunt, D.L., Haynes, R.B., Hanna, S.E. & Smith, K. (1998). Effects of computer-based clinical decision support systems on physician performance and patient outcomes: a systematic review. *JAMA, 280*(15).

Johnsons, M.E., Langton, K.B., Haynes, R.B. & Mathieu, A. (1994). Effects of computer based clinical decision support systems on clinician performance and patient outcome. *Ann Intern Med., 120*(2).

Kohn, L.T., Corrigan, J. & Donaldson. M.S. (2000). *To err is human: Building a safer health system.* Washington, DC: National Academic Press.

Lobach, D.F. & Hammond, E. (1997). Computerized decision support based on a clinical practice guideline improves compliance with care standards. *Am J Med., 102*(1).

Ludley, R.S. & Lusted, L.B. (1959). Reasoning foundations of medical diagnosis. *Science*.

MedExpert-Cadiag (2004). MedExpert WWW - A medical knowledge base server - HELP. Retrieved March 31 2004, from *http://medexpert.imc.akh-wien.ac.at/cadiag_info.html*

MedExpert-DxPlain (2004). MedExpert WWW - A medical knowledge base server - HELP. Retrieved March 31 2004, from *http://medexpert.imc.akh-wien.ac.at/dxplain_info.html*

MedExpert-HELP (2004). MedExpert WWW - A medical knowledge base server - HELP. Retrieved March 31, 2004, from *http://medexpert.imc.akh-wien.ac.at/help_info.html*

MedExpert-Prodigy (2004). MedExpert WWW - A medical knowledge base server - HELP. Retrieved March 31, 2004, from *http://medexpert.imc.akh-wien.ac.at/prodigy_info.html*

Murch, R. & Johnson, T. (1999). *Intelligent software agents*. Upper Saddle River, NJ: Prentice Hall.

Office of Technology Assessment. (1995). *Bringing health care online: The role of information technologies*. Washington, DC: US Government Printing Office.

Open Clinical-Glare (2004). Open clinical knowledge management for medical care - Guideline modelling methods and technologies. Retrieved March 31, 2004, from *http://www.openclinical.com/gmm_glare.html*

Patel, V.L, Kaufman, D.R. & Arocha, J.F. (2000). Emerging paradigms of cognition in medical decision making. *Journal of Biomedical Informatics*.

Rahavan, S., Ladik, V. & Meyer, K.B. (2005). Developing decision support for dialysis treatment of chronic kidney failure. *IEEE Transactions on Information Technology in BioMedicine*.

Seyfang, A., Miksch, S. & Marcos, M. (2002). Combining diagnosis and treatment using Asbru. *International Journal of Medical Informatics, 68*, 1-3.

Shiffman, R.N, Karras, B.T, Agrawal, A., Chen, R., Marenco, L. & Nath, S. (2000). GEM: A proposal for a more comprehensive guideline document model using XML. *JAMIA 2000, 7*(5).

Simon, H. (1971). *The new science of management*. Englewood Cliffs, NJ: Prentice Hall.

Stohr, E.A & Konsynski, B.R. (1992). *Information systems and decision processes*. IEEE Computer Society Press.

Thomas, K.W., Dayton, C.S. & Peterson, M.W. (1999). Evaluation of Internet based clinical decision support systems. *Jounal of Medical Internet Research, 1*(2), 6.

Turban, E. & Aronson, J.E. (2000). *Decision support systems and intelligent systems*. Upper Saddle River, NJ: Prentice Hall.

Wong, H.J., Legnini, M.W., Whitmore, H.H. & Taylor, R.S. (2000). The diffusion of decision support systems in healthcare: Are we there yet? *Journal of Healthcare Management, 45*(4).

Wyatt, J. & Spiegelhalter, D. (1991). Field trials of medical decision-aids: Potential problems and solutions. In P.D. Clayton (Ed.) *Proceedings of the Fifteenth Annual Symposium on Computer Applications in Medical Care.* Washington: American Medical Informatics Association. 3-7.

Section III

Knowledge Management in Action: Clinical Cases and Application

Chapter XII

Feasibility of Joint Working in the Exchange and Sharing of Caller Information Between Ambulance, Fire and Police Services of Barfordshire

Steve Clarke, The University of Hull, UK

Brian Lehaney, Coventry University, UK

Huw Evans, University of Hull, UK

Abstract

This was a practical intervention in the UK, the objective of which was to undertake an examination of the current arrangements between Barfordshire Fire, Police and Ambulance Services for the sharing and exchange of caller information, taking into account technological potential and constraints, organisational issues, and geographical factors. The process followed was highly participative. The initial event was an open space session followed by later sessions exploring information technology and human resource issues. For these two later events, interactive planning and critical systems heuristics were used. The most important outcome was that, whilst the extent to which the five organisations involved shared information and knowledge was very variable, there were no perceived barriers to this happening. Such sharing, despite considerable structural and cultural barriers, was seen to be feasible both organisationally and technically. The study further highlighted a need to more closely integrate operational and strategic planning in this area and to make more explicit use of known and tested methodologies to better enable participative dialogue.

Introduction

This chapter is based on a study into the feasibility of sharing caller information undertaken by the authors during 2000/2001. It involved Barfordshire Ambulance and Paramedic Service NHS Trust, Fire and Rescue Service, and Police Service, all in the UK, as participants. Whilst it might be construed that the only "healthcare group" included was the Ambulance and Paramedic Service, our study indicated that much of the work of all emergency services falls into or is related to healthcare. For example, fire crews and police frequently act as "first responders" to accidents in which they give support to ambulance and paramedic staff. Fire and police officers receive first aid training, and in some parts of the UK police are beginning to be trained to use defibrillators. The boundary between these three services as regards issues of healthcare is becoming, it would seem, ever more blurred.

The study began with all five services named above as participants. However, in practice, the participation of the Fire and Rescue Service was restricted to mostly Principal Officer level, and the project team has augmented information from that source by visits to similar organisations and IT suppliers, and by the collection of secondary data. There were good reasons for this lower level of involvement on the part of the Fire service, and these will be addressed later in the chapter.

Further, the outcome of the May 2000 Home Office Report (The Future of Fire Service Control Rooms and Communications in England and Wales, HM Fire Service Inspectorate, 2000), which came in the middle of the study, marked the effective withdrawal of Barfordshire Fire and Rescue Service from operational involvement in the study.

Specification of the Study

The objective, agreed between the consultants and a project board representing the emergency services, was to undertake an examination of the current arrangements between Barfordshire Fire, Police and Ambulance Services for the sharing and exchange of caller information. Specifically, we were charged with the task of exploring the feasibility of enhancing sharing and exchange of caller information between the above, taking into account technological potential and constraints; organisational issues; and

Copyright © 2005, Idea Group Inc. Copying or distributing in print or electronic forms without written permission of Idea Group Inc. is prohibited.

geographical factors. In fact, the study embraced basic data storage, information retrieval, and the use and sharing of knowledge bases which had evolved in each of the organisations: in many ways, the knowledge issues proved to be the most interesting and challenging.

We were given a framework within which to operate, which was defined in relation to enhanced customer service, "best value", economies of scale, and more effective working. The output required was a "draft report for further consideration, detailing the options for Inter-County Joint Emergency Service sharing and exchange of caller information".

Specifically, the specification of the study can be defined as below.

Problem Content, Scope and Objectives

- An examination of the current arrangements between Barfordshire Police, Ambulance and Paramedic Service, and Fire & Rescue Service for the sharing and exchange of caller information between the services.
- An exploration into the feasibility of enhancing the sharing and exchange of caller information between the above, taking into account: technological potential and constraints; organisational issues; geographical factors.

Project Outcomes

- A feasibility study looking at joint working in the exchange and sharing of caller information, and yielding the potential benefits of: enhanced customer service, "best value", economies of scale, and more effective working.

Outline Project Deliverables and/or Desired Outcomes

A Draft Report for further consideration, detailing the options for Inter-County Joint Emergency Service sharing and exchange of caller information.

The Approach

Purpose

The project was conducted within the Best Value framework (AuditCommission 1998). Best Value was a U.K. Government initiative of the time, a key purpose of which was to improve public services, especially at a local level. A series of three participative sessions were held to inform the study and to contribute to the "consult" (AuditCommission 2000) element of Best Value. The events would also contribute to informing the 'challenge' element of the Best Value framework and potentially enable people to question the rationale for the current fragmented emergency service call handling and information sharing processes.

It was felt that a participative approach involving people from all the organisations involved would begin to breakdown any organisational cultural barriers and begin to develop a common understanding of how the different organisations worked.

Process

The participative sessions were held on three separate days, one to develop the issues about the organisations working together and two others to gain information about technical (IT) issues and human resources issues respectively.

The initial event was an open space (Owen, 1997) session where approximately 20 people from each of the Ambulance and Police Services attended. Two later sessions concentrated on the technical and HR aspects of greater collaboration and attendees were fewer in number and limited to the experts in the areas being explored, namely information technology and human resources. For these two later events, interactive planning (Ackoff 1981) and critical systems heuristics (Ulrich 1983) were used. All the organisations involved except the Fire services were represented. These remaining organisations found elements of common ground and began to explore the technical and HR issues.

The events took the form of a chaired discussion testing ideas and assumptions about collaboration between the organisations. The experts shared technical information about their IT systems and began to explore the potential for working together. Lists of issues raised were produced by each group which informed the study.

Findings

The participative sessions highlighted the problem of engagement for the organisations that would have been masked had they not been held. The project board formed with executive officers from each of the emergency services appeared to exhibit a broad consensus about the value of exploring the potential for collaborative work. However, that commitment was tested demonstrably when it came to sending people to the participative sessions. In terms of participation, non-attendance speaks volumes about the commitment of participants within a partnership. The issues about these emergency services collaborating in call handling or information sharing were exposed as being primarily human centred, cultural and managerial and less about the technical issues.

Similar problems have been experienced in another area of the United Kingdom where the emergency services were actively working towards a shared control room, but the cultural and human centred issues have got in the way of implementation.

Findings of Feasibility Study

The comments in this section should be seen in the context of access to the Fire Service being only at Principal Officer level. Access to Fire control rooms in Barfordshire has not been possible. A joint Ambulance and Fire Service control room has been visited at Warwickshire, and the visits to three joint control (Fire, Police and Ambulance) sites in Wiltshire, Gloucestershire and Cleveland provided further information, although none of these three sites was operational at the time of the study.

Organisational Feasibility

The services share their knowledge on a daily, operational basis, via telephone between control rooms, and person to person communication at incident locations. There is, however, no sharing of information on an organised basis. NB: At the time of the study, the Home Office was championing the implementation of a new combined radio and other data communication system called Airwave (www.airwaveservice.co.uk).

Enhancing the sharing of information implies improved enabling technologies for such sharing, including the new public service radio communications platform, and mobile data/automatic vehicle location.

1. Improved enabling technologies for information sharing.
    a. Airwave or an equivalent communication platform [note that there is strong evidence—per Jack Straw (Home Secretary): see Police Information Technology Organisation Web site (www.pito.org.uk)—that all Police forces will be committed to Airwave; as I write, all forces have agreed to take it]. More information is needed to evaluate Airwave, and plan other technologies and the necessary organisational procedures around them.
    b. Mobile data, linked with automatic vehicle location, is a major opportunity for data sharing between agencies. IT applications providing learning systems based on these technologies are available now.

All of this adds up to a major management tool for the deployment of resources based on mapping systems, and according to known and targeted response levels.

2. Information sharing at other than operational levels.

There is evidence of sharing information at other than operational levels, of which the Inter County Joint Emergency Services Group in Barfordshire is an example. However, an overall explicit, articulated strategic focus, shared by all, is lacking. The situation with Airwave, which admittedly is driven by Government and therefore not entirely within the control of even the three services on a national basis, highlights this. A given service (e.g., the Police) is expected to "buy in" to Airwave: but none of the following questions are systematically addressed: where does this fit with communication strategies overall; what is the long term view; what about mobile data; what are the operational considerations?

PITO (Police Information Technology Organisation) is charged by government with designing common IS and IT requirements for implementation across all Police services in England and Wales. This brings added complications concerning the ability of individual forces to change IS and IT in line with government requirements, given the need to improve some systems immediately and the competing pressure to conform to a PITO standard that might be several years away from readiness. This situation is further complicated by the narrow development of Police IS and IT, and probably Ambulance and Fire, in not designing systems that integrate the three services. A notable exception being Cleveland Police where they have employed integrated IS for a number of years.

Structure Issues

The organisational structure of Fire, Police and Ambulance Services has been investigated through visits to sites and discussions with relevant personnel.

A number of issues related to structure became clearer as a result of this study. For example, one overriding issue was the different levels of complexity within each of the services in handling caller information. In general terms, the Fire Service seems to have the most clearly defined procedures, which are easiest to follow. The Ambulance Service control systems exhibit more complexity, and a greater need for overall management, whilst mostly occurring under resource pressure. The Police Service has a much wider remit, with a wider variety and greater number of calls, requiring considerably more decision-making at the point of call receipt.

Consequently, the two Police Services have to "manage" demand in order to meet the needs of their clients. Whilst the Fire Service deals only with emergency calls, and the Ambulance Service has clear categories of calls to which it has to respond, and which it prioritises. This gives the Ambulance Service a particular problem, in that it has to manage the knowledge relating to its own activities or those of the Fire and Police Services, on, at times, a minute-by-minute basis. As an example, an emergency call to the Fire Service is always responded to. The vehicles and teams to send are, for the most part, allocated according to set rules and procedures. These rules and procedures are such that they can be incorporated within an automated system. Except in major emergencies, control of the incident, from inception of the instruction to eventual return to station, can be left largely with the operational fire teams attending the incident—central co-ordination is much less of an issue than with Ambulance or Police.

Ambulance Service

The Ambulance Service for Barfordshire has a control centre in Barford. The centre uses a computerised system for call taking and despatch, included within which (currently available or planned) are mapping, mobile data and automated vehicle location systems

(AVLS). The system uses a single radio system. Ambulance and Paramedic Service command and control is supplied by a proprietary system.

The Ambulance and Paramedic Service call taking and despatch system is an example of a system often under severe resource pressure, frequently resulting in a need to carefully schedule vehicles to meet the current demand. Any joint working which can enhance Ambulance and Paramedic Service performance is estimated to yield immediate benefits in terms of response times, which are subject to National targets.

Police Service

The Police Service for Barfordshire has two central control centres (or "information centres"). One, in Barford, has recently been refurbished, and monitors all calls across the County. A computer system is used, together with UHF and VHF radio systems (now replaced). There is no AVLS or mapping, and no mobile data. The other, in a neighbouring county town, is in the process of rolling out the Airwave implementation. The same computer system as that in Barford is used (although the procedures for using it differ), together with UHF and VHF radio systems (now replaced). There is no AVLS or mapping, and no systematic application of mobile data [e.g., the Armed Response Vehicles (ARVs) use laptops to link in to central systems remotely].

Police Services have to "manage" demand in order to meet the needs of their clients. Forces have adopted systems of "graded response" which they apply to calls for service. For example, Grade 1—immediate response, grade 2—respond but at normal traffic speeds, grade 3—by appointment, grade 4—no police attendance required. As mentioned earlier, whilst the Fire Service deals only with emergency calls, and the Ambulance Service has clear categories of calls to which it has to respond, and which it prioritises, the range of calls which the Police Service has to respond to is much more varied. A distinction is made between emergency and non-emergency calls, but the classification of a call as emergency or non-emergency is not always clear cut. This is an issue which impacts future development, and to which I will return later in the report.

Fire Service

In the case of the Fire Service, there have been limited opportunities to assess the operation, but from the evidence gathered it is possible to make the following preliminary assessment of the emergency systems:

1. An emergency call to the Fire Service is always responded to.
2. The vehicles and teams to send are, for the most part, allocated according to set rules and procedures. These rules and procedures are such that they can be incorporated within an automated system.
3. Calls can be dealt with by a call taker/dispatcher, who, on receipt of the call, is able to generate an automated instruction, resulting in hard copy (e.g., computer printout) information at the relevant Fire station.

4. Except in major emergencies, control of the incident, from inception of the instruction in number 3, above, to eventual return to station, can be left largely with the operational fire teams attending the incident—central coordination is much less of an issue than with Ambulance or Police.

Structure: Summary

The organisational structures of the Fire, Police and Ambulance Services are different, as are the working practices. However, this should not be seen as preventing information sharing, but rather as the framework(s) within which such sharing must take place. For example, if the technical systems are so disparate as to be functionally incompatible, sharing might be achieved by providing a reporting structure, or an intermediary file structure, which stores and produces the necessary information in a common format, as determined, for example, by an information needs analysis.

In effect, information need must be determined, supported by effective ways of deciding how much information should be passed from service to service, and making that information available.

Human/Cultural Issues

The human and cultural issues are significant and may be summarised as a need to "manage change". Detailed issues which have surfaced as in need of consideration include:

- Differences in practice between different agencies, and different members of the same agency.
- The physical co-location of call takers and dispatchers, from all services, is seen to be an important factor in handling emergency calls.
- The need to determine, in any joint working scenario, who the employer would be.
- A need to focus on improvement in service rather than just reducing costs.
- A need to involve all levels of the organisation, and to feed back outcomes and decisions.
- Ensure adequate training.
- Ensure that there is a long term strategy for caller information handling.
- General concerns of re-location / redundancies associated with any change programme.
- Each organisation has its own conditions of service, staff associations etc.

Cultural issues are most problematic in the Fire Service, where there is direct evidence of considerable resistance to change which is a matter of public record available in reports and on web sites (see: The Future of Fire Service Control Rooms and Communications in England and Wales, HM Fire Service Inspectorate, 2000; Cleveland FBU Web Site on www.firerescue.co.uk), and in the required adherence to the status quo in each of the joint control sites visited.

The Ambulance service, whilst often overstretched, has a desire to collaborate with other services; the Police service appears willing to consider any relevant initiatives.

Technical Feasibility

The technical systems are so disparate as to be functionally incompatible: from information gathered, it would appear that the only directly compatible systems among the services studied are the IT and radio systems used by both Police Forces.

Generally, however, technology is not seen to be an obstacle. This view was particularly reinforced by visits to suppliers of equipment, visits to other sites, and technology session attended by Barfordshire Police, and the Ambulance and Paramedic Service. The issues raised as important include, for example, a recognition that not all information can be computerised—much resides as knowledge, implicit or explicit, with the people undertaking the work.

The issues raised as important are:

- A need to provide real-time shared information.
- A need to look at automated ways of dealing with non emergency information (e.g., use the Internet).
- A recognition that not all information can be shared (e.g., ethical and legal considerations).
- A recognition that not all information can be computerised—much resides with the people undertaking the work.
- The importance of ensuring only one source of accurate information—this is not currently certain even with single agency working.
- Physical proximity is important culturally, but unnecessary technically.
- Organisational structure and procedures need to be matched to defined business needs.
- Focus on people rather than systems.
- Ensure that there is a long term strategy which can be consistently followed.
- Systems need to be flexible to adapt to changing circumstances. People too often end up working to fit systems.

- Recommendations of regulative or co-ordinating organisations (e.g., Police Information Technology Organisation) need to be taken into account.

There are, however, technical barriers. Government intervention is a significant driver where new technology is concerned, and brings its own problems. Currently identified areas of concern include procurement lifecycles (e.g., Airwave appears to be a fifteen year commitment which seems to be taking five years or more to get in place), a lack of any planned commonality of systems, and an overall lack of an explicit, articulated strategy shared by all.

Risks: Structural, Organisational and Technical

Project Risks

A Risk Log compiled in January 2000 represented an assessment of the risks seen to be most threatening at that time. Events have shown this to be both accurate and of significant value.

Most of these risks were realised to a greater or lesser degree in respect of the Fire Service. The outcome has been to confirm the ability of the Police and Ambulance Service to work more closely together, whilst the Fire Service has effectively withdrawn from any operational involvement in the study. This is not to apportion any blame, but merely to state the facts, a point which is confirmed by an assessment of Risk No. 2, "Staff Association Opposition": a visit to the Cleveland FBU Web Site (www.firerescue.co.uk) will confirm some of the problems here, as will reading the recent Home Office Report (The Future of Fire Service Control Rooms and Communications in England and Wales, HM Fire Service Inspectorate, 2000).

Future Development Risks

The key risks identified are listed below.

- The potential failure of any technical solution.
- Information security and data protection issues.
- Ethical and legal issues which prevent information sharing.
- A call centre solution to non-emergency issues, and an emergency control centre solution may prove incompatible.
- Business continuity (e.g., back up systems).
- Staff dissatisfaction, expressed in terms of staff turnover, stress etc.
- Lack of resources to meet the identified requirements.

- Who is responsible when things go wrong? Currently the three separate services deal with claims individually.
- Has anyone else done this—can we learn from others? (Evidence collated from the USA, Australia and Canada during the course of this study suggests that we can).

Other Issues to be Addressed

Security

1. Information security standards need to be set across participating organisations, embracing technological and non-technological issues.
2. Data protection requirements must be met.
3. Other legal obligations (not specifically determined as part of this study) must be complied with.

Ethical/Legal

1. For ethical and legal reasons, not all information can be shared.
2. Any technological and organisational "solutions" need to enable only the sharing that is allowed.

Technology

The general view, supported by the participative session on technology, is that "anything is possible technologically", but developments will be constrained by money, time, and other resources.

Technology needs to be treated as an enabler, supporting the information needs of the business.

There is an implied need to combine call-centre and emergency control technologies.

Call centres: There are national and local issues here. Visits to the Call Centre Conference and Barfordshire Connect have helped with preliminary understanding of these, and this needs to be continued.

Emergency Control: Two main suppliers of control systems, Intergraph and Fortek, have been visited. Again, this initial work needs to be built on.

As outlined in the "risks" above, call-centre technology may be relevant to non-emergency calls; control-centre technology and procedures to emergency calls. Integrating the two within one system is a challenge.

Conclusions

These are given on the basis that the Fire Services of Barfordshire have currently withdrawn from operational studies. It is expected that they will maintain involvement in any longer term developments. As this is a feasibility study, it is assumed that further studies and initiatives will be undertaken to build on these findings.

- All caller information should be considered within any further studies: that is, not only "emergency calls".

- Operational issues: verbal information is shared between the services, but data is not shared in any systematic way. In sharing information between agencies, little use is made of technology other than mobile phones. Enhancement to this requires a view of the benefits to be derived from planned and interdependent use of emerging technologies, examples of which are Airwave, mobile data, sophisticated information systems, and automatic vehicle location.

- Evaluation and planning for the implementation of new systems needs to be embedded in the working relationships between the three organisations. Beginning this with a joint planning initiative between participating agencies for the implementation of Airwave might be a useful pilot, alongside which could be the development of joint strategies to implement mobile data, AVLS, and any other foreseen developments.

- Strategic issues: there needs to be a clearly articulated and disseminated strategy for sharing information within and between the three services. This must be developed within the overall strategic objectives of each partner organisation, and needs to be shared by all involved. Commitment to this is needed from the highest levels of the participating organisations.

- Organisational and Technological Structures: the different organisational and technological structures of the three services, and even differences within one service (e.g., Barfordshire Police), will have to be taken into account for information sharing. The differences need to be investigated and analysed in detail, and any implementation plans drawn up must cater for these.

- Different terms and conditions of control room staff, including pay scales, are a potential source of problems.

- Care should be taken not to "over distribute" call handling. There is strong evidence that physical proximity between those involved in caller information is valuable. This view is supported by the finding that much of the information to be shared cannot be held on technical systems, but is people-centred. However, set against this is the technical factor that what is required/feasible could be achieved without co-locating

- Culture: the Ambulance and Police Services exhibit different "cultures". However, current evidence suggests that sharing information, and even more complete joint working, would be achievable with careful planning.

- Government: Airwave is an example of the problems resulting from Government intervention. All three services have to work within this framework, and the problems and opportunities presented must be considered.
- Best Value: the most pressing requirement to more fully meet Best Value criteria is Consultation. A plan for both internal and external consultation needs to be agreed and implemented by the Ambulance and Police Services.

Immediately Feasible Ways Forward

The findings of this study regarding the Fire Service, together with the May 2000 Home Office Report (The Future of Fire Service Control Rooms and Communications in England and Wales, HM Fire Service Inspectorate, 2000), have led to the conclusion that any ways forward for combined emergency service activities should at present be planned to include the operations of only Barfordshire Police and the Ambulance and Paramedic Service NHS Trust.

Other Issues Raised by the Study

- The need to plan a "common" single tier operation across Barfordshire Police, which currently has a single tier environment, with all operations monitored from the central information room in Barford, but also maintains five divisional operations rooms at strategic locations throughout the County. The need for a "common" single tier operation has become even more important with the abandonment of the STEP (Single Tier Environment) project in Barfordshire. (See above—this has moved—there is a county control room now and single tier to fit with Airwave)
- A need to co-ordinate the move to Airwave across all services, but particularly Police and Ambulance.
- Liaison with Barfordshire Connect.
- The need to more explicitly address the long term, strategic issues.
- The need to explore the potential for outsourcing, if only to meet Best Value criteria.
- A requirement for measures of success which include qualitative criteria.

Postscript: Some Reflections

The chapter is completed with some reflections on the study.

The issues addressed in this study turned out to be predominantly human centred, and the approach to these to be primarily methodological. Three methodologies were used to surface the views of participants: open space, interactive planning, and critical systems heuristics.

Open space allowed a mixed group of participants the freedom to engage in open dialogue toward common goals. In the early stages it was felt that this might prove sufficient, but further methodological support proved to be needed to address technical issues and in an attempt to undermine the power which was demonstrably preventing progress. The application in particular of critical systems heuristics, focusing on normative issues, helped surface power and control at senior levels. Some of the partners were clearly committed to seeking ways to work more closely together whilst others were not, or did not feel able to demonstrate commitment to the study. The effective withdrawal of two organisations gave a clear indication to the remaining three that any collaboration was unlikely with them. The experience also demonstrated the need for positive commitment by leaders to any initiative for it to succeed.

The leadership of the two problematic organisations might have sought to open dialogue internally by becoming immersed in the discussions about the feasibility of collaborative work and thus provide a richer picture of the possibilities. The comments from the graffiti board for the open space event are revealing:

- "Day was dominated by police personnel – better if other agencies had attended"
- "Too many small groups all discussing all issues ultimately"
- "How about public participation"

The event had been restricted to people from the organisations involved and had not sought to involve users or stakeholders, a point picked up by at least one participant. The attendee list possibly represented the commitment of each organisation to actively working together.

The message that could be drawn was that the Police service showed high interest in the concept, the Ambulance service maintained an interest and stated they were constrained by staffing problems, whilst the Fire service effectively failed to engage with the other emergency services.

This application of open space also highlights the assumptions about freedom to express ideas and thoughts along with assumptions about communicative competence and power and status, that were found in the previous three case studies. However, this example demonstrates that the process made transparent the level of commitment of two participating organisations.

The application of interactive planning was essentially a chaired discussion group that addressed issues raised by the chair and those that others chose to contribute. The group at the two sessions was composed of a small number of experts in the areas of IT and HR drawn together specifically to address the perceived problems of the different organisations working together. As such there was no suggestion that they were designed to be representative or open to wider attendance. They were strongly controlled and tightly focused, intended only to identify potential problems rather than seeking any decisions or solutions to problems. Again the events were designed to inform the consultants exploration of the issues.

This final case sought to apply a mix of approaches within one case intervention to fine tune the overall method, addressing the issues raised in cases 1-4. Critical Systems Heuristics (CSH) was chosen for its ability to deal with power, Open Space (OS) for its open, collaborative, unstructured qualities, and Interactive Planning (IP) to address a particular need within the information management of the project.

In the event, much was learned from this intervention, but shortcomings remain. CSH proved hugely valuable in *surfacing* issues of power and coercion, but less adept at *dealing with them*. OS seemed to work well with those organisations which chose to take part, but two organisations withdrew! Interactive Planning did its job well, but was a fairly tightly constrained part of the project. In addition, the need for positive leadership commitment became abundantly clear.

References

Ackoff, R. L. (1981). *Creating the corporate future*. New York: Wiley.

Audit Commission (1998). *Better by far: Best value management paper*. London: Audit Commission.

Audit Commission (2000). Listen Up! Effective community consultation - Briefing.

Owen, H. (1997). *Open space technology (A user's guide)*. San Francisco: Berrett-Koehler.

Ulrich, W. (1983). *Critical heuristics of social planning: A new approach to practical philosophy*. Berne: Haupt.

Endnote

- Barfordshire is a pseudonym, and represents two adjoining counties containing two police, two fire and one ambulance service.

Chapter XIII

Organizing for Knowledge Management:
The Cancer Information Service as an Exemplar

J. David Johnson, University of Kentucky, USA

Abstract

The Cancer Information Service is a knowledge management organization, charged with delivering information to the public concerning cancer. This chapter describes how societal trends in consumer/client information behavior impact clinical knowledge management. It then details how the CIS is organized to serve clients and how it can interface with clinical practice by providing referral, by enhancing health literacy, by providing a second opinion, and by giving crucial background, assurance to clients from neutral third party. The CIS serves as a critical knowledge broker, synthesizing and translating information for clients before, during, and after their interactions with clinical practices; thus enabling health professionals to focus on their unique functions.

Introduction

The Cancer Information Service (CIS) is essentially a knowledge management (KM) organization, manifestly charged with delivering up-to-date information to the public related to scientific advances concerning cancer. Its latent purpose, increasingly important in a consumer driven medical environment, is to insure the rapid diffusion of state-of-the-art medical care. It is an award-winning national information and education network, which has been the voice of the National Cancer Institute (NCI) for more than 30 years in the US. While the CIS has extensive outreach programs dedicated to reaching the medically underserved, it is probably best known for its telephone service that has a widely available 800 number (1-800-4-CANCER). We will use the CIS as an exemplar in this chapter of issues related to a national information infrastructure that supports clinical knowledge management.

Because of the critical role of broader societal trends we will turn to a discussion of them before describing in more detail the basic services and organizational structure of the CIS and its potential interfaces with clinical KM. Many health organizations have realized that there are strategic advantages, especially in enhancing quality, maintaining market share, and developing innovations, in promoting information technologies. Improving information management, associated analytic skills, and knowledge utilization should be a top priority of clinical practice (Johnson, 1997). It has become commonplace for almost all hospitals and managed care providers to have very active information programs for their clients allowing those in clinical settings to concentrate on their central, unique missions. Government information providers can also act as information services providing knowledge before, during, and after client interactions with clinical organizations. Health professionals can partner with KM services that recognize the public's demand for information and the various difficulties involved in reaching the people who need information. Indeed, the CIS focuses on the classic KM functions of retrieving and applying knowledge, combining it, and finally distributing/selling it.

This chapter's objectives are to answer the following questions:

1. How do societal trends in consumer/client information behavior impact clinical KM?
2. How the CIS is organized to serve clients?
3. How it can interface with clinical practice?
    a. By providing referral
    b. By enhancing health literacy
    c. By providing a second opinion
    d. By giving crucial background, assurance to clients from neutral third party
4. How can the CIS serve as an answer to information explosion?
    a. For client it acts as synthesizer, translator who can relieve clinical settings of this task
    b. Through client it directly acts to disseminate information to improve practice

Background

Knowledge Management

KM has been loosely defined as a collection of organizational practices related to generating, capturing, storing, disseminating, and applying knowledge (MacMorrow, 2001; Nonaka & Takeuchi, 1995). It is strongly related to information technology, organizational learning, intellectual capital, adaptive change, identification of information needs, development of information products, and decision support (Choo, 1998; Fouche, 1999). They are so intimately related, in fact, that it is often difficult to say where one approach stops and another begins. We will primarily view KM as a system for processing information. This is certainly the organizing thrust of the CIS. The CIS obtains the knowledge it translates to the public from the NCI (Figure 1). It is organized to provide consistent, quality information translated in a manner that can result in meaningful responses on the part of callers (e.g., course of treatment) that transforms basic information into knowledge (and perhaps in some cases even into wisdom and ultimately action). Thus, the KM service the CIS provides makes information purposeful and relevant to individuals that need it, who are often in dire circumstances.

Knowledge itself runs the gamut from data, to information, to wisdom, with a variety of distinctions made in the literature between these terms (e.g., Boahene & Ditsa, 2003). Special weight in the context of the CIS is given to a consensus surrounding refereed scientific findings that can be translated to the public to improve cancer prevention, detection, and treatment. What is knowledge is often a matter of intense negotiation between various stakeholders as evidenced by the recent controversy over the most appropriate ages for screening mammography (HHS, 2002). Accordingly, the KM structure of the CIS is fundamentally a responsive one, very dependent on changes in the larger biomedical environment. It does not have the luxury of operating in a command environment, it must respond to the diversity, complexity, and sense-making abilities (health literacy) of its audience, placing it clearly in more post modern conceptions of KM (Dervin, 1998). So, the CIS must respond to the information environment of an information seeking public, with information specialists acting as the public's human interface in transforming information to knowledge.

Public Information Seeking and Clinical Practice

Information seekers are often frustrated in their attempts to seek information, especially from doctors in clinical settings. A major focus of health communication in recent years has been on the long-standing problems with patient-doctor communication (Thompson, 2003). The amount of information physicians give to patients is correlated with patient satisfaction with healthcare and with compliance (Street, 1990). People are more

Figure 1. Knowledge flow

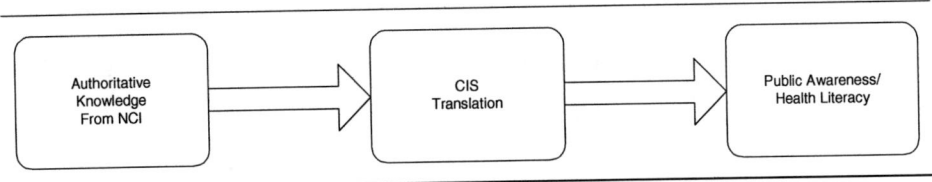

likely to follow behavioral recommendations when their understanding is high. Lack of communication between doctor and patient has been found to relate to decreased ability to recall information given by the caregiver, patient dissatisfaction, decreased adherence to prescribed regimens, and other forms of non-compliance (Brown, Stewart, & Ryan, 2003).

Cancer-related information seeking is often a great challenge to individuals. They have to overcome their tendency to deny the possibility of bad news and the distasteful problems associated with cancer. They also have to be willing to believe that their individual actions can make a difference, that by seeking information they gain some control and mastery over their problems. They also have to overcome the limits of their education and prior experience. They have to possess skills as information seekers, An awareness of databases, familiarity with the Internet, an ability to weigh conflicting sources of information and to make judgments about their credibility. In short, any one of the factors in this long-linked chain could severely impede, if not halt, the information seeking process. Recognizing this, individuals are increasingly relying on knowledge brokers, such as the CIS, as intermediaries to conduct their searches for them (Johnson, 1997).

There are a number of ways that use of knowledge brokers can complement clinical practice. First, individuals who want to be fully prepared before they visit the doctor often consult the Internet (Fox & Rainie, 2002; Taylor & Leitman, 2002). Indeed, Lowrey and Anderson (2002) suggest prior web use impacts respondent's perception of physicians. Second, there appears to be an interesting split among Internet users, with as many as 60 percent of users reporting that while they look for information, they only rely on it if their doctors tell them to (Taylor & Leitman, 2002). While the Internet makes a wealth of information available for particular purposes, it is often difficult for the novitiate to weigh the credibility of the information, a critical service that a knowledge broker, such as a clinical professional, can provide. This suggests that a precursor to a better patient-doctor dialogue would be to increase the public's knowledge base and to provide alternative, but complementary information sources, by shaping client's information fields. By shaping and influencing the external sources a patient will consult both before and after visits, clinical practices can at one in the same time reduce their own burden for explaining (or defending) their approach and increase the likelihood of client compliance.

Why is Cancer-Related Information Seeking Important?

Information is an important first step in health behavior change (Freimuth, Stein, & Kean, 1989). The consequents of information carrier exposure and seeking are many, including information gain, affective support, emotional adjustment, social adjustment, attitude change, knowledge change, behavior maintenance, a feeling of greater control over events, reduction of uncertainty, and compliance with medical advice (Johnson, 1997). Information seeking can be defined simply as the purposive acquisition of information from selected information carriers. Cancer-related information seeking has become increasingly important over the last decade. Not too long ago, information related to cancer was the exclusive preserve of doctors and other health professionals. Today in the U.S., not only is the diagnosis shared, but individuals have free access to an often bewildering wealth of information. With this access has come an increasing shift of responsibility (some might say burden) to the individual to convert information into knowledge, in the process making decisions concerning his/her treatment and adjustment to cancer (Johnson, Andrews, Case, & Allard, in press).

Individual Responsibility/Client/Consumer Movement

Increasingly the responsibility for health-related matters is passing to the individual, partly because of legal decisions that have entitled patients to full information (Johnson, 1997). The consumer movement in health is in part actively encouraged by hospitals, insurance providers, and employing organizations (Duncan, 1994) who want to encourage health consumers to 'shop' for the best product at the most affordable price (Hibbard & Weeks, 1987). Facilitating and enhancing this consumer movement have been explosive developments in information technologies, which make more specialized media sources available, permitting increased choice in information carriers, and increased connectivity with other interested parties (Case, Johnson, Andrews, Allard, & Kelly, 2004; Duncan, 1994).

Large numbers of patients do not receive state of the art treatments (Freimuth et al., 1989), partly because physicians cannot keep up with the information explosion (Duncan, 1994; NCI, 2003). The overload of information on health professionals today forces decentralization of responsibilities, with increasing responsibility passing to individuals if they are going to receive up-to-date treatment. The physician is no longer the exclusive source of medical knowledge; they must be cognizant of the welter of information available and their role in this complex system (Parrott, 2003). Recognition of the limits of health professionals also requires individuals to be able to confirm and corroborate information by using multiple sources. In fact, patients often call the CIS to verify information they receive elsewhere (Freimuth et al., 1989).

Cancer patients tend to want much more information than healthcare providers can give to them, even if willing (Johnson, 1997). Information that to a client is necessary for coping with cancer, may be seen by doctors as an intrusion into their prerogatives. Exacerbating this problem is the fact that doctors and patients may not share similar outcome goals. Traditionally, doctors have viewed the ideal patient as one who came to them recognizing their authority and who were willing to comply totally (with enthusiasm) with recommended therapies (Hibbard & Weeks, 1987).

Perhaps the most threatening aspect of enhanced information seeking for health professionals is their loss of control. Many doctors have legitimate concerns about self-diagnosis and patients possessing just enough information to be dangerous (Broadway & Christensen, 1993) and the general preferences of consumers may cause them to avoid unpopular, albeit effective, invasive procedures (Greer, 1994). Still, the more control that health professionals have, the less effective they may ultimately be, especially in terms of insuring that clients act according to consensus views of treatment (Johnson, 1997).

Solutions and Recommendations

One approach to managing these problems is for those in clinical practice to either create or to partner with KM services that recognize the public's demand for information and the various difficulties involved in reaching the people who need information in the most timely fashion. There are a number of indications that programmatically the best channels for providing cancer-related knowledge are those channels that constitute a hybrid of the positive properties of both mediated and interpersonal channels.

Information and referral centers can take many forms, such as hotlines, switchboards, and units within organizations (e.g., nurse's medical help lines) where individuals can go to get answers to pressing concerns from knowledge brokers. They serve three primary functions: educating and assisting people in making wise choices in sources and topics for searches; making information acquisition less costly; and being adaptable to a range of users (Doctor, 1992). These services have been found to offer considerable help and assistance to callers (Marcus, Woodworth, & Strickland, 1993).

Telephone information and referral services like the CIS represent a unique hybrid of mediated and interpersonal channels, since they disseminate authoritative written information, as well as verbal responses to personal queries (Freimuth et al., 1989). The hybrid nature of telephone services is important, since it can overcome some of the weaknesses of other channels. They have the additional advantage of homophily of source, a crucial factor in effective communication (see Rogers, 2003), since the calls are handled by individuals of closer status and background to potential callers than are physicians. It has been suggested that the CIS provides an important link between symptomatic people and health services, since a substantial proportion of callers follow up with more information seeking, passing on information to others, or consultations with health professionals (Altman, 1985).

The advantages of telephone services as a channel include they: are free, are available without appointment and forms, offer a high level of empathic understanding, offer greater client control, permit anonymity for both parties, bridge geographic barriers, provide immediate responses, the client can take greater risks in expressing feelings, convenience, cost effectiveness, and personalized attention (Johnson, 1997). All these factors are reflected in respondents rating CIS as the highest quality source of cancer information (Mettlin et al., 1980) and the extremely high rate of user satisfaction in subsequent surveys (Morra, Bettinghaus, Marcus, Mazan, Nealon, & Van Nevel, 1993; Thomson & Maat, 1998; Ward, Duffy, Sciandra, & Karlins, 1988).

Knowledge Management Services of CIS

We need every weapon against cancer, and information can be a powerful, lifesaving tool. ... A call is made, a question is answered. NCI reaches out through the CIS, and the CIS is the voice of the NCI (Broder, 1993, pp. vii).

... one phone call, one conversation, can save a life. This is the true essence of the service and the most rewarding aspect of the program (Morra, Van Nevel, O'D.Nealon, Mazan, & Thomsen, 1993, p. 7)

In broad sweep (see Figure 1), the CIS has traditionally been the disseminator/translator of consensus based scientific information from NCI to broader segments of the public, responding to demand characteristics of an increasingly knowledgeable, responsible public. We can easily organize a description of the CIS around the major functions of KM organizations: transforming information into knowledge; identifying and verifying knowledge, capture/securing knowledge; organizing knowledge; retrieving and applying knowledge, combining it, creating it, and finally distributing/selling it (Liebowitz, 2000).

The CIS is one section of the Health Communication and Informatics Research Branch (HCIRB) of the Behavioral Research Program (BRP) of the Division of Cancer Control and Population Services (DCCPS). The DCCPS is a major research division of NCI that also has research programs focusing on Epidemiology and Genetics, Applied Research, and Surveillance Research. The BRP contains additional branches relating to Applied Cancer Screening, Basic BioBehavioral Research, Health Promotion, and Tobacco Control. (For more information see http://dccps.nci.nih.gov). The "HCIRB seeks to advance communication and information science across the cancer continuum-prevention, detection, treatment, control, survivorship, and end of life" (http://dccps.nci.nih.gov/hcirb).

The CIS was implemented in 1975 by the NCI to disseminate accurate, up-to-date information about cancer to the American public, primarily by telephone (Ward et al., 1988; Morra, Van Nevel et al., 1993) and is currently governed by the principles of "responsiveness, tailoring information to audience needs, and proactively sharing information" (http://cis.nci.nih.gov/about/orgprof.html). The CIS was one of the first federally-funded health-related telephone information systems in the US (Marcus et al., 1993).

Information is available, free of charge, in both English and Spanish to anyone who calls 1-800-4-CANCER. In recent years the CIS has added a special telephone service to promotion smoking cessation (1-877-44U-QUIT), as well as services for the hearing impaired, and has instituted a range of web-based services. The CIS responds to nearly 400,000 calls annually and many more requests for assistance on its LiveHelp website feature (http://cis.nci.nih.gov/about/underserved.html). In total, in 2002 the CIS handled over 1.4 million requests for service (NCI, 2003). The impetus underlying the creation of the CIS was the assumption that all cancer patients should receive the best care. To accomplish that end it was felt that free and easy access of consumers to credible information was critical (Morra, Van Nevel et al., 1993).

KM Roles in the CIS

To accomplish its KM work the CIS has developed several specialized, functional roles in its 14 Regional Offices (ROs). The major roles in each of the RO's include, Project Directors (PD), Telephone Service Managers (TSM), and Partnership Program Managers (PPM) and subordinate Partnership Coordinators. These roles differ both in their position requirements and in their organizational status level, with PDs the day-to-day managers for the regional CIS offices (Morra et al., 1993).

TSMs are in charge of managing the Information Specialists (IS) and the referral resources. The IS within the CIS, who serve as knowledge brokers, have a unique cluster of skills: though the CIS is not a help/counseling "hotline," callers are often very anxious; information specialists must be able to communicate highly technical information clearly to callers who come from all demographic groups and who differ considerably in their levels of knowledge (Davis & Fleisher, 1998; Morra, 1998). Performance standards for telephone calls are set nationally and are monitored by an extensive formal evaluation effort unique for telephone services (Kessler, Fintor, Muha, Wun, Annett, & Mazan, 1993; Morra, 1998). IS are clearly constrained to be information providers within strict protocols, to not interfere with existing medical relationships, to refrain from counseling, and to provide callers with quality assurances of information they provide (e.g., "this is consensus scientific information") (NCI, 1996).

The CIS Partnership Program is designed to reach out to people, especially the medically underserved and minority communities, who do not traditionally seek health information, providing equal access to cancer information from a trusted source (http://cis.nci.nih.gov/about/orgprof.html). PPMs are responsible for disseminating health messages through networking with other organizations such as local American Cancer Societies, state health departments, and so on. In doing this they serve a strategic multiplier function for the CIS, increasing the health literacy of clients.

Designing for KM

The traditional CIS telephone service is literally the tip of the iceberg of a very sophisticated KM system that has considerable strategic advantages because of its commitment to translating the highest quality, consensus based scientific information to the public. Enhanced information seeking possibilities for the public are created by new technologies, representing an information architecture, in three areas: data storage, data transport, and data transformation (Cash, Eccles, Nohria, & Nolan, 1994). While we will discuss these components separately, increasingly it is their blending and integration that is creating exciting new opportunities for information seeking (Case et al., 2004).

Networks, like the Internet, usually combine enhanced telecommunication capabilities with software that allows linkage and exchange of information between users (one of which is often a database). In the health arena these benefits and possibilities are captured under the heading of health informatics. Consumer health information is perhaps the fastest growing area of this specialty (MacDougall & Brittain, 1994).

Telecommunication systems such as fiber optic cables and satellite systems provide the hardware that links individuals and provides enhanced access to systems. Telecommunication systems maintain communication channels (e.g., e-mail) through which information is accessed and reported. They can specifically enhance the information seeking of patients by creating new channels for sending and receiving information (e.g., Live Aid feature on CIS's Web site), helping them in filtering information, reducing their dependence on others, leveraging their times to concentrate on the most important tasks, and enhancing their ability for dealing with complexity. One example of enhanced data transport lies in the benefits of telemedicine: increased access to information, increased consistency in medical decision making, matching diagnostic and management options to patient needs, increased quality of care, more interpretable outcomes, increased efficiency, increased efficacy, decreased costs, and a more uniform structure for healthcare (Turner, 2003).

Essentially databases provide a means for storing, organizing, and retrieving information. Modern conceptions of storage have broadened this function considerably to include verification and quality control of information entering a storage system. Doctors have historically mistrusted medical information systems because they may not capture the subtlety and nuance that only their long experience and training can bring to a situation (Shuman, 1988). They also do not provide much assistance to health professionals in areas where there is low consensus knowledge (Brittain, 1985). Unfortunately, a number of scientific controversies in well established areas have emerged in recent years that even peer review may not successfully address.

To their credit, the authoritativeness of the information they provide has always been a paramount concern for government databases. In the cancer area the NCI's Physician Data Query (PDQ) has been paying systematic attention to these issues for two decades (Hibbard, Martin, & Thurin, 1995). PDQ was originally designed to address the knowledge gap between primary care physicians and specialists. In a 1989 survey of primary care practitioners two-thirds felt that the volume of the medical literature was unmanageable and 78 percent reported that they had difficulty screening out irrelevant information from it (Hibbard et al., 1995). "This knowledge gap is responsible for the prolonged use of outmoded forms of cancer treatment resulting in unnecessarily high rates of cancer morbidity and mortality" (Kreps, Hibbard, & DeVita, 1988, pp. 362).

Fundamental to PDQ is the recognition that the rapid dissemination of health information is critical to successful treatment, since at the time it was created in the early 1980's approximately 85 percent of cancer patients were treated by primary care doctors (Kreps et al., 1988). In its early days one-third of the PDQ usage came from the CIS (Kreps et al., 1988), reflecting a high level of usage by the lay public that has continued to this day.

PDQ seeks to provide a current peer-reviewed synthesis of the state-of-the-art clinical information related to cancer (Hibbard et al., 1995). PDQ contains several components. Cancer information summaries in both health professional and patient versions on adult treatment, pediatric treatment, supportive care, screening and prevention, genetics, and supportive care. A registry of clinical trials and directories of physicians, professionals who provide genetic services, and organizations that provide cancer care is also available.

A critical issue facing all databases is how old, irrelevant information is culled from any storage system. A not so apparent problem of public databases, like many of those available on the Internet, is the potential lack of timeliness of the information. The cancer information file of PDQ is reviewed and updated monthly by six Editorial Boards of cancer experts in different areas. These Editorial Boards have clear guidelines on levels of evidence for information to be considered for the database. CIS IS receive extensive training on cancer-related issues and the use of PDQ (Davis & Fleisher, 1998; Fleisher, Woodworth, Morra, Baum, Darrow, Davis, Slevin-Perocchia, Stengle, & Ward, 1998).

More generally, HHS through the NIH's National Library of Medicine and the Office of Disease Prevention and Health Promotion's National Health Information Center, have devoted considerable resources and given thought to building a national information infrastructure in the U.S. The integration of data storage and transport with sophisticated software offers unique opportunities for solutions that transcend the limits of individual information processing, especially that of novices. Combining databases and telecommunications with software creates telematics that allows for the possibility of increasingly sophisticated searches for information and analysis/interpretation of it once it is compiled. The CIS IS, acting as a knowledge broker, serves a critical function in translating information into knowledge for increasingly literate health consumers.

Future Trends

Consortia and Clinical Knowledge Management

The future of KM in clinical settings is likely to be a turbulent one. It is clear, however, that those in clinical settings will be asked to do more with less. Thus, increasingly organizations will be looking for partners to provide services they cannot provide—in this role the CIS can be a key strategic partner for clinical practice allowing those in clinical settings to concentrate on their central, unique missions. Government information providers can act as central repositories for information services that would be provided before, during, and after client interactions with clinical organizations. They can do this in the *ad hoc*, random basis currently characteristic of a client's information environment (pattern a in Figure 2), or they can form explicit partnerships (paths b, c, and d) with clinical practice in consortia.

Consortia are particularly interesting settings in which to examine these issues because of the voluntary nature of relationships within them, which often creates a situation that is a mix of system/altruism and market/self-interest. A consortium can be defined simply as a collection of entities (e.g., companies, occupational specialties, community members) brought together by their interest in working collaboratively to accomplish something of mutual value that is beyond the resources of any one member (Fleisher et al., 1998; Johnson, 2004).

Organizations often find that they are either strapped for resources or are pursuing projects of such magnitude that they must pool their resources to pursue innovations (Hakansson & Sharma, 1996); developing cooperative relationships with other entities promotes the possibility of resource sharing and greater efficiencies, especially in community (Dearing, 2003) and governmental settings (Dorsey, 2003; Parrott & Steiner, 2003). Fundamentally, consortiums are formed so that their members can accomplish more than they could do on their own. Given the interest in new organizational forms, heightened competition, fractured communities, and declining resources available to any one group, this topic has captured the attention of researchers in a wide range of disciplines (Hakansson & Sharma, 1996; Johnson, 2004).

The focus of these consortiums should be on providing consistent health-related messages and information (resulting from relationships described by b, c, and d in Figure 2) that through their repetition from multiple trustworthy institutions increases the probabilities that clients will comply with the best medical advice. The CIS's existing PPM provides the critical foundations for the development of such consortia (Thomsen & Maat, 1998). Indeed, increasing health literacy by encouraging autonomous information seekers should be a goal of our healthcare system (Parrott, 2003). While it is well known that individuals often consult a variety of others before presenting themselves in clinical settings (Johnson, 1997), outside of HMO and organizational contexts, there have been few systematic attempts to shape the nature of these prior consultations.

If these prior information searches happen in a relatively uncontrolled, random, parallel manner (pattern a), expectations (e.g., treatment options, diagnosis, drug regimens) may be established that will be unfulfilled in the clinical encounter. The emergence of the Internet as an omnibus source for information has apparently changed the nature of opinion leadership; both more authoritative (e.g., medical journals) and more interpersonal (e.g., cancer support groups) sources are readily available and accessible online (Case et al., 2004). This is a part of a broader trend that Shapiro and Shapiro (1999) refer to as "disintermediation"—the capability of the Internet to allow the general public to bypass experts in their quest for information, products and services.

Figure 2. Relationships between CIS and Clinical Practice

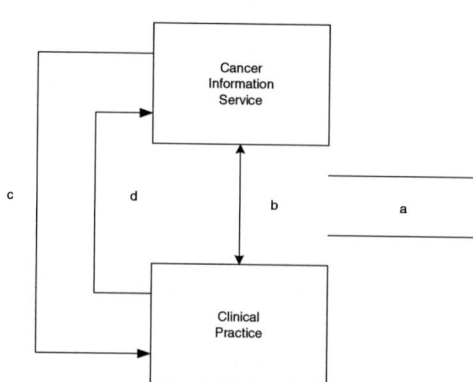

Contemporary views of health communication are more likely to stress a dialogic view of interaction, with both parties initiating and attending to messages in turn (paths b, c, and d). Health professionals' most important role in these more contemporary perspectives is as stimulus or cue to action, defining the agenda of the most important issues that an audience needs to face (path d). For those clients who have had no prior contact with a clinical organization, the CIS can provide referral services (path c in Figure 2) for clients predisposed to act. About 20 percent of callers to Midwest Regional Office of the CIS last year asked for referrals to health professionals, treatment facilities, and/or community services (http://www.karmonos.org/cis/telephone.html). Clinical entities should be familiar with this referral process and the kinds of information provided to insure a smooth encounter with referred clients.

Proactive prior referral and/or exposure to an information source such as the CIS can ease the burden on clinicians by transmitting basic information to clients that then does not have to be repeated in an office visit (path c). Increasingly the focus of health communication campaigns is on getting people to seek more information on health topics (Parrott, 2003; Salmon & Atkin, 2003). Creating rich information fields through such practices as 'self-serving' to information from data bases, for example, should make for a more informed consumer, who is likely to consume less time of health professionals "being brought up to speed" on the basics of his/her disease and its treatment. As we have seen, this enhanced level of health literacy can result in various improved outcomes for clients.

Similarly, as part of a visit clinicians can encourage individuals to use services like the CIS to secure a second, complementary opinion (path d). These services can provide crucial background, support, and assurance to clients from neutral third party. For clients the CIS IS act as a knowledge broker synthesizing and translating information, thus relieving clinical settings of this task. The CIS can perform many of the same functions that call centers and customer relations staff can provide for information technology services. Similarly, it has taken on the increasingly critical role of referral to clinical trials for new and/or experimental treatments (http://cancer.gov/cancerinfo/pdq/cancerdatabase), something that should be done in concert with clinical practice (path b).

While the CIS, and similar government services, have often been viewed as competitors or interlopers into clinical practice, its access to NCI's research infrastructure, the development of authoritative guidelines (PDQ), and its role in translating this information to the public, demonstrate it can be a key contributor to a more effective healthcare delivery system (NCI, 2003). In the past, through clients, it has directly acted to disseminate information to improve practice. The ultimate outcome of effective KM is the rapid adoption or creation of appropriate innovations that can be successfully implemented within a particular organization's context. In the case of the CIS there was an even broader mandate, a latent purpose to insure the rapid diffusion of state-of-the-art medical care to the public. By explicitly recognizing, managing, and supporting this role, clinical entities can more effectively manage their limited resources to insure that consonant information is provided to clients that is more likely to result in positive healthcare outcomes. While often services like the CIS have been viewed as a competitor, it may be better to view it as a complementary/supportive service, especially for government driven healthcare systems.

Conclusions

In broad sweep (see Figure 1), the CIS has traditionally been the disseminator/translator of consensus based scientific information from NCI to broader segments of the public. Societal trends in consumer/client information behavior make this a critical role in today's broadly envisioned clinical practice environment. While the CIS is organized to serve clients directly, it can also interface with clinical practice by providing referrals, enhancing health literacy, providing second opinions and crucial background, assurance to clients from neutral third party. Thus, the CIS is a partial answer to information explosion: for client it acts as synthesizer, translator who can relieve clinical settings of this task and through clients it directly acts to disseminate information to improve clinical practice.

This chapter highlights how information providers can interface with clinical practice to form a strategic partnership that advances health outcomes for the public. All this suggests the increasing importance of information as a strategic asset that should be systematically incorporated in the planning of health professionals. Health institutions need to recognize the potential benefit of marketing unique corporate knowledge and expertise to other information seekers and to consider their unique role in a complex information seeking environment. Health professionals also need to lobby the government to maintain critical information infrastructures.

While more and more information can be produced more efficiently, there is a concomitant increase in the costs of consuming (e.g., interpreting, analyzing) this information (More, 1990). Thus, proactively working to shape a client's information fields by insuring concordant information is provided may forestall increasing problems for clinical practice in the rapidly growing information jungle.

Acknowledgments

I would like to thank Dr. Sally Johnson for reviewing an earlier version of this manuscript.

References

Altman, D.G. (1985). Utilization of a telephone cancer information program by symptomatic people. *Journal of Community Health, 10,* 156-171.

Boahene, M. & Ditsa, G. (2003). Conceptual confusions in knowledge management and knowledge management systems: Clarifications for better KMS development. In E. Coakes (Ed.), *Knowledge management: Current issues and challenges* (pp. 12-24). Hershey, PA: IRM Press.

Brittain, J.M. (1985). Introduction. In J. M. Brittain (Ed.), *Consensus and penalties for ignorance in the medical sciences: Implications for information transfer* (pp. 5-10). London: British Library Board.

Broadway, M.D. & Christensen, S. B. (1993). Medical and health information needs in a small community. *Public Libraries*, September/October, 253-256.

Broder, S. (1993). Foreward. *Journal of the National Cancer Institute*, Monograph 14, vii.

Brown, J.B., Stewart, M. & Ryan, B.L. (2003). Outcomes of patient-provider interaction. In T. L. Thompson, A. M. Dorsey, K. I. Miller, & R. Parrott (Eds.), *Handbook of health communication* (pp. 141-161). Mahwah, NJ: Lawrence Erlbaum.

Case, D., Johnson, J.D., Andrews, J.E., Allard, S. & Kelly, K.M. (2004). From two-step flow to the Internet: The changing array of sources for genetics information seeking. *Journal of the American Society for Information Science and Technology*, 55, 660-669.

Cash, J.I., Jr., Eccles, R.G., Nohria, N. & Nolan, R.L. (1994). *Building the information-age organization: Structure, control, and information technologies*. Boston: Irwin.

Choo, W.C. (1998). *The knowing organization: How organizations use information to construct meaning, create knowledge, and make decisions*. New York: Oxford University Press.

Davis, S.E. & Fleisher, L. (1998). Treatment and clinical trials decision making: The impact of the Cancer Information Service, Part 5. *Journal of Health Communication*, 3, 71-86.

Dearing, J.W. (2003). The state of the art and the state of the science in community organizing. In T. L. Thompson, A. M. Dorsey, K. I. Miller, & R. Parrott (Eds.), *Handbook of health communication* (pp. 207-220). Mahwah, NJ: Lawrence Erlbaum.

Dervin, B. (1998). Sense-making theory and practice: An overview of user interests in knowledge seeking and use. *Journal of Knowledge Management*, 2, 36-46.

Doctor, R.D. (1992). Social equity and information technologies: Moving toward information democracy. In M. E. Williams (Ed.), *Annual review of information science and technology* (pp. 44-96). Medford, NJ: Learned Information.

Dorsey, A.M. (2003). Lessons and challenges from the field. In T. L. Thompson, A. M. Dorsey, K. I. Miller, & R. Parrott (Eds.), *Handbook of health communication* (pp. 607-608). Mahwah, NJ: Lawrence Erlbaum.

Duncan, K.A. (1994). *Health information and health reform: Understanding the need for a national health information system*. San Francisco: Jossey-Bass.

Fleisher, L., Woodworth, M., Morra, M., Baum, S. Darrow, S. Davis, S., Slevin-Perocchia, R., Stengle, W. & Ward, J.A. (1998). Balancing research and service: The experience of the Cancer Information Service. *Preventive Medicine*, 27, S84-92.

Fouche, B. (1999). Knowledge networks: Emerging knowledge work infrastructures to support innovation and knowledge management. *ICSTI Forum*, 32.

Fox, S. & Rainie, L. (2002). How internet users decide what information to trust when they or their loved ones are sick. Retrieved from the Pew Research Center Web site *http://www.pewinternet.org/*

Freimuth, V.S., Stein, J.A., & Kean, T.J. (1989). *Searching for health information: The Cancer Information Service Model*. Philadelphia: University of Pennsylvania Press.

Greer, A. L. (1994). You can always tell a doctor ... . In L. Sechrest, T.E. Backer, E.M. Rogers, T.F. Campbell & M.L. Grady (Eds.), *Effective dissemination of clinical and health information* (pp. 9-18). Rockville, MD: Agency for Health Care Policy Research, AHCPR Pub. No. 95-0015.

Hakansson, H. & Sharma, D. D. (1996). Strategic alliances in a network perspective. In D. Iacobucci (Ed.), *Networks in marketing* (pp. 108-124). Thousand Oaks, CA: SAGE.

HHS (February 21, 2002). Press release: HHS affirms value of mammography for detecting cancer.

Hibbard, J.H. & Weeks, E.C. (1987). Consumerism in health care. *Medical Care, 25*, 1019-1032.

Hibbard, S.M., Martin, N.B. & Thurn, A.L. (1995). NCI's Cancer Information Systems: Bringing medical knowledge to clinicians. *Oncology, 9*, 302-306.

Johnson, J.D. (1997). *Cancer-related information seeking*. Cresskill, NJ: Hampton Press.

Johnson, J.D. (2004). The emergence, maintenance, and dissolution of structural hole brokerage within consortia. *Communication Theory, 14*, 212-236.

Johnson, J.D., Andrews, J.A., Case, D.O. & Allard, S. (in press). Genomics-The perfect information seeking research problem. *Journal of Health Communication*.

Kessler, L., Fintor, L., Muha, C., Wun, L., Annett, D. & Mazan, K.D. (1993). The Cancer Information Service Telephone Evaluation and Reporting System (CISTERS): A new tool for assessing quality assurance. *Journal of the National Cancer Institute, Monograph, 14*, 617-65.

Kreps, G. L., Hibbard, S. M., & DeVita, V. T. (1988). The role of the physician data query on-line cancer system in health information dissemination. In B. D. Ruben (Ed.), *Information and behavior* (Vol. 2, pp. 1,6-7). New Brunswick, NJ: Transaction Books.

Liebowitz, J. (2000). *Building organizational intelligence: A knowledge management primer*. New York: CRC Press.

Lowery, W. & Anderson, W.B. (2002). The impact of Web use on the public perception of physicians. Paper presented to the annual convention of the *Association for Education in Journalism and Mass Communication*, Miami Beach, FL.

MacDougall, J. & Brittain, J.M. (1994). Health informatics. *Annual Review of Information Science and Technology, 29*, 183-217.

MacMorrow, N. (2001). Knowledge management: An introduction. *Annual Review of Information Science and Technology, 35*, 381-422.

Marcus, A.C., Woodworth, M.A. & Strickland, C.J. (1993). The Cancer Information Service as a laboratory for research: The first 15 years. *Journal of the National Cancer Institute, Monograph 14*, 67-79.

More, E. (1990). Information systems: People issues. *Journal of Information Science, 16,* 311-320.

Morra, M., Bettinghaus, E.P., Marcus, A.C., Mazan, K.D., Nealon, E. & Van Nevel, J.P. (1993). The first 15 years: What has been learned about the Cancer Information Service and the implications for the future. *Journal of the National Cancer Institute, Monograph, 14,* 177-185.

Morra, M.E. (1998). Editorial. *Journal of Health Communication, 3,* V-IX.

Morra, M.E., Van Nevel, J.P., O'D.Nealon, E., Mazan, K.D., & Thomsen, C. (1993). History of the Cancer Information Service. *Journal of the National Cancer Institute, Monograph 14,* 7-34.

NCI (1996). *CIS policy and procedures manual.* Bethesda, MD: National Cancer Institute.

NCI (2003). *The nation's investment in cancer research.* Rockville, MD: NIH.

Nonaka, I. & Takeuchi, H. (1995). *The knowledge-creating company: How Japanese companies create the dynamics of innovation.* New York: Oxford University Press.

Parrott, R. (2003). Media issues. In T. L. Thompson, A. M. Dorsey, K. I. Miller & R. Parrott (Eds.), *Handbook of health communication* (pp. 445-448). Mahwah, NJ: Lawrence Erlbaum.

Parrott, R. & Steiner, C. (2003). Lessons learned about academic and public health collaborations in the conduct of community-based research. In T. L. Thompson, A. M. Dorsey, K. I. Miller & R. Parrott (Eds.), *Handbook of health communication* (pp. 637-649). Mahwah, NJ: Lawrence Erlbaum.

Rogers, E.M. (2003). *Diffusion of innovations* (5th edition). New York: Free Press.

Salmon, C.T. & Atkin, C. (2003). Using media campaigns for health promotion. In T. L. Thompson, A. M. Dorsey, K. I. Miller & R. Parrott (Eds.), *Handbook of health communication* (pp. 449-472). Mahwah, NJ: Lawrence Erlbaum.

Schuman, T.M. (1988). Hospital computerization and the politics of medical decision-making. *Research in the sociology of work, 4,* 261-287.

Shapiro, A.L. & Shapiro, R.C. (1999). *The control revolution: How the Internet is putting individuals in charge and changing the world we know.* New York: Public Affairs.

Street, R. L., Jr. (1990). Communication in medical consultations: A review essay. *Quarterly Journal of Speech, 76,* 315-332.

Taylor, H. & Leitman, R. (May, 2002). Four-nation survey shows widespread but different levels of Internet use for health purposes. *Health Care News, 2,* Retrieved from http://www.harrisinteractive.com/newsletters_healthcare.asp

Thomsen, C.A. & Maat, J. T. (1998). Evaluating the Cancer Information Service: A model for health communications. Part 1. *Journal of Health Communication, 3,* 1-14.

Thompson, T.L. (2003). Provider-patient interaction issues. In T. L. Thompson, A. M. Dorsey, K. I. Miller & R. Parrott (Eds.), *Handbook of health communication* (pp. 91-93). Mahwah, NJ: Lawrence Erlbaum.

Turner, J.W. (2003). Telemedicine: Expanding health care into virtual environments. In T. L. Thompson, A. M. Dorsey, K. I. Miller & R. Parrott (Eds.), *Handbook of health communication* (pp. 515-535). Mahwah, NJ: Lawrence Erlbaum.

Ward, J.A.D., Duffy, K., Sciandra, R. & Karlins, S. (1988). What the public wants to know about cancer: The Cancer Information Service. *The Cancer Bulletin, 40,* 384-389.

Chapter XIV

Clinical Decision Support Systems:
Basic Principles and Applications in Diagnosis and Therapy

Spyretta Golemati, National Technical University of Athens, Greece

Stavroula Mougiakakou, National Technical University of Athens, Greece

John Stoitsis, National Technical University of Athens, Greece

Ioannis Valavanis, National Technical University of Athens, Greece

Konstantina S. Nikita, National Technical University of Athens, Greece

Abstract

This chapter introduces the basic principles of Clinical Decision Support (CDS) systems. CDS systems aim to codify and strategically manage biomedical knowledge to handle challenges in clinical practice using mathematical modelling tools, medical data processing techniques and Artificial Intelligence (AI) methods. CDS systems cover a wide range of applications, from diagnosis support to modelling the possibility of occurrence of various diseases or the efficiency of alternative therapeutic schemes, using not only individual patient data but also data on risk factors and efficiency of available therapeutic schemes stored in databases. Computer-Aided Diagnosis (CAD) systems can enhance the diagnostic capabilities of physicians and reduce the time required for accurate diagnosis. Modern Therapeutic Decision Support (TDS) systems make use of advanced modelling techniques and available patient data to optimise and individualise patient treatment. CDS systems aim to improve the overall health of the population by improving the quality of healthcare services, as well as by controlling the cost-effectiveness of medical examinations and treatment.

Introduction

Advances in the areas of computer science and Artificial Intelligence (AI) allow the development of computer systems that support clinical diagnostic or therapeutic decisions based on individualised patient data (Berner & Ball, 1998; Shortliffe, Perrault, Wiederhold, & Fagan, 1990). Clinical Decision Support (CDS) systems aim to codify and strategically manage biomedical knowledge to handle challenges in clinical practice using mathematical modelling tools, medical data processing techniques and AI methods (Bankman, 2000). CDS systems cover a wide range of applications, from diagnosis support to modelling the possibility of occurrence of various diseases or the efficiency of alternative therapeutic schemes, using not only individual patient data but also data on risk factors and efficiency of available therapeutic schemes stored in databases.

To diagnose a disease, a physician is usually based on the clinical history and physical examination of the patient, visual inspection of medical images, as well as the results of laboratory tests. In some cases, confirmation of the diagnosis is particularly difficult because it requires specialisation and experience, or even the application of interventional methodologies (e.g., biopsy). Interpretation of medical images (e.g., Computed Tomography, Magnetic Resonance Imaging, Ultrasound, etc.) usually performed by radiologists, is often limited due to the non-systematic search patterns of humans, the presence of structure noise (camouflaging normal anatomical background) in the image, and the presentation of complex disease states requiring the integration of vast amounts of image data and clinical information. Computer-Aided Diagnosis (CAD), defined as a diagnosis made by a physician who uses the output from a computerised analysis of medical data as a "second opinion" in detecting lesions, assessing disease severity, and making diagnostic decisions, is expected to enhance the diagnostic capabilities of physicians and reduce the time required for accurate diagnosis.

The first CAD systems were developed in the early 1950s and were based on production rules (Shortliffe, 1976) and decision frames (Engelmore & Morgan, 1988). More complex systems were later developed, including blackboard systems (Engelmore & Morgan, 1988) to extract a decision, Bayes models (Spiegelhalter, Myles, Jones, & Abrams, 1999) and Artificial Neural Networks (ANNs) (Haykin, 1999). Recently, a number of CAD systems have been implemented to address a series of diagnostic problems. CAD systems are usually based on biosignals, including the electrocardiogram (ECG), electroencephalogram (EEG), electromyogram (EMG) or medical images from a number of modalities, including radiography, CT, MRI, and US imaging.

In therapy, the selection of the optimal therapeutic scheme for a specific patient is a complex procedure that requires sound judgement based on clinical expertise, and knowledge of patient values and preferences, in addition to evidence from research.

Usually, the procedure for the selection of the therapeutic scheme is enhanced by the use of simple statistical tools applied to empirical data. In general, decision making about therapy is typically based on recent and older information about the patient and the disease, whereas information or prediction about the potential evolution of the specific patient disease or response to therapy is not available. Recent advances in hardware and software allow the development of modern Therapeutic Decision Support (TDS) systems, which make use of advanced simulation techniques and available patient data to optimise and individualise patient treatment, including diet, drug treatment, or radiotherapy treatment. In addition to this, CDS systems may be used to generate warning messages in unsafe situations, provide information about abnormal values of laboratory tests, present complex research results, and predict morbidity and mortality based on epidemiological data.

The aim of this chapter is to describe the main principles of CDS systems, including diagnostic and therapeutic decision support, and to present examples of such systems developed in the Biomedical Simulations and Imaging Laboratory, Faculty of Electrical and Computer Engineering, National Technical University of Athens.

Computer-Aided Diagnosis

CAD systems aim to enhance the ability to detect pathological structures in medical examinations and to support evaluation of pathological findings during the diagnostic procedure. In the following, the fundamental theory for the development of CAD systems is presented along with two examples of CAD systems.

Basic Principles

The typical structure of a CAD system is shown in Figure 1. Diagnostic tests, which are determined by the clinical protocol, include imaging tests (CT, MRI, US, etc.), laboratory tests and tests for recording tissue electrical activity (e.g., ECG, EEG, EMG, etc.). Input in the form of plain text (e.g., text describing the clinical symptoms of a patient) may also be used in a CAD system. The main modules of a CAD system include:

- Data pre-processing
- Definition of regions of interest (ROIs)
- Extraction and selection of characteristic features
- Classification

Figure 1. Typical structure of a CAD system

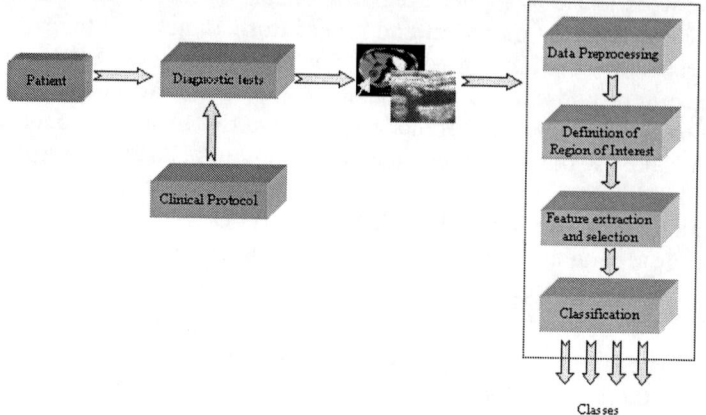

Data Pre-Processing

The data pre-processing module aims to improve the quality of data through the application of methods for denoising (application of mean filters, median filters, etc.), resampling (linear interpolation, spline interpolation, etc.) and, in the case of images, enhancing image contrast (histogram equalisation, wavelet transform, etc.) (Bankman, 2000).

Data pre-processing is critical, especially when the quality of data is low. The computational cost of data pre-processing depends on the size of the data, but it is significantly lower than other processing tasks, including automatic segmentation and feature extraction.

Definition of Regions of Interest

Within the Region-of-Interest-(ROI) definition module, possible pathological structures appearing in patient biosignals recordings or medical imaging data are defined. Regions corresponding to lesions (e.g., tumours, cysts, etc.) can be detected from medical images. Definition of ROIs can be performed using manual and semi-automatic methodologies where the user interacts with the system in order to define a possible pathological region or fully automated methodologies where suspect regions are detected with appropriate digital image processing techniques (Bankman, 2000).

An example of semi-automatic ROI definition is seeded region growing. The user selects a point on the structure of interest and, subsequently, new points are added through an iterative recursive procedure using similarity and spatial connectivity criteria. Furthermore, a number of techniques have been implemented for automatic definition (segmentation) of known anatomic structures from medical images. Many of these structures rely on a priori shape information of the organ or structure of interest to segment it out (Cootes, Edwards, & Taylor, 2001). Region-based methods, for example, fuzzy connectedness (Udupa & Samarsekera, 1996), use some homogeneity statistics coupled with low-level image features, like intensity, texture, histograms, and gradient to assign pixels to objects. If two pixels are similar in value and connected to each other in some sense, they are assigned to the same object. These approaches do not consider any shape information. Hybrid segmentation techniques have also been proposed by some researchers (Chen & Metaxas, 2000). These approaches seek to exploit the local neighbourhood information of region-based techniques, and the shape and higher level information of boundary-based techniques.

Extraction and Selection of Characteristic Features

"Characteristic features" are a group (vector) of quantitative indices extracted after processing the patient data. The aim of feature extraction and selection is to reliably discriminate between pathological and physiological structures or to characterise a pathological structure (e.g., benign or malignant tumour). Characteristic features may be derived from measurements of mean image intensity or image texture, shape analysis, and so on.

It is often desirable to reduce the dimension of the vector of characteristics extracted from patient data (Jain, Duin, & Jianchang, 2000). The selection of the most robust characteristics allows maximisation of classification accuracy and minimisation of system complexity.

Ideally, a method for feature selection would examine all 2^N (N: the number of characteristic features) different combinations of characteristic features and would find the optimal combination that satisfies the previous criteria. The disadvantage of this exhaustive search technique is its high computational cost which may be prohibitive even for relatively small values of N. Statistical methods, including ANalysis Of VAriance (ANOVA), have been used for feature selection in a number of applications. ANOVA compares the means of two or more groups of data and returns the probability for the null hypothesis that the means of the groups are equal (Furlong, Lovelace & Lovelace, 2000). Other techniques, based on heuristic or random search methods, attempt to compensate computational complexity with classification accuracy. Feature selection is usually achieved through the following steps:

- Production of a candidate feature subgroup
- Evaluation of the eligibility of the candidate subgroup through an appropriate evaluation function
- Procedure termination after satisfaction of a predefined criterion
- Confirmation of the validity of the selected feature subgroup

Methodologies such as Genetic Algorithms (GAs) may be applied to feature selection problems. GAs are adaptive heuristic search methods which may be used to solve complex pattern recognition and classification problems (Goldberg, 1989). They are based on the evolutionary ideas of natural selection and genetic processes of biological organisms. As the natural populations evolve according to the principles of natural selection and "survival of the fittest" first laid down by Charles Darwin, so by simulating this process GAs are able to evolve solutions to real-world problems, if they have been suitably encoded (Holland, 1975). Feature selection through GA-based search along with an ANN classifier has been used to classify "difficult-to-diagnose" microcalcifications from mammography (Dhawan, Chitre, Kaiser-Bonasso, & Moskowitz, 1996). The use of similar procedures has also been reported for the recognition of skin tumours and endothelial cells (Yamani, Khiani & Farag, 1997).

Classification

The role of the classification module is to decide about the inclusion of the feature vectors appearing at its input into a category or class, among a set of predefined classes. The development and implementation of the classification module may be based either on supervised or unsupervised training methods.

In supervised training, a set of characteristic vectors is considered, known as training set. For each vector in the training set the class to which it belongs, is known a priori. The vectors in the training set are used to train the classifier, which can subsequently classify any vector appearing to its input into one of the available classes (Duda, Hart & Stork, 2000).

In unsupervised training a set of characteristic vectors is examined and the class each vector belongs to is unknown. Classification of vectors in the available categories is done automatically during the formation of the classifier. Subsequently, as before, each characteristic vector is classified in one of the categories (Duda, et al., 2000).

Supervised Classification

The most common supervised classifiers include Bayes classifiers, nearest neighbour classifiers, decision trees, and artificial neural networks (Jain et al., 2000). In Bayes classifiers, each feature belongs to the class with the highest a posteriori estimated probability. In nearest neighbour classifiers, features are classified based on their relative similarity. In decision trees, classification is the result of a sequence of logical steps.

One of the most popular supervised classification methods is based on the use of ANNs, which are computational modelling tools that have recently emerged and found extensive acceptance in many disciplines for modelling complex real-world problems. ANNs may be defined as structures comprised of densely interconnected adaptive simple processing elements (called artificial neurons or nodes) that are capable of performing massively parallel computations for data processing and knowledge representation (Schalkoff, 1997). ANNs are applied to problems that cannot be described with rules or mathematical formulas.

The fundamental element of each ANN is the neuron. Each neuron is the synapse of its inputs, i.e. the sum of the products of the inputs with appropriate weighing coefficients.

The inputs can be external signals or the outputs of other neurons. Subsequently, the sum of the inputs with the corresponding weights is transformed with a linear or a non-linear transform function to produce the neuron output. Neurons are organised in layers. The neurons of the input layer have only one input. The outputs of the neurons of the input layer are conveyed to neurons of intermediate layers. The neurons of the output layer produce the final output of the ANN.

The ability to train an ANN consists in the adaption of weighs during the training process. The most commonly used procedure for training ANNs is the feedforward error-backpropagation learning algorithm (Haykin, 1999). According to this method, each vector in the training set appears at the ANN input and is propagated through the intermediate layers toward the output layer. The response of the neurons at the output layer is compared with the desired response of each vector and the error is calculated. Subsequently, the error is backpropagated and used to re-estimate the weight of each neuron. The procedure is repeated for all vectors of the training set. An *epoch* is completed after all vectors have been interrogated. At the end of each successful epoch the total error, for all vectors of the training set, is reduced. The procedure continues until a criterion for the termination of the training is satisfied.

Unsupervised Classification

The most common unsupervised classification method is C-means. A characteristic vector, known as the class centre, is created for each class. Subsequently, each vector is assigned to the class for which the euclidean distance from the class centre is minimal. After all characteristic vectors are classified, the class centre is re-estimated. The procedure is repeated until a convergence criterion is satisfied (e.g., there is no variation in vector classification in two successive iterations, the reduction in the total euclidean distance of the vectors from the class centres is lower than a predefined threshold). The advantages of the method include low complexity and easy implementation. However, the method is sensitive to the initial estimation of the class centres and, as a result, convergence may appear at a local minimum of the euclidean distance of the vectors from the class centres.

A variation of the C-means method is the fuzzy C-means method (Bezdek, 1981). The difference in the two methods lies in the fact that in C-means method each vector is strictly assigned only to one class, whereas in fuzzy C-means method each vector belongs to all classes. The percentage with which each vector belongs to a class is the member function (Zadeh, 1965).

Although ANNs are mainly supervised classifiers, unsupervised ANNs can be used to cluster characteristic vectors. The most representative network of this category is the Self-Organising Map (SOM) ANN (Kohonen, 1989). Characteristic vectors appear at the ANN input and are related to the output neurons. The weights between input and output neurons are recursively varied (training phase) until a termination criterion is satisfied. SOMs are sensitive to the selection of the initial weights, because a sub-optimal solution may be the result of improper initial weight selection. Furthermore, convergence is determined by a number of parameters including the training rate and the neuron neighbourhood where training takes place.

Unsupervised classification also includes hierarchical clustering (King, 1967), clustering using genetic algorithms (Goldberg, 1989), and so on.

Diagnosis of Carotid Atherosclerosis from B-Mode Ultrasound Images

B-mode ultrasound imaging of the carotid artery is widely used in the diagnosis of carotid atherosclerosis because it allows non-invasive assessment of the degree of stenosis and of plaque morphology. Currently, disease severity and selection of patients to be considered for endarterectomy, that is, surgical removal of plaque, is based on previous occurrence of clinical symptoms (e.g., stroke) and the degree of stenosis caused by the plaque. However, there is evidence that atheromatous plaques with relatively low stenosis degree may produce symptoms and that the majority of asymptomatic patients with highly stenotic atherosclerotic plaques remain asymptomatic (Inzitari, Eliasziw, Gates, Sharpe Chan, Meldrum, et al., 2000). To assist diagnosis of carotid atherosclerosis, quantitative indices characterising plaque severity may be estimated from the analysis of digitised ultrasound images.

More specifically, image texture analysis may be used to quantify plaque echogenicity, and movement of the carotid artery wall and plaque may be estimated from temporal image sequences and used as an index of plaque strain. In addition to this, automatic classification of atheromatous plaques into symptomatic or asymptomatic using AI techniques is expected to support decision-making about management of patients.

Plaque texture may be estimated with a number of techniques and be used as input to a CAD system. First-order statistics have been used to characterise texture in the following three distinct areas of a typical B-mode ultrasound image of a diseased carotid artery (Figure 2): atheromatous plaque, blood and surrounding tissue (Golemati, Tegos, Sassano, Nikita, & Nicolaides, 2004). First order statistical features were significantly different in different areas of the ultrasound image. In addition to this, the fractal dimension was estimated using the kth nearest neighbour method in 10 symptomatic and nine asymptomatic carotid atheromatous plaques at different phases of the cardiac cycle, namely systole, diastole and average during the cardiac cycle. As shown in Table 1, the fractal dimension was found significantly higher in the symptomatic group (Asvestas, Golemati, Matsopoulos, Nikita & Nicolaides, 2002). ANOVA (ANalysis Of Variance) statistics demonstrated that the phase of the cardiac cycle had no systematic effect on the calculation of the fractal dimension. The fuzzy c-means algorithm was applied to classify the fractal dimension into two classes. All features corresponding to symptomatic subjects were assigned to Class 1, whereas the features of eight out of nine asymptomatic subjects were assigned to Class 2. Furthermore, Laws texture energy features have been estimated in 54 symptomatic and 54 asymptomatic plaques. Energy features were significantly different between symptomatic and asymptomatic plaques (Mougiakakou, Golemati, Gousias, Nikita, & Nicolaides, 2003b).

Figure 2. Example of a typical B-mode ultrasound image of a diseased carotid artery. The atheromatous plaque was outlined by an expert. The rectangles correspond to ROIs in blood and the surrounding muscle tissue.

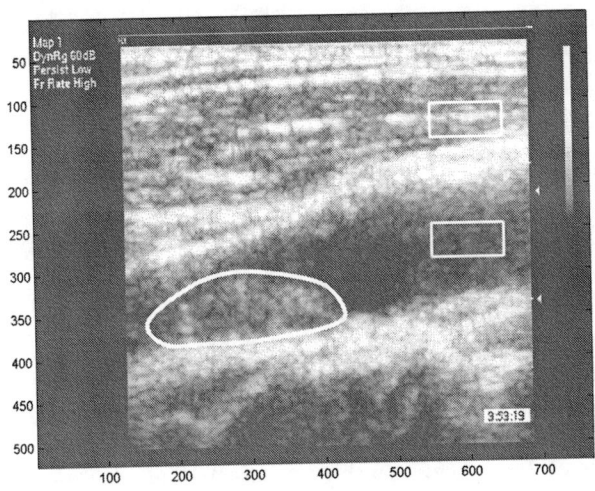

Automatic classification of carotid plaques into symptomatic or asymptomatic based on texture features was achieved with a CAD system, which consists of three modules: a feature extraction module, where texture features are estimated based on Laws' texture energy and first-order statistics, a dimensionality reduction module, where the number of features is reduced using ANOVA statistics, and a classifier module with an ANN, trained via a novel hybrid method using GAs, to recognize the type of atheromatous plaques (Mougiakakou, Golemati, Gousias, Nikita, & Nicolaides, 2003b). The hybrid method was able to select the most robust features, to automatically adjust the ANN architecture, and to optimise the classification performance. The developed CAD system achieved a total classification performance of 99 percent.

The motion of the carotid atheromatous plaque relative to the adjacent wall may be related to the risk of cerebral events. A quantitative method for motion estimation was applied to analyse arterial wall movement from sequences of two-dimensional B-mode ultrasound images (Golemati, Sassano, Lever, Bharath, Dhanjil, & Nicolaides, 2003). Image speckle patterns were tracked between successive frames using the correlation coefficient as the

Table 1. Fractal dimension (mean ± standard deviation) of symptomatic and asymptomatic carotid atheromatous plaques

	Diastole	Systole	Average during the cardiac cycle
Symptomatic	2.320±0.041	2.313±0.078	2.324±0.049
Asymptomatic	2.159±0.111	2.185±0.048	2.183±0.061
p-value	0.000558	0.000493	0.0000294

Table 2. Motion and texture features of symptomatic and asymptomatic carotid plaques

Motion/Texture features	Symptomatic	Asymptomatic	p-value
Maximal surface velocity, MSV (mm/s)	1.84±1.00	0.75±0.73	0.01
Maximal relative surface velocity, MRSV (mm/s)	2.85±1.69	0.52±0.53	0.0009
Standard deviation – Mask $L^T E$	17.46±4.01	21.76±5.41	0.04
Histogram width – Mask $L^T E$	38.3±8.74	46.11±12.95	0.03

matching criterion. The results showed expected cyclical motion in the radial direction and some axial movement of the arterial wall. The method can be used to study further the axial motion of the carotid artery wall and plaque and, thus, provide useful insight into the mechanisms of atherosclerosis.

The combination of texture and motion information may be valuable for accurate diagnosis of vascular disease. ANALYSIS, a modular software system designed to assist interpretation of medical images may be used to analyse texture and motion from B-mode ultrasound images of the carotid artery (Stoitsis, Golemati, Nikita, & Nicolaides, 2004). Texture analysis allows the estimation of first-order statistics, second-order statistics (Haralick's method), Laws' texture energy, and the fractal dimension using the differential box-counting method.

Motion is estimated using block-matching and region tracking for rectangular ROIs along the surface of the plaque. The following indices of motion can be estimated for each plaque: (a) Maximal Surface Velocity (MSV) and (b) Maximal Relative Surface Velocity (MRSV), defined as the maximum of differences between maximal and minimal surface velocities throughout the sequence. Motion and texture features were estimated in 10 symptomatic and nine asymptomatic carotid atheromatous plaques. After dimensionality reduction, the two strongest texture features (lowest p-value) were retained, namely standard deviation after convolution with Laws mask $L^T E$ and histogram width after convolution with Laws mask $L^T E$. Their average values (±standard deviations) are shown in Table 2 together with the values of the motion features.

Differential Diagnosis of Focal Liver Lesions from CT Images

One of the most common and robust imaging techniques for the detection of hepatic lesions is CT (Taylor & Ros, 1998). Although the quality of CT images has been significantly improved during the last years, some cases are difficult to accurately diagnose. In these cases, the diagnosis has to be confirmed by administration of contrast agents or invasive procedures (biopsies). In order to assist clinicians in diagnosis, and reduce the number of required biopsies, CAD systems can be employed for the characterisation and classification of liver tissue. A number of approaches have been proposed based on different image characteristics, such as texture features, from ultrasound B-scan and CT images. Texture analysis of liver CT images based on the Spatial Gray Level Dependence Matrix (SGLDM), the Gray Level Run Length Method (GLRLM) and the Gray Level Difference Method (GLDM) has been used by various researchers to discriminate normal from malignant hepatic tissue. Texture features estimated from the SGLDM have been applied to a probabilistic Neural Network (NN) for the characterisation of hepatic tissue (hepatoma and haemangioma) from CT images (Chen, Chung, Chen, Tsa, & Chang, 1998). Although a lot of effort has been devoted to liver tissue characterisation, the developed systems are usually limited to two or three classes of liver tissue and do not gain from the interaction of different texture characterization methods or the combination of different classifiers.

Two different approaches have been used to design a CAD system that characterises and automatically classifies hepatic lesions from non-enhanced CT images. ROIs corresponding to normal liver, hepatic cysts, haemangiomas, and hepatocellular carcinomas (Figure 3) were delineated by an experienced radiologist and were used as input to the systems. Of a total of 147 ROIs that were interrogated, 76 corresponded to healthy controls, 19 to cysts, 28 to haemangiomas, and 24 to hepatocellular carcinomas.

In the first approach, the average gray level and 48 texture characteristics derived from the SGLDM (Gletsos, Mougiakakou, Matsopoulos, Nikita, Nikita & Kelekis, 2003) consisted the feature vector providing input to a classifier module which was made up of three sequentially placed feed-forward ANNs. The first ANN was used to discriminate normal from pathological liver tissue. The pathological liver tissue regions were characterised by the second ANN as cyst or "other disease". The third ANN classified "other disease" into haemangioma or hepatocellular carcinoma. Three feature selection techniques have been applied to each individual ANN: the sequential forward selection, the sequential floating forward selection, and a GA for feature selection. The comparative study of the above dimensionality reduction methods showed that GAs result in lower dimensionality feature vectors and improved classification performance.

In another study, characterisation and classification of liver tissue into one of the previous classes was achieved through a CAD system based on the use of various texture features and ensembles of classifiers (Mougiakakou et al., 2003a) in an attempt to define an optimally performing CAD system architecture. The effect of feature selection on the resulting classification performance was also studied through the application of a GA-based feature selection technique to feature vectors with dimensionality exceeding a predefined threshold. For each ROI, five distinct sets of texture features were extracted using the following methods: first order statistics, SGLDM, GLDM, Laws' texture energy measures, and the fractal dimension. A set of classifiers whose individual predictions are combined in some way (typically by voting) to classify new examples consists an ensemble of classifiers.

The attraction that this topic exerts on machine learning and diagnostic decision support researchers is based on the premise that ensembles are often much more accurate than the individual classifiers that make them up. Two different ensembles of classifiers were constructed and compared. The first one consists of five multi-layer perceptron ANNs, each using as input either one of the computed texture feature sets or its reduced version after GA-based feature selection. The second ensemble of classifiers was generated by combining five different type primary classifiers, namely one multi-layer perceptron ANN, one probabilistic ANN, and three k-Nearest Neighbour (k-NN) classifiers. The primary classifiers of the second ensemble used identical input vectors, which resulted from the combination of the five texture feature sets, either directly or after proper GA-

Figure 3. Examples of typical CT images of liver showing different types of liver tissue: (a) normal liver tissue, (b) hepatic cyst, (c) haemangioma, and (d) hepatocellular carcinoma

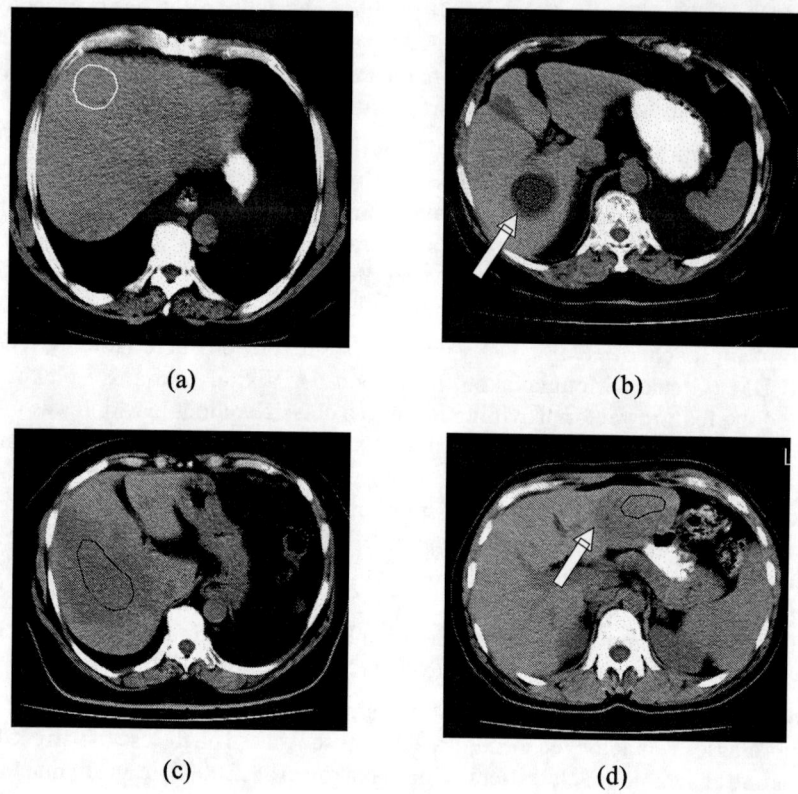

based feature selection. The final decision of each ensemble of classifiers was extracted by applying appropriate voting schemes across the outputs of the primary classifiers. The optimally performing CAD system architecture was chosen, based on the achieved classification performances in a testing data set. The highest individual classification performances in the testing set were 90.63 percent using the dimensionality-reduced Laws' texture energy features, followed by an 87.5 percent performance using first-order statistical features.

Therapeutic Decision Support

TDS systems aim to optimise and individualise patient treatment, including diet, drug treatment or radiotherapy treatment. In the following, the fundamental theory for the development of TDS systems is presented along with an example of a TDS system.

Basic Principles

TDS systems are based on the combined use of available patient or literature data and simulation models (Figure 4). More specifically, individualised patient data (medical history, laboratory tests, medical imaging data) along with information about the disease available in databases (DBs) provide input to appropriate simulation models. Simulation is then performed for alternative therapeutic schemes leading to the estimation of an optimal therapeutic scheme for the specific patient.

TDS systems can be based on the use of simulation models of biological and physiological procedures and/or stochastical analysis of available data. The most widely used methodologies include Compartmental Models (CMs), stochastical simulation methods, cellular automata methods, ANNs and AI methods (Brown & Rothery, 1993).

The specific goal of CMs is to represent complicated physiological systems with relatively simple mathematical models. In compartmental analysis, systems that are continuous and essentially non-homogeneous are replaced with a series of discrete spatial regions, termed compartments, considered to be homogeneous (Jacquez, 1985). A system can be defined by a class of dynamic models widely used in quantitative studies of the kinetics of materials in physiological systems. Materials are considered to be either exogenous (such as a drug or tracer) or endogenous (such as a substrate like glucose or an enzyme or hormone like insulin). Kinetics refers to time-variant processes, such as production, distribution, transport, utilisation, and substrate-hormone control interactions.

A compartment is an amount of material or spatial region that behaves as though it is well mixed and kinetically homogeneous. The concept of well mixed is related to uniformity of information. This means that any samples taken from the compartment at the same time will have identical properties and are equally representative of the system. Kinetic homogeneity means that each particle within a chamber has the same probability of taking any exit pathway. A CM then is defined as a finite number of compartments with specific interconnections between them, each representing a flux of material which physiologically represents transport from one location to another and/or a chemical transformation (Cobelli & Foster, 1998). The definition of a compartment is actually a theoretical

Figure 4. Typical structure of a TDS system

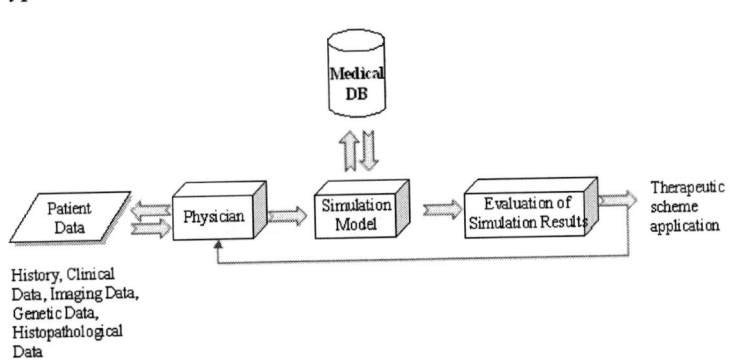

construct which could combine material from several physical spaces within a system. Consequently, the ability to equate a compartment to a physical space depends on the system being studied and associated model assumptions.

Stochastical models involve random variables that are functions of time and include probabilistic considerations. For a given set of initial conditions, a stochastic model yields a different solution each and every time (Brown & Rothery, 1993).

Cellular automata are widely used to solve diffusion problems in physiological systems. A grid of points is used to define a number of geometrical cells. Each cell corresponds to a variable with a limited number of possible values. The set of variables is renewed, at regular intervals, using the same renewal rule for every point. The rules are local in space and time, that is, their value in a point at time t depends only on the values of neighbouring points at time $t-1$ (Deutsch & Dormann, 2003).

ANNs can be used to simulate complex physiological/biological systems, such as the glucose metabolism, presented in the next paragraph. Their wide use in the simulation of non-linear systems is due to their ability to extract hidden data information, and to efficiently approximate any function of many variables.

Therapeutic Decision Support for Insulin Dependent Diabetes Mellitus

Diabetes Mellitus (DM) is a clinical condition caused by a disturbance in the metabolism of glucose introduced through nutrition. This disturbance is due to insufficient secretion or unsatisfactory action of the hormone insulin, which regulates glucose metabolism and

Figure 5. Simplified model of glucose metabolism

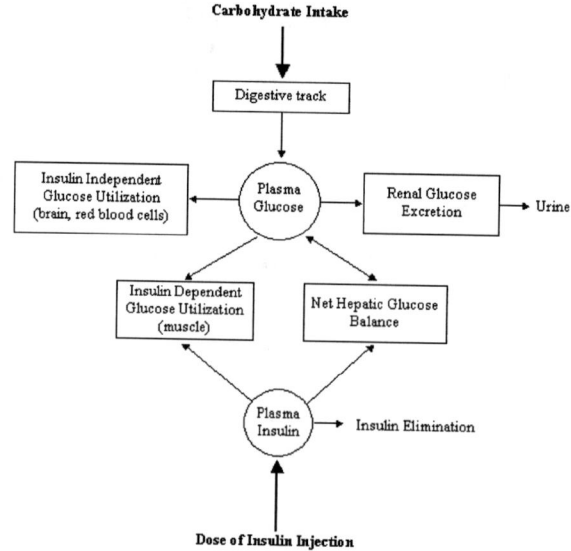

is normally secreted by the pancreas. In Insulin Dependent Diabetes Mellitus (IDDM) , also known as Type 1 DM, insulin is completely absent. DM causes extensive disorders in the body and thus affects the normal function of several organs (renal insufficiency, progressive vision reduction, infarction, etc.). These complications can be prevented through regular glucose control, that is, daily regular measurement of blood glucose concentration, and appropriate insulin treatment to maintain blood glucose within normal levels. The appropriate insulin regime has to be carefully adjusted, since blood glucose profile is rapidly, non predictably changing due to its dependence not only on the internal mechanism of glucose-insulin metabolism but also on a number of life style factors, like carbohydrate content of meals, physical activity and exercise, stress, other diseases, and so on. A simplified model of the most important internal and external factors associated with glucose metabolism, is presented in Figure 5.

During the last years, many TDS systems have been developed to assist not only the diabetes patients to handle their blood glucose levels, but also the physicians to understand the metabolic mechanism of their diabetes patients (Carson, 1998; Lehmann, 1997). These systems can store and display information about measured blood glucose concentration, food and insulin intake, physical activity, and other diseases. The efficiency of the aforementioned systems is significantly enhanced in case they comprise a simulation of glucose metabolism in order to predict short-term blood glucose levels, visualize the effect of the life-style parameters in the blood glucose profile, and make recommendations about the appropriate dose and time of insulin injections. In this context, many research approaches have been reported using expert systems, time series analysis, Mathematical Models (MM), and causal probabilistic networks.

The acceptance of the aforementioned TDS systems by diabetes patients was limited because these systems take into account only a limited number of the factors associated with glucose metabolism, and they are not easily individualised to accurately simulate metabolic processes for a specific Type 1 diabetes patient. In view of the above difficulties, the use of ANNs for the implementation of diabetes management decision support systems has been proposed. ANNs have been recently employed (Mougiakakou & Nikita, 2000) to advise on insulin regime and dose adjustment for Type 1 diabetes

Figure 6. Outline of TDS system for short-term prediction of blood glucose levels in Insulin Dependent Diabetes Mellitus

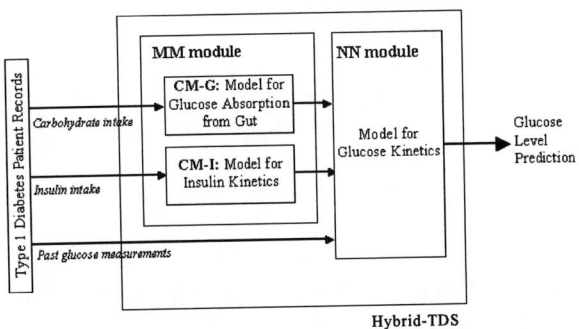

Table 3. Root mean square (RMS) errors along with mean absolute differences (MAD) and standard deviations (SD) between measured and estimated by the hybrid-TDS and the simple-TDS blood glucose levels (mg/dl) for all datasets

	Training Set		Validation Set		Testing Set	
	Hybrid-TDS	Simple-TDS	Hybrid-TDS	Simple-TDS	Hybrid-TDS	Simple-TDS
RMS	45.1	46.5	34.3	44.0	33.0	40.9
MAD (SD)	35.7 (27.7)	36.3 (29.1)	28.1 (20.1)	36.1 (25.6)	24.5 (22.2)	31.6 (26.5)

patients using information about their glucose levels, previous insulin intake, and hypoglycaemia symptoms.

In the following, a hybrid patient-specific TDS system is presented able to provide short-term blood glucose profile prediction for Type 1 diabetes patients employing data from their day-to-day life. The system is based on the combination of CMs and ANN techniques in an attempt to accurately model glucose metabolism. Figure 6 shows an outline of the TDS system. The system development and performance is demonstrated using input data from a Type 1 diabetes patient, stored in a database. The data contain information about blood glucose levels, insulin intake, and description of food intake.

The data are passed to two separate compartmental systems, which produce estimations about (i) the effect of insulin intake on blood insulin concentration and (ii) the effect of carbohydrate intake on blood glucose absorption from the gut, both with a time resolution of 15 minutes. The outputs of the two compartmental systems are passed to an ANN, able to handle delayed inputs, in order to predict subsequent blood glucose levels. The patient data are divided into three disjoint datasets: a) a training set, consisting of 257 glucose measurements recorded during the first 55 days, b) a validation set, consisting of 27 measurements recorded during the next seven days, and c) a testing set, consisting of 26 measurements recorded during the last seven days. Comparison of the proposed hybrid TDS system with a simple TDS system consisting of a single ANN which makes direct use of patient data shows that the use of CMs, to model the internal mechanisms of response to insulin and food intake, results in a significant improvement in the accuracy of blood glucose concentration prediction. Table 3 shows the Root Mean Square (RMS) errors along with Mean Absolute Differences (MAD) and Standard Deviations (SD) between measured and estimated by the hybrid-TDS and the Simple-TDS blood glucose levels for all datasets. As we can see, the accuracy of the hybrid-TDS system in predicting blood glucose levels is improved compared to that of the simple-TDS system. The proposed system can assist Type 1 diabetes patients to handle their blood glucose profile and recognize dangerous metabolic states, using only the basic information that is necessarily written in their diary.

Conclusions

In this chapter, the fundamental theory of CDS systems was described together with useful applications in diagnosis and therapy. The systems described in this chapter were designed to offer user-friendly interfaces, short response times, and low cost, enabling their introduction to routine clinical practice. The primary goal of CDS systems development, as for any branch of biomedical research, is to improve the overall health of the population. CDS systems may contribute to this by improving the quality of healthcare services, as well as by controlling the cost-effectiveness of medical examinations and treatment.

The ultimate acceptance of CDS systems will depend not only on the performance of the computerised method alone, but also on how well the human performs the task when the computer output is used as an aid and on the ability to integrate the computerised analysis method into routine clinical practice (Hunt, Haynes, Hanna & Smith, 1998). Issues, such as a friendly user-interface, a short system response time and low cost, are critical for the daily routine use of CDS systems.

Obviously, the development of CDS systems requires close collaboration of two scientific areas: medicine and computer science. This collaboration aims to codify knowledge and define the logical procedures used by the physician to reach a conclusion. As a result, the engineer must "extract" knowledge from the physician and reproduce it appropriately. This is particularly difficult because the physician's decisions are the result of a complex procedure combining special knowledge and experience.

Future trends and challenges in the area of CDS systems include the creation of links to patient electronic medical records. A universally-agreed upon medical vocabulary, so that the entries in the medical record have well-defined meanings, is important for this purpose. In addition to this, studies that evaluate the performance of CDS systems in clinical practice, in conjunction with demonstrations of cost-effectiveness, are a critical stage in further developing CDS systems. Users should be responsible for carefully monitoring the introduction of any new system carefully (Delaney, Fitzmaurice, Riaz & Hobbs, 1999).

Acknowledgments

The development of the CDS systems described in this chapter has been based on data provided by 2nd Department of Radiology, Medical School, University of Athens (Prof. D.Kelekis, Dr A.Nikita), and St. Mary's Hospital Medical School, Imperial College of Science, Technology and Medicine (Professor A.N.Nicolaides).

References

Asvestas, P., Golemati, S., Matsopoulos, G. K., Nikita K. S. & Nicolaides, A.N. (2002). Fractal dimension estimation of carotid atherosclerotic plaque from B-mode ultrasound: A pilot study. *Ultrasound Med Biol, 28*, 1129-1136.

Bankman, I. (2000). *Handbook of medical imaging: Processing and analysis*. Academic Press.

Berner, E.S. & Ball, M.J. (1998). *Clinical decision support systems: Theory and practice*. Springer-Verlag.

Bezdek, J.C. (1981). *Pattern recognition with fuzzy objective function algorithms*. New York: Plenum Press.

Brown, D. & Rothery, P. (1993). *Models in biology: Mathematics, statistics and computing*. John Wiley & Sons.

Carson, E.R. (1997). Decision support systems in diabetes: A systems perspective. *Computer Methods Programs Biomed, 56*, 77-91.

Chen, E.L., Chung, P.C., Chen, C.L., Tsa, H.M. & Chang, C.I. (1998). An automatic diagnostic system for CT liver image classification. *IEEE Trans Biomed Eng, 45*, 783-794.

Chen, T. & Metaxas, D. (2000). Image segmentation based on the integration of Markov random fields and deformable models. *Proceedings of the International Conference on Medical Image Computing and Computer-Assisted Intervention* (pp. 256-265).

Cobelli, C. & Foster, D.M. (1998). Compartmental models: Theory and practice using the SAAM II software system. *Advances in Experimental Medicine and Biology, 445*, 79-101.

Cootes, T.F., Edwards, G.J. & Taylor, C.J. (2001). Active appearance models. *IEEE Trans Pattern Analysis Machine Intell, 23*, 681-685.

Delaney, B.C., Fitzmaurice, D.A., Riaz, A. & Hobbs, F.D.R. (1999). Can computerised decision support systems deliver improved quality in primary care? *BMJ, 319*, 1-3.

Deutsch, A. & Dormann, S. (2003). *Cellular automaton modelling of biological pattern formation*. Boston: Birkhauser.

Dhawan, A.P., Chitre, Y., Kaiser-Bonasso, C. & Moskowitz, M. (1996). Analysis of mammographic microcalcifications using gray-level image structure features. *IEEE Trans Med Imag, 15*, 246-259.

Duda, R.O., Hart, P.E., & Stork, D.G. (2000). *Pattern classification* (2nd ed.). Wiley-Interscience.

Engelmore, R. & Morgan, T. (1988). *Issues in the development of a blackboard-based schema system for image understanding*. Reading, MA: Addison-Wesley.

Furlong, N., Lovelace, E. & Lovelace, K. (2000). *Research methods and statistics: An integrated approach*. Orlando, FL: Harcourt Brace & Company.

Gletsos, M., Mougiakakou, S., Matsopoulos, G., Nikita, K.S., Nikita A., & Kelekis, D. (2003). A computer-aided diagnostic system to characterize CT focal liver lesions: Design and optimisation of a neural network classifier. *IEEE Trans Inform Technology Biomed, 7,* 153-162.

Goldberg, D.E. (1989). *Genetic algorithms in search, optimization and machine learning.* Redwood City, CA: Addison-Wesley.

Golemati, S., Sassano, A., Lever, M.J., Bharath, A.A., Dhanjil, S. & Nicolaides, A.N. (2003). Carotid artery wall motion estimated from B-mode ultrasound using region tracking and block-matching. *Ultrasound Med Biol, 29,* 387-399.

Golemati, S., Tegos, T.J., Sassano, A., Nikita, K.S. & Nicolaides, A.N. (2004). Echogenicity of B-mode sonographic images of the carotid artery – Work-in-progress. *Journal of Ultrasound Med., 23,* 659-669.

Haykin, S. (1999). *Neural networks: A comprehensive foundation.* Prentice-Hall.

Holland, J.H. (1975). *Adaptation in natural and artificial systems.* Cambridge, MA: MIT Press.

Hunt, D., Haynes, R.B., Hanna, S. & Smith, K. (1998). Effects of computer-based clinical decision support systems on physician performance and patient outcomes. *Journal of American Medical Association, 280,* 1339-1346.

Inzitari, D., Eliasziw, M., Gates, P., Sharpe B.L., Chan, R.K., Meldrum, H.E., et al. (2000). The causes and risk of stroke in patients with asymptomatic internal carotid artery stenosis. *New England Journal of Medicine, 342,* 1693-1700.

Jacquez, J.A. (1985). *Compartmental analysis in biology and medicine* (2nd ed.). Ann Arbor: University of Michigan Press.

Jain, A.K., Duin, R.P.W. & Jianchang, M. (2000). Statistical pattern recognition: A review. *IEEE Trans Pattern Analysis and Machine Intelligence, 22,* 4-37.

King, B. (1967). Step-wise clustering procedures. *Journal of Am. Stat. Assoc., 69,* 86-101.

Kohonen, T. (1989). *Self-organization and associative memory* (3rd ed.). Springer Information Sciences Series. New York: Springer-Verlag.

Lehmann, E.D. (1997). Application of computers in clinical diabetes care. *Diabetes, Nutrition and Metabolism: Clinical and Experimental, 10,* 45-59.

Mougiakakou, S.G. & Nikita, K.S. (2000). A neural network approach for insulin regime and dose adjustment in type 1 diabetes. *Diabetes Technology and Therapeutics, 2,* 381-389.

Mougiakakou, S., Golemati, S., Gousias, I., Nikita, K.S. & Nicolaides, A.N. (2003b). Computer-aided diagnosis of carotid atherosclerosis using Laws' texture features and a hybrid trained neural network. *The 25th International Conference of the IEEE-EMBS,* Cancun, Mexico, September 17-23.

Mougiakakou, S., Valavanis, I., Nikita, K.S., Nikita, A. & Kelekis, D. (2003a). Characterization of CT liver lesions based on texture features and a multiple neural network classification scheme. *The 25th International Conference of the IEEE-EMBS,* Cancun, Mexico, September 17-23.

Schalkoff, R.J. (1997). *Artificial neural networks.* New York: McGraw-Hill.

Shortliffe, E.H. (1976). *Computer-based medical consultation: MYCIN.* New York: Elsevier.

Shortliffe, E.H., Perrault, L., Wiederhold, G. & Fagan, L. (1990). *Medical informatics – Computer applications in healthcare and biomedicine.* Wokingham: Addison-Wesley.

Spiegelhalter, D.J., Myles, J.J., Jones, D.R. & Abrams, K.R. (1999). An introduction to Bayesian methods in health technology. *British Medical Journal, 319,* 508-512.

Stoitsis, J., Golemati, S., Nikita, K.S. & Nicolaides, A.N. (2004). A modular software system to assist interpretation of medical images – Application to vascular ultrasound images. *IEEE Workshop on Imaging Systems and Techniques,* Stresa-Lago Maggiore, Italy, May 14.

Taylor, H.M. & Ros, P.R. (1998). Hepatic imaging: An overview. *Radiol Clin North Am, 36,* 237-245.

Udupa, J.K. & Samarsekera, S. (1996). Fuzzy connectedness and object definition: Theory, algorithms, and applications in image segmentation. *Graphical Models Image Processing, 58,* 246-261.

Yamani, S.M., Khiani, K.J. & Farag, A.A. (1997). Application of neural networks and genetic algorithms in the classification of endothelial cells. *Pattern Recognition Letters, 18,* 1205-1210.

Zadeh, L.A. (1965). Fuzzy sets. *Inf. Control, 8,* 338-353.

Chapter XV

Towards Knowledge Intensive Inter-Organizational Systems in Healthcare

Teemu Paavola, LifeIT Plc, Finland and
Helsinki University of Technology, Finland

Pekka Turunen, University of Kuopio, Finland

Jari Vuori, University of Kuopio, Finland

Abstract

The aim of this chapter is to share recent findings and understanding on how information systems can be better adopted to support new ways of work and improve productivity in public funded healthcare. The limits of transferring explicit and tacit knowledge are discussed and moreover, the chapter elaborates barriers to the widespread use of knowledge management tools among clinicians. The impacts of an inter-organizational system used for remote consultation between secondary and primary care providers are examined. Furthermore, the authors suggest that issues related to clinical knowledge management such as the varying information and knowledge processing needs of clinicians from various medical expertise domains should be examined carefully when developing new clinical information systems.

Copyright © 2005, Idea Group Inc. Copying or distributing in print or electronic forms without written permission of Idea Group Inc. is prohibited.

Introduction

It is reasonable to expect information technology to bring benefits to healthcare organizations just as it does to any other business. But recent studies demonstrate that the introduction of information technology does not in itself improve employee effectiveness in healthcare (Littlejohns, Wyatt, & Garvican, 2003). Moreover, it has proved particularly difficult to evaluate information systems in healthcare. This may involve the "evaluation paradox": we refuse to use a new technology until an evaluation study of its use has proved it useful. Particularly true in the healthcare sector, this cautious approach can be seen as a virtue, considering that hospitals fortunately were never taken in by all the information technology hype; but on the other hand the healthcare sector is regrettably slow in adopting even the best new practices. Although successful adoption of inter-organizational systems in healthcare still lacks a substantial body of research we argue that a key issue to be addressed may well be the natural logic of information and knowledge processing in various medical areas.

The term "telemedicine" covers the application of information technology to medical care chains. Conventional operations can always be pushed up a notch with state-of-the-art technology. In many cases it would also be more efficient to reorganize the operations altogether. For example, an e-mail application is at its most useful when its users are considered as active users of information rather than passive recipients. An organization can learn new ways of profiting from this new communications channel, which lies at the medium level according to the information richness theory introduced by Daft and Lengel (1986). A good example may be found in the healthcare sector, where new technology has managed to loosen the shackles of conventional thinking.

The electronic consultation model in medical care emerged in the early 1990s, beginning with an experiment at the internal medicine department of a hospital in southern Finland, where referrals were returned with care instructions in cases where lab results coupled with the information on the referral were sufficient for an accurate diagnosis. If the specialist receiving the referral considered that the case had not been examined sufficiently, the referral would be returned with instructions as to further tests or examinations. The new model evolved naturally from this practice. The information entered on a referral is often sufficiently comprehensive to enable remote consultation, and under the new model about half the patients referred to the hospital could be treated at their local health center on the basis of instructions provided by specialists through an information network.

There are many phenomena at play in shaping the practices of the healthcare sector. Identifying and allowing for these phenomena may be the key to successful telemedicine projects, indeed even more important than the technology itself. In the case of the new model described above, the innovation was supported by studies in the field: over half of all referrals in Finland and, for example, in Britain contain enough information for making an accurate diagnosis. So why are these patients being sent to hospital if there is already enough information for a diagnosis to be made at the health center? Convention, insecurity, financial incentives and similar factors have been cited as possible reasons.

This chapter aims to share recent findings and understanding on how information systems can be better adopted to support new ways of work and improve productivity in public funded healthcare. The effects of an integrated electronic referral system used for remote consultation between secondary and primary care providers is examined in a case study of two healthcare units in southern Finland. The study demonstrates how costly investments in videoconferencing in orthopedics yielded lesser benefits than the cheaper investment in e-mail-type application in internal medicine. Evidently internal medicine relies on fixed-format information, whereas orthopedics is more dependent on direct sensory inputs and tacit knowledge (Sternberg & Hovarth, 1999). Consequently, the natural logic of information and knowledge processing needs to be examined carefully before investing in information technology.

Development Toward Knowledge Intensive Systems in Healthcare

The information system presented in this chapter is representing information systems of new generation. The amount and meaning of these knowledge intensive systems will be highly expanding in the field of healthcare in the future. The progress of information system presented here is concluded based on theoretical frameworks of information systems in different disciplines. Phases of information system development in organizations have been dived from two to five phases by information system science researches. Most explanatory model seems to be Friedman and Cornford's (1989) intensive study (Checkland & Holwell, 1998; Korpela, 1994).

Friedman and Cornford argue that information system development can be seen as the interaction of changes in the core technology with changes in applications, linked by activities which mediate between them. These three elements are acting upon agents of change. The agents of change are technical development, changes in applications, market pressures and internal pressures, such as management culture (Friedman & Cornford, 1989).

After given this analysis, researchers see the history of development of the information systems in terms of three phases, each defined by a "critical factor" that has limited the development of computerization during that period. These phases are 1) hardware, 2) software and 3) user relation constraints. According to Friemdan and Cornford (1989) the first phase was approximately until the mid-1960s, the second one lasted from the mid-1960s until the early 1980s and the third one was early in the 1980s at least until their research was published in 1989.

In the year 1989, Friedman and Cornford speculated that a fourth phase might be organization environment constrain. The fourth phase could be also divided into subphases. At the first issue of developing systems is to make large-scale organizational systems and after these problems have been solved the main concern is to build inter-organizational systems. These inter-organizational systems are often customer orientated information systems (Friedman & Cornford, 1989).

Not every organization is going causally through these four phases. The development of information systems in the organization depends also on time when computerization has started in the particular organization or on other issues such as size of organization and ability to change organization (Checkland & Holwell, 1998; Friedman & Cornford, 1989).

At the moment, the state of healthcare information systems resembles that found in other fields in the early 1990s. As a rule, the field seems to be 10 years behind other areas (Ragupathi, 1997). However, the development of healthcare information systems is not exactly linear. Some innovations are put into use in healthcare faster than others (that are not seen very useful in healthcare). For example, Intranet technologies have been applied fast in healthcare organizations at least for non-medical purposes. Also (medical) knowledge intensive solutions are implemented fast in the field of healthcare (see Figure 1) (Turunen, Forsström, & Tähkäpää, 1999). The use of knowledge intensive systems have seen already in other fields, thus this same development is expected to happen also in healthcare.

General development of information systems in the 1990s as well Friedman's genuine work is clearly concluding that healthcare information systems will develop into the direction of customer orientated systems. This progress is in some cases passing through organizational environment constraints. It is possible that organizational phase is skipped over, if the particular organization is innovative, small or organic ones (cf. Courtright, Fairhurst, & Rogers, 1989). The more organic organization is relation to its environment the more capable it is to react the demands of customers and adapt the new technology. Naturally, the rapid development is also helped by new Internet technology.

Because of information intensive medical work (see right side of Figure 1) these new applications are in the healthcare strongly knowledge intensive ones. As a matter of fact, medicine can be considered as a knowledge-based business, which in practice means that experienced doctors use about two million pieces of information to manage their patients

Figure 1. An example of diffusion time of IT innovations and medicine into the field of health care information systems (HCIS)

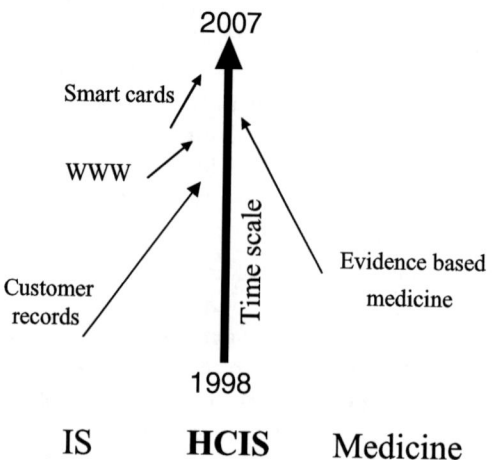

(Pauker, Gorry, Kassirer, & Schwartz, 1976; Smith, 1996; Wyatt, 1991). About a third of doctors' time is spent recording and combining information and a third of the costs of a healthcare provider are spent on personal and professional communication (Hersch & Lunin, 1995). Still most of the information doctors use when seeing patients is kept unrecorded in their heads and unfortunately some of this information is out of date or wrong. Thus the interest in applying networking media, like Internet in medical practice has arisen from the ever greater demand of meeting the needs of patients by drawing on the knowledge accumulated by medicine over 5,000 years.

New information may not have penetrated and the information may not be there to deal with patients with uncommon problems (Smith, 1996). These problems have become more serious as the rate of change in medical knowledge has accelerated. There are new scientific findings every day and at the moment the medical knowledge is estimated to increase fourfold during a professional lifetime (Heathfield & Louw, 1999; Wyatt, 1991). This inevitably means that doctors cannot practice high quality medicine without constantly updating their knowledge and finding information to help them with particular patients. The historical arrangement, in which doctors individually held the responsibility for ensuring an adequate supply of knowledge to guide their practice, is now failing to meet the knowledge needs of modern healthcare (Smith, 1996; Tannenbaum, 1994). Therefore diverse organizations, also health authorities, spread a vast amount of information to doctors.

It seems that knowledge intensive inter-organizational systems are the latest result of evolution in the field of healthcare. Healthcare organizations will adopt these kinds of systems sooner or later. Thus, it is important to understand the use of these knowledge intensive inter-organizational systems.

Managing Explicit Knowledge: Towards Optimal Allocation of Supply and Demand

The demand for healthcare services in both quantitative and qualitative terms is growing constantly. The aging of population will cause more pressure on healthcare organizations in several countries. The demand for healthcare services is greater, almost double among the old when comparing to average adults (Bethea & Balazs, 1997). Managing supply and demand between the primary parties, doctors and patients, seems to be a simple constellation of resources and requests. However, when managing supply and demand within the public funded healthcare system, the constellation changes into allocation of information resources of varying proficiencies and cost structures. By tradition, the tool used for this allocation has been the referring process.

The traditional referring process has its roots in the late 19th century England. There were three medical proficiencies operating in hospitals: surgeons, doctors and pharmacists. The third group was later called general practitioners, and before the turn of century they were pressured to leave hospital premises. As a result, citizens had no longer free passage to hospital care without seeing a general practitioner first.

Reengineering the One-Way Information Process: A Case Study

In the following example we concentrate to analyze the factors behind the startling change that took place when paper referrals were replaced with electronic referrals. Our example is based on a study that took place in southern Finland focusing on cross-functional processes within two healthcare units (Harno, Paavola, Carlson, & Viikinkoski, 2000). The process chosen for the study was simple: a health center general practitioner referring a patient to a specialist at a regional hospital. While one unit began to utilize information system when referring a patient to hospital, the other send referrals traditionally by ordinary mail. Parts of the study are reported in TQM and Human Factors conference, Sweden (Lillrank, Paavola, Harno, & Holopainen, 1999).

Both healthcare units, Peijas Hospital and Hyvinkää Hospital, consist of a hospital and several health centers serving 150,000 people. Both healthcare units use similar internal administrative processes, however, Peijas has an Intranet-type system connecting the hospital with health centers while the information process between the parties in unit Hyvinkää is done traditionally with paper documents. All internal medicine cases that were referred from health centers to the hospitals during the study are included.

The process of making diagnosis goes in short according to following. First, in health center a general practitioner begins to diagnose by taking the anamnesis, that is, a preliminary case history of a medical or psychiatric patient. Patient explains the matter in his or her own words and general practitioner uses interviewing technique to lead the conversation in order to confirm or dispatch any presentiments of diagnosis that well forth in mind. Based on the anamnesis and verbal communication general practitioner can also decide to take laboratory tests in order to make the diagnosis more accurate. The diagnosis can be then defined and the treatment process can begin.

In roughly 95 percent of visits, the health center can handle the whole case. The remaining five percent of cases require specialist's resources for diagnosis, tests, treatment or all of these. In such cases a referral is sent to a hospital. The behavior behind the sending of a referral can be explained by several factors that may all have an effect to tip the scale in favor of referring (Coulter, Noone & Goldacre, 1989).

At the hospital, after examining an incoming referral, the specialist decides its level of urgency, again in one out of the three categories. Within the given time period the patient will be invited to hospital for consulting. The other option would be to send the referral back to the general practitioner if the specialist decides the case does not warrant specialist care. In practice, however, almost all referrals lead to a hospital visit. During our study the specialists working with traditional paper documents estimated that one out of five cases could have been treated at a health center with the help of some consultation. However, they felt that inviting a patient in for consulting that takes normally 20 minutes was less troublesome and took less time than writing and mailing instructions back to the general practitioner. Further, there were no economic incentives to take this trouble, since the local governments paid hospitals by the number of patients that were actually called in. In sum, the traditional process had neither procedures nor incentives for optimal resource allocation.

The Peijas hospital was built in the early 1990s. The health centers in the corresponding healthcare unit Peijas were linked to the new hospital with a tailor-made Intranet-type system. Initially it was assumed that the system would simply replace paper forms and mail, and allow electronic filing. The effects were assumed to be speedier traffic of information. There had been some plans to use the new information route also to integrate secondary and primary care providers with an electronic referral system linked to an electronic medical record. However, it was not known what the new process would look like, or even less, if the procedures were going to work at all. The precautions were justified since the effects of an integrated electronic referral system used for remote consultations had not been studied.

For a few years the system was utilized as such, a speedway for one-way referral traffic giving the predictable, but rather minor impact on the performance of the whole system. Only after a few specialists in the internal medicine area started, on their own initiative, to utilize the system for remote consultation with promising results, the other specialists in the same area joined in. Eventually the local governments had to revise their economic system to hospitals to include remote consultations as well as actual visits as basis for remuneration. Thus, a new business process supported by remote consulting had emerged in internal medicine.

The basic assumption when the integrated referral system was installed was that some 20 percent of referral cases could be treated at a health center with the help of remote consultation. In practice, the amount was more than double. The ratio of referrals that were returned to the health center grew to 50 percent of all referrals.

The consequences from this were, first, that the number of patients who were asked to actually visit a specialist was reduced to half thus reducing the burden on specialist resources. According to the doctors involved, this happened in 75 percent of the cases without any observable problems to fulfill of the appropriate treatment. Second, it was discovered that the rate of remote consultations given by specialist varied depending on referral urgency. Only 10 percent of the most urgent, priority I referrals lead to the use of remote consultation, while the percentage was 30 for category II, and over 50 for the category III referrals. Consequently, the major cycle time reduction came to the least urgent cases. Third, the number of health center patients whose problems got the attention of a specialist through remote consultation, roughly doubled that is from five to ten patient visits out of hundred. This was due to the learning experience to utilize cooperation with specialists in borderline cases. As a fourth effect it can be added the hard to measure effects that frequent exchange of experiences had on learning on health center side.

Encouraged by the positive results achieved in internal medicine, the possibility of expanding the system into other disciplines was discussed at Peijas hospital. Orthopedic surgery was used as a comparative case, since its business process is in many ways different. Internal medicine is, by and large, an intellectual process where specialists make conclusions from pieces of information of symptoms, medical histories and laboratory tests. The information used is reasonably well structured and treatment is often medication. Orthopedic specialist's were not inclined to use e-mail consultations, therefore a videoconferencing system was installed allowing the specialist's to study x-ray images and the patient located at health center while interacting with the general practitioner. In half of the cases videoconferencing was found to work satisfactorily, however, it did not reduce the number of patients that were eventually sent to the hospital. Thus, the relatively cheap e-mail system in internal medicine yielded a better return on investment than the costly videoconferencing system in orthopedic surgery.

Managing Implicit Knowledge in Medicine: Challenge for Shared Knowledge

The limits of transferring knowledge as well as managing it only the basis of rational view of man has proved to be inconsistent with the reality of medical experts in many ways (Ferlie, Wood, & Fitzgerald, 1999; Patel, Arocha, & Kaufman, 1999). Medical professionals produce a great amount of knowledge, which a doctor has but is unable to articulate or quantify easily. In frame of transferring technology managing knowledge refers more explicit than implicit knowledge. To be more precise, most of medical experts transfer more data than practical know-how to each other. The problems with sharing knowledge among a specific districts of medicine rises from the fact that every each of doctor attach a different meanings to data on the basis of their experiences of previous patients. However, studies of social construction, learning and tacit knowledge gives us a opportunity to approach the dilemma in promising way (Argyris, 1999; Berger & Luckmann, 1966; Dixon, 1994; Koskinen, 2001; Polanyi, 1966; Polanyi & Prosch, 1975; Sternberg & Horvarth, 1999).

It is commonly accepted scholars that are two basic types of knowledge: explicit knowledge that can be verbalized, such as knowledge of facts and concepts; and implicit knowledge that cannot be made verbal, such as intuition and knowledge of procedures (Patel et al., 1999). Even if the conventional synonym of implicit knowledge is tacit knowledge, some of scholars use the concept of tacit meaning structures as well (cf. Dixon, 1994). Tacit knowledge, by definition, refers to the inarticulate aspects or meaning structures that cannot be taught explicitly and therefore are only learnt via direct experience (Dixon, 1994; Patel et al., 1999). According to Polanyi and Prosch (1975) textbooks of medicine are so much empty talk in the absence of personal, that tacit knowledge of their subject matter. Therefore evaluating explicit knowledge (e.g., education) is pretty easy to do, but to evaluate its impact on tacit meaning structures of orthopedic specialist's is almost impossible.

In management literature the meaning of intuition for leaders has not been studied very much even if the importance of it for management was found already in 1930's (Barnard, 1938). However, for example Patel et al. (1999) outlined a systematic approach to the study of medical expertise in order to understand the role that the acquisition of tacit knowledge plays in competent and expert performance. According to them intensive care decision-making is characterized by a rapid, serial evaluation of options leafing to immediate action. In this real-time decision-making, the reasoning is schema-driven in a forward direction toward action with minimal interference or justification (ibid.). Practically this means that doctors are tend to attach meaning to data more on the basis of their experience than logical interpretation of facts. On the other hand we may say that the scientific knowledge is in part socially constructed in different ways in a case of general practitioner who is sending data compared with medical expertise who is interpreting it (cf. Ferlie et al., 1999). Therefore managing knowledge should be always understood in the terms of socially construction reality among a certain group of professionals.

The importance of tacit knowledge in decision-making is proved to be more important than explicit knowledge (Leprohon & Patel, 1995). In one study (ibid.) decision-making accuracy was significantly higher in nurses with 10 years or more of experience than in nurses with less experience, which is consistent with that we know about acquisition of expertise in their domains.

Managing the tacit knowledge may easily sound the answer for the problems of electronic referring system. However, the tacit knowledge is also the issue of cooperation between medical professionals. To some extent there will be always deliberate amount of 'tacit knowledge' in healthcare organizations. Clinical professionals will retain a monopoly of knowledge as far as professionals are forced to compete each other in terms of budget and resources (cf. Ferlie et al., 1999).

Barriers to open communication between different medical experts is understandable when the goal of medicine is not integrated (cf. primary care vs. special care, science vs. clinical care in university hospitals). Findings of Smith and Preston (1996) are pretty common: senior management have problems regarding the interface between junior doctors and other professional groups. An elite of professional workers secures easily a high degree of autonomy for itself on the basis of possession of expertise (Ferlie et al., 1999). However, tacit knowledge is said to be the primary basis for effective management, and the basis for its deterioration (Argyris, 1999; Nonaka & Takeuchi, 1995). Understanding the management of knowledge in the frame of rational man leads us easily situation where the benefits of electronic referring system are limited (cf. Breite, Koskinen, Pihlanto, & Vanharanta, 1999).

Knowledge Management Issues for Clinical Applications

When utilizing applications of information technology in product-line orientated organization aspects such as the level of utilization becomes important. For example if only every third salesman in a company types his or hers orders in a given database the benefits of using an enterprise resource planning system barely exists at the shop floor. In knowledge intensive organizations, on the other hand, the quality and usefulness of the information content seems to play a leading role.

Information collected and stored in product-line orientated organizations is normally fixed-format information, for example, ordering codes that typically consist of numbers and letter acronyms. When storing qualitative data in knowledge intensive organizations the case becomes more complex. Storing personal notes is a trivial case, since one understands all the matters that are relevant and connected with the stored information. On the other hand storing qualitative information for other people's later interpretation is a greater challenge. One must presume the form that the data will be later most useful in; that is, one must understand the context the data will be later utilized in. In order to fulfill the validity and relevance requirements of users the rationale behind the information must be somehow included in the stored data. Until we learn to do that we must grapple with the quest of various oracles and gurus to explain us the true nature of knowledge management.

Slawson and Shaughnessy (1997) determine the usefulness of medical information according to the following equation (Figure 2). Since in practice the fixed consulting hours of doctors set limits to the dispersion or flexibility of the work amount, we can argue that the more validate and relevant information the practitioners get within constant time to access it the better.

A cautious approach to introduce knowledge management issues in clinical applications can be seen as a virtue. Sackett et al. (1997) argue that still 80 per cent of all treatments are not based on scientific proof but on believes. Validity can be determined as the probability of the information being true. For example the fact that an article is published in an academic journal does not guarantee that the information is true. One such article presented findings from a research that demonstrated how sodium fluoride effectively prevents contracting osteoporosis. Later another more validate research findings proved that there truly was causality between these two—only the effect sodium fluoride had on the illness was just the opposite (Slawson & Shaughnessy, 1997).

One barrier to the widespread use of knowledge management tools is the psychological need of clinicians. Weinberg, Ullian, Richards, and Cooper (1981) found that 81 American doctors in a distant geographical area consulted 23 experts, logging 11 calls to experts each month. Six of these experts received over 90 percent of the calls, about 90 each a month. Doctors consult experts, because it is a quick, cheap, and easy method. The other doctors can also provide the psychological benefits that are not available from books, journals, and computers. On the other hand, the doctors "answering" the questions may sometimes not be much more knowledgeable than the doctors seeking answers.

Also a review article by Smith in 1996 which looked at the information needs of general practitioners made several similar conclusions, one of which was that doctors are looking for guidance, psychological support, affirmation, commiseration, sympathy, judgement, and feedback. Smith (1996) argues that this aspect of doctors' information needs is

Figure 2. Usefulness of medical information

$$\text{USEFULNESS OF MEDICAL INFORMATION} = \frac{\text{RELEVANCE} \times \text{VALIDITY}}{\text{WORK}}$$

particularly poorly explored, and yet it may well be the most important need and the biggest stumbling block to a technical solution. The electronic consultation and knowledge management tools, such as decision-support systems, could support each other in various ways. Doctors could use the knowledge system for retrieving required information before referring to the specialist. They could for example check what are the necessary laboratory tests that should be done before referring. This would save specialists' time in electronic consultation as wouldn't need to separately ask for those tests. In addition that would be convenient for the patient.

Conclusions

Within the healthcare sector, different types of process can be identified with different demands on the quality of information and knowledge required. The amount of reference information needed to make a diagnosis differs from one specialization to the next in a hospital. In some fields, the information required by a doctor consists mainly of basic data, observations recorded in numbers and text, whereas in other fields an image or video clip is considered crucial for drawing conclusions. While the significance of what is known as tacit information is important in some fields, in the majority of cases it is explicit information that is required; and it is in care chains relying heavily on explicit information that the potential of information technology can be put to the best use.

We examined the impacts on adoption of a new inter-organizational system in healthcare. The consequences from using the Intranet-type referral system in the example case were, first, that the number of patients who were asked to actually visit a specialist was reduced to half thus reducing the burden on specialist resources. This was due to the replying option in the system. According to the doctors involved, this happened in 75 percent of the cases without any observable problems to fulfill the appropriate treatment. Secondly, the rate of consultations was varied depending on referral urgency. The major time reduction came to the least several cases. Thirdly, 5 out of 100 general practitioners' patients normally treated also by specialists was doubled to 10. However, there were differences between the benefits in different medical areas according to nature of information related to education of different specialists. This should be considered more carefully in the future, when new information systems will be planned and developed.

Evidently in internal medicine, specialists make conclusions from pieces of information of symptoms, medical histories and laboratory tests. The information used is explicit, reasonably well structured and treatment is often medication. Intranet referral system both increased the utilization of specialist consultation and decreased the need for secondary care services by transferring information and knowledge to primary care. The system allowed more patients to be treated at lower expense. Because all patients were thoroughly examined beforehand, the numbers of repeat visits as well as direct costs remained lower.

In orthopedic surgery the situation appeared to be different, as the diagnostic process is based more on tacit knowledge and the specialist frequently rely on direct observations. In orthopedics the specialists were not inclined to use e-mail consultations, therefore a videoconferencing system was installed allowing the specialists to study x-ray images and the patient located at health center while interacting with the general practitioner. In half of the cases videoconferencing was found to work satisfactorily, however, it did not reduce the number of patients that were eventually sent to the hospital. The relatively cheap e-mail solution in internal medicine yielded a better return on investment than the costly videoconferencing system in orthopedics. Consequently, the natural logic of information and knowledge processing in various areas needs to be examined carefully before investing in new information technology solutions.

Although it has been said that it is possible to ease sharing of tacit knowledge by using computer environment that is simulating real world (see Breite et al., 1999), it seems to be expensive task in healthcare. To optimize relatively small resources it should be invested more for developing sharing knowledge systems in those medical areas (like in internal medicine) that are using explicit information. In those medical areas that are using more like implicit information (like in surgery) development knowledge sharing systems should be delayed until we have effective mass-products that are supporting multimedia features. However, even then the development of those systems that are supporting open communication between different specialists and more generally between different professional groups should be encouraged. This would ease the change of working culture into the direction of non-hierarchical knowledge sharing.

References

Argyris, C. (1999). Tacit knowledge and management. In R.J. Sternberg & J.A. Horvath (Eds.), *Tacit knowledge in professional practice: Researcher and practiner perspectives* (pp. 123-140). London: LEA.

Berger, P. & Luckmann, T. (1966). *The social construction of reality*. New York: Penguin.

Breite, R., Koskinen, K., Pihlanto, P. & Vanharanta, H. (1999). To what extent does the tacit knowledge embodied in a technology product limit its electronic commerce potential? *Proceedings of IRIS22,* August 7-10, Keuruu, Finland.

Buchan, I., Heathfield, H., Kennedy, T. & Bundred, P. (1996). Decision support for primary care using the Path.Finder System. *British Journal of Healthcare Computing, 13*(6), 20-21.

Checkland, P. & Holwell, S. (1998). *Information, systems and information systems.* Chichester: John Wiley & Sons.

Coulter, A., Noone, A. & Goldacre, M. (1989). General practitioners' referrals to specialist outpatient clinics I. Why general practitioners refers patients to specialist outpatient clinics. *British Medical Journal, 299*(6694), 304-306.

Courtright, J.A., Fairhurst, G.T. & Rogers, L.E. (1989). Interaction patterns in organic and mechanistic systems. *Academy of Management Journal, 32*(4), 773-802.

Daft, R. & Lengel, R. (1986). Organizational information requirements, media richness and structural design. *Management Science, 32*(5), 554-571.

Dixon, N. (1994). *The organizational learning cycle. How we can learn collectively.* London: McGraw-Hill.

Duff, L.A. & Casey, A. (1999). Using informatics to help implement clinical guidelines. *Health Informatics Journal, 5*(2), 90-97.

Ferlie, E., Wood, M. & Fitzgerald, L. (1999). Some limits to evidence-based medicine: A case study from elective orthopaedics. *Quality in Health Care, 8*(2), 99-107.

Friedman, A.L. & Cornford, D.S. (1989). *Computer systems development: History, organization and implementation.* Chichester, UK: John Wiley & Sons.

Haines, A. & Jones, R. (1994). Implementing findings of research. *British Medical Journal, 308*(6942), 1488-1492.

Hanka, R., O'Brien, C., Heathfield, H. & Buchan, I. E. (1999). WAX ActiveLibrary: A tool to manage information overload. *Topics in Health Informatics Management, 20*(2), 69-82.

Harno, K., Paavola, T., Carlson, C. & Viikinkoski, P (2000). Patient referral by telemedicine: effectiveness and cost analysis of an Intranet system. *Journal of Telemedicine and Telecare, 6*(6), 320-329.

Heathfield, H. & Louw, G. (1999). New challenges for clinical informatics: knowledge management tools. *Health Informatics Journal, 5*(2), 67-73.

Hersch, W.R. & Lunin, L.F. (1995). Perspectives on medical informatics: information technology in health care. Introduction and overview. *Journal of American Social Infrastructure Science, 46,* 726-727.

Korpela, M. (1994). *Nigerian practice in computer systems development. A multidisciplinary theoretical framework, applied to health informatics.* Doctoral thesis. Helsinki University of Technology, Department of Computer Science Reports 1994: TKO–A31. Helsinki.

Lillrank, P., Paavola, T., Harno, K. & Holopainen, S. (1999). The impact of information and communication technology on optimal resource allocation in healthcare. *Proceedings of the International Conference on TQM and Human Factors QERGO,* Linköping, Sweden (pp. 169-174).

Littlejohns, P., Wyatt, J. & Garvican, L. (2003). Evaluating computerised health information systems: hard lessons still to be learnt. *British Medical Journal, 326*(7394), 860-863.

Nonaka, I. & Takeuchi, H. (1995). *The knowledge-creating company.* New York: Oxford University Press.

O'Brien, C. & Cambouropoulos, P. (2000). Combating information overload: A six-month pilot evaluation of a knowledge management system in general practice. *British Journal of General Practice, 50*(455), 489-490.

Patel, V.L., Arocha, J.F. & Kaufman, D.R. (1999). Expertise and tacit knowledge in medicine. In R.J. Sternberg & J.A. Horvath (Eds.) *Tacit knowledge in professional practice. Researcher and practitioner perspectives* (pp. 75-100). London: LEA.

Pauker, S.G., Gorry, G.A., Kassirer, J.P. & Schwartz, W.B. (1976). Towards the simulation of clinical cognition. Taking a present illness by computer. *American Journal on Medicine, 60*(7), 981-996.

Polanyi, M. (1966). *The tacit dimension.* New York: Doubleday & Co.

Polanyi, M. & Prosch, H. (1975). *Meaning.* Chicago: The University of Chicago Press.

Ragupathi, W. (1997). Health care information systems. *Communications of ACM, 40*(8), 81-82.

Sackett, D.L., Richardson, W.S., Rosenberg, W. & Haynes, R.B. (1997). *Evidence-based medicine: How to practice and teach EBM.* London: Churchill Livingstone.

Slawson, D.C. & Shaughnessy, A.F. (1997). Obtaining useful information from expert based sources. *British Medical Journal, 314*(7085), 947-949.

Smith, A.J. & Preston, D. (1996). Communications between professional groups in an NHS trust hospital. *Journal of Management in Medicine, 10*(2), 31-39.

Smith, R. (1996). What clinical information do doctors need? *British Medical Journal, 313*(7064), 1062-1068.

Sternberg, R.J. & Horvath, J.A. (1999). *Tacit knowledge in professional practice: Researcher and practioner perspectives.* London: LEA.

Tannenbaum, S.J. (1994). Knowing and acting in medical practice: The epistemological politics of outcomes research. *Journal of Health Politics, Policy and Law, 19*(1), 27-44.

Turunen, P., Forsström, J. & Tähkäpää, J. (1999). Visions of Health Care Information Systems in Finland by the year 2007 – At the point of intersection between different cultures. In C. Carlsson & R. Suomi (Eds.), *The State of the Art of Information Systems Applications in 2007.* TUCS Publications No. 16, University of Turku, Finland.

Weinberg, A.D., Ullian, L., Richards, W.D. & Cooper, C. (1981). Informal advice- and information-seeking between physicians. *Journal of Medical Education, 56*(3), 174-180.

Wyatt, J. (1991). Use and sources of medical knowledge. *Lancet, 338*(8779), 1368-1373.

Chapter XVI

An Overview of Efforts to Bring Clinical Knowledge to the Point of Care

Dean F. Sittig, Medical Informatics Department,
Kaiser Permanente Northwest, USA, Care Management Institute,
Kaiser Permanente, USA and Oregon Health & Sciences University, USA

Abstract

By bringing people the right information in the right format at the right time and place, state of the art clinical information systems with imbedded clinical knowledge can help people make the right clinical decisions. This chapter provides an overview of the efforts to develop systems capable of delivering such information at the point of care. The first section focuses on "library-type" applications that enable a clinician to look-up information in an electronic document. The second section describes a myriad of "real-time clinical decision support systems." These systems generally deliver clinical guidance at the point of care within the clinical information system (CIS). The third section describes several "hybrid" systems, which combine aspects of real-time clinical decision support systems with library-type information. Finally, section four provides a brief look at various attempts to bring clinical knowledge, in the form of computable guidelines, to the point of care.

"To be effective, (clinical decision support) tools need to be grounded in the patient's record, must use standard medical vocabularies, should have clear semantics, must facilitate knowledge maintenance and sharing, and need to

be sufficiently expressive to explicitly capture the design rational (process and outcome intentions) of the guideline's author, while leaving flexibility at application time to the attending physician and their own preferred methods." (Shahar, 2001)

Introduction

By bringing people the right information in the right format at the right time and place, informatics helps people make the right clinical decisions. Cumulatively, these better decisions improve health outcomes, such as quality, safety, and the cost-effectiveness of care. This improvement has been a mantra of informatics at least since the landmark article by Matheson and Cooper in 1982. This chapter provides an overview of the efforts over the years to develop systems capable of delivering such information at the point of care. Such an overview should help illustrate both the opportunities and challenges that lie ahead as we struggle to develop the next generation of real-time clinical decision support systems for use at the point of care.

Vision to Achieve

The ultimate goal is to provide patient-specific, evidence-based, clinical diagnostic and therapeutic guidance to clinicians at the point of care; this guidance should be available within the clinical information system (CIS) that defines their current workflow. In addition, we must have the tools necessary to enable clinicians, without specialized programming knowledge, to enter, review, and maintain all the clinical knowledge required to generate this advice. Finally, we must have the ability to rapidly change the clinical knowledge, test it, and make it available to clinicians without having to wait for a regular CIS updating schedule.

Questions That Must Be Answered Prior to Creating Such Systems

What information or knowledge is required to help the clinician make the right decision to achieve the desired health outcome?

Who will be the information's recipient (e.g., physician, nurse, pharmacist, or even a specific individual such as the patient's primary care physician)?

When in the patient care process is the intervention applied, for example, before, during (which can be broken down into sub-activities such as order entry or progress note creation) or after the patient encounter?

How is the intervention triggered and delivered? For example, does the system or the clinician initiate it? How much patient-specific data (if any) is needed to trigger system-initiated interventions? How much is the intervention output customized to the clinical workflow stage, the clinician, and or the patient? For system-initiated interventions, how can the threshold be set to minimize nuisance alerts? How intrusive should the information be (Krall, 2001)?

Will the clinicians find the information useful (Krall, 2002a, 2002b)? What can we do to minimize the number of false positive alerts?

Where will the clinician be when receiving the intervention, for example, with the patient at the bedside or in the office? What should happen if the information becomes available at some future point when the clinician is no longer with the patient to whom the information pertains?

Which medium will be used to convey the message, for example, e-mail inbox, wireless and/or handheld device, pager, CIS/CPOE screen, or printed pre-visit encounter sheet?

Is there a demonstrable return on investment (ROI) that is due exclusively to the clinical decision support intervention or feature?

Background

Where Do We Stand?

Numerous attempts have been made to bring various forms of clinical information to the clinician at the point of care. One way to solve this problem is to develop computer applications that stand alone, although often network accessible, and are available to the clinician upon his/her request. Another way is to incorporate the clinical knowledge directly into the clinical information system used by clinicians while giving care. Once it is there, the CIS system can automatically prompt the clinician or the clinician can request help. Recently, a third model for system development has been proposed that enables the clinician to request help from an outside source. Often several patient-specific data items are included in the request. Using such a system, a clinician is still completely in-charge of making the request for information and the information can be automatically configured based on a sub-set of patient information (Cimino, 1996). In addition, it is now possible to embed medication or procedure codes along with instructions that tell the CIS how to handle this information that allows a clinician to place an order directly into the patient's electronic medical record from the externally available information resource (Tang & Young, 2000).

The following diagram (Figure 1) is an attempt to illustrate both the key types of information or knowledge that investigators have focused on along with their mode of interaction [i.e., directly to the clinician or through the clinical information system (CIS)]. The boxes (yellow) represent the type of knowledge and the labels on the links (green) represent some of the key projects or concepts vendors have focused on this particular

Figure 1. An overview of the numerous efforts researchers have made to bring pertinent clinical knowledge to clinicians at the point of care

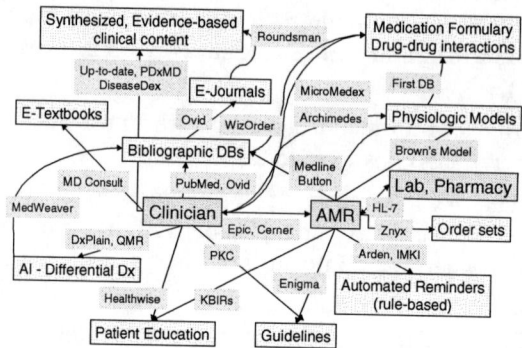

form of clinical decision support. The aim of this figure is to highlight the myriad attempts that have been made to develop clinical knowledge management applications and to help everyone understand how different clinical knowledge resources and applications are both related in terms of what they are trying to accomplish and different in the resources they utilize. Although this diagram is fairly complex, it is only a small, imperfect and incomplete representation of the entire clinical decision support landscape.

The following sections briefly review each of the projects or concepts. The first section focuses on "library-type" applications that enable a clinician to look up information in an electronic document. The second section describes a myriad of "real-time clinical decision support systems". These systems generally deliver clinical guidance to clinicians at the point of care within the CIS. The third section describes several "hybrid" systems, which combine aspects of knowledge-based clinical decision support systems with library-type information. Finally, section four looks at various attempts to bring clinical knowledge in the form of computable guidelines to the point of care.

Library-Type Applications: Front-Ends to Applications That Directly Interact with Clinicians

Bibliographic Databases (DBs)

Biomedical bibliographic databases contain on the order of millions of records, each representing a unique scientific journal article that has been published. Each record typically contains the title of the article, the authors, their affiliation(s), and the abstract of the article. In addition, several metadata tags may have been applied by human indexers to improve the likelihood that each document will be retrieved when and only when it is relevant to the user's query.

For the last 30 years, the National Library of Medicine has maintained the MedLine database, the most common bibliographic database used in clinical medicine. During that period, various attempts have been made to develop easy to use and reliable interfaces to this vast resource including Grateful Med (Cahan, 1989) and COACH (Kingsland, 1993). Currently, PubMed is the most widely used of all these interfaces. PubMed relies on a sophisticated free-text query processor to map freetext user queries to MeSH terms, when appropriate, and returns a highly relevant set of documents.

In addition to PubMed, several commercial vendors have created proprietary interfaces to the MedLine database in an attempt to improve either the recall or precision of the user's queries. For example, Ovid has developed an interesting MeSH mapper and query expander that has gathered outstanding reviews from highly trained librarians. Knowledge Finder has developed a fuzzy mapping algorithm that has also generated some good reviews. Unfortunately, none of these systems consistently enables clinicians to retrieve more than half of all the relevant articles on any particular topic (Hersh, 1998). In addition to these variations on a search interface, several projects have used automated differential diagnosis generators, such as DxPlain, as an interface to the bibliographic DBs. Finally, the Science Citation Index uses the reference list at the end of every scientific article published to generate linked lists of references. Such a scheme can also identify particularly noteworthy articles since these articles are referenced many more times. Interestingly, the Google search engine uses this same concept to generate its index of relevant web sites. It does this by keeping track of the number and quality of referring web sites rather than references at the end of the article (Brin, 1998).

Other Clinical Reference Materials

In addition to access to the bibliographic literature, clinicians could use other clinical reference information. For example, various systems exist to provide clinicians with access to the complete text and figures contained in textbooks (e.g., MD Consult) or journals (e.g., Ovid) in electronic form. In addition, several commercial companies have begun developing synthesized, evidence-based clinical summaries of the diagnosis and treatment of common clinical conditions [e.g., Up-to-Date, CliniAnswers, DiseaseDex, PDxMD (see http://www.informatics-review.com/KnowMan/Examples.html for other examples)].

Real-Time Clinical Decision Support Systems

Real-time clinical decision support systems (RCDSS) are fundamentally different from "library" applications in that they interact with the clinician through the CIS. Rather than relying on the user to interpret the text and then make a decision, RCDSS combine the patient's clinical data with the clinical knowledge to help the clinician reach a decision. At least six distinct methods can provide this type of RCDSS:

1. Medication formularies and drug-drug interaction checkers provide routine checks of all orders entered into the EMR. They can recommend alternative medications based on the formulary and check for potential drug-drug interactions automatically following entry of a medication order assuming they have access to the patient's list of current medications. FirstDataBank, Micromedex, and Multum are examples of commercial providers of drug-drug interaction checking databases.

2. Automated rule-based reminders are used most commonly to remind clinicians to perform routine health maintenance procedures. The Arden syntax is currently the only "standard" means of encoding these rule-based systems (Arden, 2003). The Institute for Medical Knowledge Interchange (IMKI), another recent entry into this field (IMKI, 2003), recently scaled back their operations while they seek additional funding and support.

3. Condition-specific order sets and charting templates help clinicians remember and facilitate the entry of all related orders or answers to questions asked of a patient with a particular condition. Zynx, a recently formed company, is developing these ready-to-use, condition-specific order sets. In addition, most of the commercially available CISs either offer ready-to-use order sets or provide tools to help clinicians construct these order sets (Franklin, Sittig, Schmiz, Spurr, Thomas, O'Connell, & Teich, 1998).

4. Integrated clinical guidelines allow a clinician to compare a specific patient's clinical data against a guideline and automatically receive a recommendation. The most advanced example of this type of application is Enigma's PREDICT application. In an early attempt to develop this sort of functionality, Epic Systems created their "Active Guidelines" application. A major distinguishing feature between the Active Guidelines (AG) effort and that of Enigma is that the AG user is still required to search for, read, interpret, and then decide which portion of the large, text-based guideline document applies to a specific patient and then select the appropriate recommendation. Enigma's PREDICT automatically does all of this work and simply presents the clinician with the appropriate suggestions. Another recent entry into this market space is Theradoc, a small company located in Salt Lake City, Utah (www.theradoc.com).

5. Complex, physiologic models incorporate our best current understanding of human physiology along with evidence-based results of randomized clinical trials to generate patient- or population-specific predictions of future events. Archimedes, a complex model of diabetes and the entire healthcare system, is currently configured to allow specially trained clinicians to access this resource (Schlessinger, 2002). Jonathan Brown, from Kaiser Permanente's Center for Health Research in Portland, has recently begun working on a method to automatically link therapeutic suggestions generated by his Global Diabetes Model to a specific patient's clinical data from an EMR or clinical data warehouse (Brown, 2000).

6. Artificial Intelligence (AI) systems have been developed to help clinicians complete their differential diagnosis list. To date these systems have not had widespread acceptance by clinicians, even though they have been shown to be at least as good and often better than un-aided clinicians (Bankowitz, 1989; Berner, 1994).

DxPlain, QMR, and Weed's Problem Knowledge Couplers (Weed, 1986) are the best examples of these types of systems.

Hybrid Applications

In addition to each of these fairly straightforward attempts to link clinical knowledge to the point of patient care, several companies have attempted to combine various clinical knowledge resources and applications to generate even more useful applications. MedWeaver (Detmer, 1997) was an early attempt to combine DxPlain, an AI-based differential diagnosis (DDx) system with a bibliographic database. The DDx in conjunction with the Unified medical Language system (UMLS) act as an automatic query expander and presents the user with information related to all the known items on the patient's differential diagnosis.

Geissbuhler and Miller (1998) developed WizOrder, a system to automatically generate potential drug-drug interaction reminders when clinicians order medication through the system. These reminders are based on the co-occurrence of clinical term frequencies[1] that are a part of the UMLS's Metathesaurus database. Rennels, Shortliffe, Stockdale, and Miller (1987) developed Roundsman, a prototype system, to automatically extract evidence-based recommendations from on-line structured journal articles.

Cimino (1996) developed the concept of the "InfoButton" which allows a clinician to automatically generate queries to a variety of clinical information resources to answer a set of generic concept-specific questions such as, "What's the treatment for X?", where X can be any clinical condition. Recently, the HL-7 Clinical Decision Support Task Force has begun efforts to develop a standard for such queries.

Computable Guidelines

Current State of the Art

Various research and commercial efforts are underway in an attempt to create systems to achieve the vision outlined above. The distinguishing feature of computable guideline systems is that they are more than a simple application of "if-then-else" rules to generate alerts and reminders at the point-of-care. All of the systems described below are attempting to go one step further with multi-step clinical algorithms at the point-of-care. The more complex algorithms attempt to track a patient's treatments and progress over time, while the alerts and reminders serve only to look at what is best at each point in time. Therefore, I did not include the efforts of the Institute for Medical Knowledge Interchange (IMKI) or the Arden group to create libraries of rules and systems to facilitate their incorporation in various clinical information systems.

The following sections overview several attempts and describe which parts of the vision they either have or are currently focusing on. I also point out the strengths and weaknesses of each approach.

Academic Research Efforts

GLEE is a system for execution of guidelines encoded in the Guideline Interchange Format v3 (GLIF3) (Wang, 2002). GLEE provides an internal event-driven execution model that can be hooked up with the clinical event monitor of a local clinical information system environment. It was developed at Columbia University and is currently only a research prototype. No tools exist for knowledge maintenance other than those proposed by the GLIF3 organization.

GLARE (GuideLine Acquisition, Representation and Execution) is a domain-independent system for the acquisition, representation and execution of clinical guidelines (Terenziani, 2001). GLARE provides expert physicians with an "intelligent" guideline acquisition interface. The interface has different types of checks to provide a consistent guideline: syntactic and semantic tests verify whether or not a guideline is well-formed. GLARE technology has been successfully tested in different clinical domains (bladder cancer, reflux esophagitis, and heart failure), at the Laboratorio di Informatica Clinica, Azienda Ospedaliera S. Giovanni Battista, Torino, Italy. Currently, this application is not available for use by people outside of the research team. See: www.openclinical.org/gmm_glare.html#aimj01

Commercial Vendors

Therapy Edge (http://www.therapyedge.com/) is a condition-specific, stand-alone, web-based, EMR with extensive clinical decision support for HIV patients. It is a stand-alone tool, that is, it has no integration with the existing CIS; it is only for HIV patients; and has no tools for knowledge based maintenance.

PRODIGY is a guideline-based decision support system in use by a large number of General Practitioners in the United Kingdom (Purves, 1999). The evaluation of PRODIGY 3 is currently underway and Phase 4 is being planned. PRODIGY I and PRODIGY II were implemented as extensions to proprietary UK electronic patient record systems. The PRODIGY system includes a proprietary guideline model, which in PRODIGY II was used to implement guidelines for the management of acute diseases. See: http://www.prodigy.nhs.uk/ClinicalGuidance/ReleasedGuidance/GuidanceBrowser.asp for a demo.

PREDICT is a decision support system for delivering evidence-based care directly to clinical practitioners desktops or handhelds. PREDICT integrates with existing clinical information systems using standard internet communication protocols, and allows clinicians to create new disease modules and manage their own guideline content. See: http://www.enigma.co.nz/framed_index.cfm?fuseaction=ourknowledge_products for more info.

eTG complete is an easy-to-use, HTML-based product, available for use on stand alone or networked PC or Mac computers. It covers over 2000 clinical topics and will be updated at regular intervals (three to four times per year). They do not even mention the possibility of integrating their work with any commercially available EMR products. See: http://www.tg.com.au/complete/tgc.htm for an online demonstration. This is just one of many examples of stand-alone, proprietary format, internet-based, guideline presentation systems. In my opinion, none of these are significantly better than what the USA's National Guideline Clearinghouse (www.guideline.gov) or the New Zealand Guidelines Group (www.nzgg.org.nz) are doing by simply creating a large database of clinical guidelines that are arranged by clinical condition.

AREZZO is a decision support technology (based on PROforma), for building and running clinical applications. AREZZO applications are designed to provide patient-specific advice, guiding the user through data collection, clinical actions and decision making. Applications can be quickly modeled and tested, using the AREZZO Composer, and instantly deployed on the Internet. AREZZO is capable of supporting the development of multi-disciplinary care pathways (customized for local circumstances); its safety-critical and task-based features support the rigor needed for clinical protocols. Currently the system is limited to supporting decisions on pain control in cancer, but this will be expanded in the future to include pain control in much more general terms (for example, arthritis, chronic pain, etc.) See: http://www.infermed.com/ds_arnocs.htm for more information.

Summary and Conclusions

Library-type applications that provide users with easy access to the latest medical knowledge in the form of journal articles or textbooks will not be going away in the near future. Even with our increasing understanding of how to build intelligent computer systems that can recognize the current patient context and suggest potentially relevant clinical information resources, the users must still have a means of getting directly to the knowledge resources so they can look up additional information. Therefore, work based on the Medline Button, as described by Cimino (1996) must continue. In fact, an effort is now underway within the HL-7 Clinical Decision Support Technical Committee to develop a standard interface that will facilitate further development of these types of applications.

The use and utility of real-time clinical decision support systems (RCDSS) will continue to grow exponentially. As more and more pressure is placed on clinicians to practice evidenced-based medicine and do it as inexpensively as possible, these RCDSSs will proliferate. A recent publication by the Health Information Management Systems Society (HIMSS) entitled, the "Clinical Decision Support Implementers' Workbook" is now freely available on the internet (Osheroff, 2004). This workbook and others should lead to increased use of various clinical decision support systems. Although little work is currently underway on the development of new hybrid systems, I believe that these types of systems will soon reappear and provide users with significant additional functionality that cannot be achieved using any other means. Finally, as healthcare delivery systems choose and implement commercially available clinical information systems, clinicians and researchers alike could gain valuable knowledge by beginning work on various clinical knowledge management projects with one or more of the vendors or research groups listed above. At the present time, this field has no clear-cut winner. In fact, no system is beyond the advanced research prototype stage. As such, there will be considerable changes in whichever solution is ultimately successful. I believe it is in everyone's best interest to be a part of the effort in defining and experimenting with these types of systems. Only then will we truly understand what is really useful and how we should proceed.

Acknowledgments

I would like to thank Jerry Osheroff, M.D. for his comments on an early draft of this paper. A preliminary version of this chapter was presented at a health informatics conference in Auckland, New Zealand (2003).

References

Arden Syntax for Medical Logic Systems. (2003). Home Page *http://cslxinfmtcs.csmc.edu/hl7/arden/*

Bankowitz, R.A., McNeil, M.A., Challinor, S.M., Parker, R.C., Kapoor, W.N. & Miller, R.A. (1989). A computer-assisted medical diagnostic consultation service. Implementation and prospective evaluation of a prototype. *Ann International Med., 110*(10), 824-32.

Berner, E.S., Webster, G.D., Shugerman, A.A., Jackson, J.R., Algina, J., Baker, A.L., Ball, E.V., Cobbs, C.G., Dennis, V.W., Frenkel, E.P., et al. (n.d.). Performance of four computer-based diagnostic systems. *New England Journal of Medicine, 330*(25), 1792-6.

Brin, S. & Page, L. (1998). The anatomy of a large-scale hypertextual {Web} search engine. *Computer Networks and ISDN Systems, 30*(1-7), 107-117.

Brown, J.B., Russell, A., Chan, W., Pedula, K., & Aickin, M. (2000). The global diabetes model: User friendly version 3.0. *Diabetes Res Clin Pract., 50*(Suppl 3), S15-46.

Cahan, M.A. (1989). GRATEFUL MED: A tool for studying searching behavior. *Med Ref Serv Q., 8*(4), 61-79

Cimino, J.J. (1996). Linking patient information systems to bibliographic resources. *Methods Inf Med., 35*(2), 122-6.

Detmer, W.M., Barnett, G.O., & Hersh, W.R. (1997). MedWeaver: Integrating decision support, literature searching, and Web exploration using the UMLS Metathesaurus. *Proceedings of AMIA Annual Fall Symposium* (pp. 490-494).

Franklin, M.J., Sittig, D.F., Schmiz, J.L., Spurr, C.D., Thomas, D., O'Connell, E.M. & Teich, J.M. (1998). Modifiable templates facilitate customization of physician order entry. *Proceedings of AMIA Symposium* (pp. 315-319).

Geissbuhler, A. & Miller, R.A. (1998). Clinical application of the UMLS in a computerized order entry and decision-support system. *Proceedings of AMIA Symposium* (pp. 320-324).

Hersh, W.R. & Hickam, D.H. (1998). How well do physicians use electronic information retrieval systems? A framework for investigation and systematic review. *JAMA, 280*(15), 1347-1352.

Institute for Medical Knowledge Implementation (IMKI) (2003). Online *http://www.imki.org/*

Kingsland, L.C., Harbourt, A.M., Syed, E.J. & Schuyler, P.L. (1993). Coach: Applying UMLS knowledge sources in an expert searcher environment. *Bull Med Libr Assoc., 81*(2), 178-183.

Krall, M.A. (2002). Clinicians' assessments of outpatient electronic medical record alert and reminder usability and usefulness requirements: A qualitative study. Master's thesis. Online *http://www.ohsu.edu/biccinformatics/people/ms/theses/2002/krall.pdf 1-104*

Krall, M.A. & Sittig, D.F. (2001). Subjective assessment of usefulness and appropriate presentation mode of alerts and reminders in the outpatient setting. *AMIA Proceedings* (pp. 334-338).

Krall, M.A. & Sittig, D.F. (2002). Clinicians' assessments of outpatient electronic medical record alert and reminder usability and usefulness requirements. *AMIA Proceedings* (pp. 400-405).

Matheson, N.W. & Cooper, J.A. (1982). Academic information in the academic health sciences center: Roles for the library in information management. *Journal of Med. Educ., 57*(10 Pt 2), 1-93.

Osheroff, J.A., Pifer, E.A., Sittig, D.F., Jenders, R.A. & Teich, J.M. (2004). Clinical decision support implementers' workbook: Health information management and systems society. Available online *http://www.himss.org/CDSworkbook*

Purves, I.N., Sugden, B., Booth, N. & Sowerby, M. (1999). The PRODIGY project: The interactive development of the release one model. *Proceedings AMIA Symposium* (pp. 359-363).

Rennels, G.D., Shortliffe, E.H., Stockdale, F.E. & Miller, P.L. (1987). A computational model of reasoning from the clinical literature. *Comput. Methods Programs Biomed., 24*(2), 139-149.

Schlessinger, L. & Eddy, D.M. (2002). Archimedes: A new model for simulating health care systems: The mathematical formulation. *Journal of Biomed Inform., 35*(1), 37-50.

Shahar, Y. (2001). Automated support to clinical guidelines and care plans: The intention-oriented view. Online *http://www.openclinical.org/docs/int/briefingpapers/shahar.pdf*

Tang, P.C. & Young, C.Y. (2000). Free full text active guidelines: Integrating Web-based guidelines with computer-based patient records. *Proceedings of the AMIA Symposium* (pp. 843-847).

Terenziani, P., Molino, G. & Torchio, M. (2001). A modular approach for representing and executing clinical guidelines. *Artificial Intelligence Med., 23*(3), 249-276.

Wang, D. & Shortliffe, E.H. (2002). GLEE: A model-driven execution system for computer-based implementation of clinical practice guidelines. *Proceedings of the AMIA Symposium* (pp. 855-859).

Weed, L.L. (1986). Knowledge coupling, medical education and patient care. *Crit Rev Med Inform., 1*(1), 55-79.

Endnote

[1] The UMLS contains the co-occurrence relationships for the number of times concepts have co-occurred as key topics within the same articles, as evidenced by the Medical Subject Headings assigned to those articles in the MEDLINE database. Co-occurrence relationships have been also computed for different ICD-9-CM diagnosis codes assigned to the same patients as reflected in a discharge summary database. In contrast to the relationships asserted within source vocabularies, the statistical relationships in the Metathesaurus can connect very different concepts, such as diseases and drugs. There are specific Metathesaurus files for the co-occurrence relationships (MRCOC.RRF and MRCOC in ORF).

Chapter XVII

Social Capital, An Important Ingredient to Effective Knowledge Sharing:
Meditute, A Case Study

Jay Whittaker, University of Ballarat, Australia

John Van Beveren, University of Ballarat, Australia

Abstract

This chapter introduces social capital as a concept useful in identifying the ingredients necessary for knowledge sharing in Healthcare. Social capital is defined and its importance as a concept for identifying the conditions necessary for knowledge exchange is discussed. Furthermore the introduction of an online database and tutorial system (Meditute) is presented as a case study to highlight the importance of social capital where information systems are used in the sharing process. The authors anticipate that the use of social capital to analyse knowledge sharing initiatives will lead to more holistic approaches. Such approaches will inform both researchers and managers as to the many factors that affect knowledge sharing.

Copyright © 2005, Idea Group Inc. Copying or distributing in print or electronic forms without written permission of Idea Group Inc. is prohibited.

Introduction and Background

Clinical staff is an extremely important asset to the emergency departments in hospitals. The knowledge they have, their competence and capabilities are vital to providing adequate and appropriate care to patients. Maintaining this knowledge and assuring that it is disseminated and renewed is an important management goal. People are the repositories of knowledge and effectively managing the human and intellectual capital within people involves providing suitable education and an organisational culture that encourages sharing the knowledge.

Knowledge sharing is dependent on social interaction, direct communication, and contact among individuals (Abou-Zeid, 2002). However the increased demands and workloads within the emergency departments have resulted in a lack of time and flexibility for staff to be included in the process of sharing knowledge. Until recently the only option for many senior clinicians to gain new or additional knowledge was through formal postgraduate education. However such programs are unpopular with senior clinicians as they tend to be lecture based with the emphasis more on teaching than learning. Formal education often lacks the "on the job" teaching, which incorporates the tacit forms of knowledge of "how we do things around here" and the context-specific cues for memory recall.

A premise for the discussion in this chapter is that although the clinical information shared is constructivist rational (based on known facts and linear logic), the access to and sharing of such information requires particular mechanisms. These mechanisms serve to create awareness that the information exists and is accessible. Furthermore the mechanisms allow access, use and ability to evaluate the content of the information

Accessible information is a transferable commodity. However the context and meanings that is often required to increase the usability of information involves sharing knowledge. Knowledge is not a transferable commodity, but rather a constructivist mechanism established from access to people, relationship building and developing shared understandings. The mechanisms will be discussed in detail as components of the social capital concept.

In March 2002, an Australian Hospital initiated a program called Meditute, which was based on a commercial system built in the United Kingdom. Meditute is a series of Web pages designed to teach small chunks of medical knowledge. It provides basic scientific information in a relevant clinical context for the learner to read and absorb, before undertaking a test of their learning by answering a series of short questions at the end of the tutorial. Meditute enables the learner to select the tutorials appropriate to their needs and complete them at their own pace through self directed learning. The intention is that the tutorials will contain more information than what is available in the textbooks. The main strength of Medititue is that the participants can write and are encouraged to write their own tutorials, which are peer reviewed before being shared with others.

The Meditute program has received grant funding from the Postgraduate Medical Council of Victoria and a donation of content from the Open University Centre Education in Medicare. While this has certainly helped kick start the program, its continued success relies on something far more important, that of social capital. In the absence of social capital, Meditute is inanimate. It is unable to command the necessary interactivity and does not convey the context and meaning that underpin its success. Without social capital it is merely a collection of facts presented without context and meaning.

Social capital is an umbrella concept that has been used to describe the "goodwill" that exists between individuals who share social relations. Goodwill refers to the sympathy, trust, and forgiveness offered to each individual by friends and acquaintances (Adler and Kwon, 2002). While "goodwill is the substance of social capital, its effects flow from the information, influence and solidarity that goodwill makes available" (Adler and Kwon, 2002, p.18). Therefore appropriate leveraging of social capital results in the fluid sharing of information and knowledge between those who share the goodwill.

In the context of Meditute, opportunities have been created for the sharing of information and clinical knowledge. However, the sustainability of Meditute as a collaborative technology and shared knowledge repository relies on nurturing motivation among potential and actual participants and providing continual resources (Lesser, 2000). Social capital is an important ingredient in effective knowledge sharing. But as discussed later, leveraging it successfully involves many challenges. The objectives of this chapter are to:

- Define social capital and emphasise its importance to understanding the conditions necessary for knowledge exchange
- Argue for the social capital perspective in clinical knowledge management
- Discuss the effects of social capital in clinical knowledge management with reference to the use of Meditute
- Present some strategies to leverage social capital to improve knowledge sharing between clinicians who participate with Meditute.

The following section defines the concepts of social capital and the importance of understanding the dimensions of social capital to identify the conditions necessary for effective knowledge exchange. Next, the relevance of the social capital perspective to clinical knowledge management is argued. With this foundation, we explore the importance of social capital to knowledge sharing in clinical settings, with suggestions for strategies to overcome barriers. We conclude by relating the social capital perspective to broader trends in knowledge management research and by offering a research and action agenda for social capital in knowledge management.

Social Capital is an Important Ingredient for Knowledge Sharing

The social capital perspective has been espoused as an appropriate framework for identifying the conditions necessary for effective knowledge sharing. In order to understand the social capital perspective and its relevance to knowledge sharing, it is important to review how the concept has evolved theoretically and has been applied practically. This section aims to briefly explore the development of social capital as a concept to emphasise its importance for effective knowledge sharing. The intention is not to provide a comprehensive account of how social capital has evolved, but rather highlight the major contributions that have formed the current concept and its application.

The notion of social capital can be traced back to classical times, where many scholars have noted and discussed the importance of involvement and participation in groups and the consequences for both the individuals and the communities (cf. List, Durkheim and Marx). Hanifan (1916, cited in Woolcock 1998) first coined the term social capital with a meaning similar to its use today. Later, Jacobs (1961) used the term social capital to describe the relational links between individuals in community social organisations that can be leveraged for the further development of both individuals and the communities (Tsai and Ghoshal, 1998).

The contemporary view of social capital is often credited to Bourdieu (1985) who defined the concept as "the aggregate of the actual or potential resources which are linked to possession of a durable network of more or less institutionalised relationships of mutual acquaintance or recognition" (Bourdieu, 1985, p. 248). Further to Bourdieu's definition of social capital, he asserted that "the profits which accrue from membership in a group are the basis of the solidarity which makes them possible" (Bourdieu, 1985, p. 249). This definition suggests that social capital consists of two components. First, the relationship that one has with others provides them access to the resources possessed by them, and second, that associates of an individual have resources, which are of value to the individual.

Although Bourdieu (1985) was the earliest scholar credited with the origin for the contemporary view of social capital, some have credited Coleman (1988) with this tribute. Coleman's essays on social capital were published in the *American Journal of Sociology*, while Bourdieu's work was published as "Provisional Notes" in the *Actes de la Recherche en Sciences Sociales*. Consequently Coleman's work gained widespread attention and provided visibility to the concept in American sociology.

Coleman emphasised the importance of social capital for the acquisition of human capital and the establishment of social norms for the protection of the community against rogue behaviours of individuals. Coleman's work on social capital was strongly influenced by Loury (1977), who mentioned tentatively the role of social capital in the creation of human capital. In his refined analysis of social capital, Coleman also acknowledges the contributions of Ben-Porath, an economist, and Nin Lin and Mark Granovetter, who are sociologists. Granovetter and his work had an enormous influence on the practical study of social capital.

Granovetter's (1973) work takes a larger perspective of the relationships between individuals in social groups. His work is referred to as embededness, which postulates that social ties among individuals enable, shape and constrain the economic actions undertaken by organisations. Coleman and Loury asserted that dense networks with strong ties among individuals are a necessary condition for the emergence of social capital (Portes, 1988). Where strong ties are close relationships between actors in a group with strong social norms.

Burt (1992) applied further the theory of embededness to the study of social capital. He investigated the relationships between employees of a firm and their external contacts.

Burt viewed social capital as pertaining to the relationships with "friends, colleagues, and more general contacts through whom you receive opportunities to use your financial and human capital" (Burt, 1992, p. 9). This definition of social capital has an external focus on the relationships between organisational members and their access to peripheral networks. Counter to Coleman, who argued for strong ties and network closure, Burt asserted that social capital emerges from "structural holes", where weak ties are established outside the network. These weak ties create a competitive advantage for an individual whose relationships span the holes. They can broker the flow of information between the groups and control the projects that bring together people from opposite sides of the hole. Structural holes separate non-redundant sources of information.

Nahapiet and Ghoshal (1997) formally linked the concept of social capital to the creation and sharing of knowledge. They defined social capital in terms of three separate but interrelated dimensions and explored they way that each facilitates the combination and exchange of knowledge. The three dimensions are referred to as: structural, relational and cognitive. The structural and relational dimensions were developed based on Granovetters distinction between structural and relational embeddedness (Tsai and Ghoshal, 1998). The cognitive dimension was developed from what Coleman described as "the public good aspect of social capital" (Coleman, 1990, p. 315). The cognitive dimension is a shared understanding of collective goals and proper ways of acting in the social system (Tsai and Goshal, 1998).

Structural Dimension

As mentioned, the structural dimension of social capital refers to the structural configuration of ties (i.e., relationships) between individuals. Such ties are used to identify individuals with desired knowledge who are in the network (Lesser and Cothrel, 2001), and enables access to the people and their knowledge resources (Hazleton and Kennan, 2000). This dimension is influenced by such things as: network density, connectivity, and referrals.

Network connectivity is the degree to which actors within a network can interact. Therefore the number of ties between members of a network is closely associated with network connectivity. Network ties function as a medium for developing shared customs, values, attitudes and standards of behaviours (Andrews, Basler, and Coller, 1999). An individual with a high number of network ties will have access to a greater number of resources and a wider variety of influences. An associated concept is the centrality of an individual. Central actors often have a high number of network contacts spanning disparate groups. It is a powerful position as they are situated at a confluence of relationships and are best situated to influence the behaviours and attitudes of others (Andrews et al., 1999).

Another structural element is network proximity, which assists in the creation of similar identities, perceptions and judgments between network members (Andrews et al., 1999). Proximal network members are characterised by their frequent and substantial contact with each other. They adopt shared behaviours to each other that could be differentiated from those of the network as a whole (Andrews et al., 1999). Essentially proximal network members attain structural equivalence, which occurs when participants share similar

Relational Dimension

The relational dimension refers to various social dynamics including trust, shared norms, interpersonal obligations and expectations. Arenius (2002, p. 64) stated that "social capital exists in a relationship between two actors if they develop personal bonds, attachment and trust." Perry, Cavaye, & Coote (2002) identified some core attributes of relationship bonds. They found that trust, commitment, equity, conflict resolution, and benevolence shaped the relationships within an interorganisational network. An individual is more likely to commit to a relationship if they perceive the existence of trust and benevolence. Moreover, a sense of equity (sharing of benefits and costs) will result in increased trust and decreased conflict. Thus, the formation of effective and stable relationships relies on these relational dimensions.

Likewise, Arenius (2002) found that the relational dimension was evident in patterns of behaviour. This includes the willingness and capacity to cooperate and coordinate. Furthermore, the relational dimension encompasses embedded network resources such as reputation, credibility and trustworthiness. These attributes are used by network members to express others' patterns of behaviour.

Cognitive Dimension

The cognitive dimension refers to the possession of a shared context. It has also been referred to as a shared thinking and language, values, experience, and culture (Nahapiet and Ghoshal, 1998; Snowden, 2002). Arenius (2002), provides an expanded definition, in which the cognitive dimension includes shared representations, interpretations, and systems of meaning embodied in shared codes, languages and narratives. A shared context can both assist in the emergence of social capital and enable access to and utilization of resources. The cognitive dimension (along with the relational dimension) is an indicator of the quality of a relationship (Arenius, 2002).

Context does not refer to a fixed set of surroundings but an extensive dynamic process (Augier, Shariq, and Thanning Vendelo, 2001). Context is created from our knowledge, experience, understanding, desires, practical interests, values, emotions and culture (Augier et al., 2001; Heyman, 1994; Malhotra, 2001; Snowden, 2002). It is both constructed through the process of interpretation (Augier et al., 2001; Heyman, 1994) and used to interpret new situations (Augier et al., 2001). The cognitive dimension is required to make sense of the knowledge exchange. It serves to reconstruct meaningful representations of the knowledge so that it retains the necessary context for the knowledge to remain useful. Shared contexts resultant of the cognitive dimension provide meaningful communication that enables the exchange and combination of knowledge between and within the subjects.

Interdependencies of the Dimensions

Although these social capital dimensions are important in their own right, they are not mutually exclusive. They are highly interrelated (Nahapiet and Ghoshal, 1997). Tsai and Ghoshal (1998), investigated the associations between the dimensions, they found:

- Social interactions (structural) generate and strengthen trusting relationships (relational). Strong ties allow individuals to develop knowledge of each other and share points of view. Therefore an individual connected by strong ties is perceived as more trustworthy.
- A trusting relationship (relational) implies that individuals share common goals and values (cognitive). Such an alignment and harmony of interests diminishes the chance of opportunistic behaviour that reduces trust. Individuals who share common goals are less likely to act in a contrary or self-interested manner.
- Social interaction (structural) enables the formation of a common set of goals and values (cognitive) among individuals. Through interaction, individuals learn and adopt the languages, values and visions of others in the social system.

Conditions for Knowledge Exchange

Johannessen, Olaisen, and Olsen (2002), found that there are four conditions that need to be fulfilled for any exchange to take place. They are:

- Accessibility: the opportunity must exist to make an exchange
- Anticipation: the parties in the exchange must anticipate the creation of value
- Motivation: parties must be motivated. This occurs when they feel that the exchange will provide them with some benefits.
- Capability: they have the capability to execute the exchange

Furthermore, both Nahapiet and Ghoshal (1997) and Tsai and Ghoshal (1998), make the significant connection between these conditions and social capital.

Structural

People are bounded rational. That is that they have limits in their problem solving and information processing abilities. An individual cannot possess all the knowledge and information they require to solve problems and perform tasks. Thus, knowing *who knows what* is important (Vinding, 2002). Individuals who are highly connected are more likely to know someone with the knowledge they desire and therefore have better access to resources. These individuals are also more likely to expand their knowledge of who knows what. A major influence on *accessibility* here is the reduction of search costs (Productivity Commission, 2003).

Relational

When ties are rich in trust, the parties are more willing to partake in knowledge exchange. Trust has been found to be an antecedent for cooperation, as it is a type of expectation that alleviates fear of opportunistic behaviour (Tsai and Ghoshal, 1998). Feelings of trust are an expression of confidence that another will adhere to social norms and meet their obligations. Norms are mechanisms that ensure parties behave in a particular manner, while obligations ensure parties will undertake certain actions or activities in the future.

The expectation that another will abide by certain norms and obligations and decline to partake in opportunistic behaviour is alternatively known as benevolence based trust. That is, in the absence of immediate returns, the knowledge donor can expect to receive some personal benefits sometime in the future. This mechanism may provide *motivation* for exchange (Nahapiet and Ghoshal, 1997).

Sometimes benefits are derived from a relationship, not from the benevolence of the partner but from their own self-interest (Productivity Commission, 2003). It is the competence of the individual which enables them to deliver value. Their ability to effectively deliver usable and timely resources defines their competence. When dealing with a competent partner an individual is likely to be confident that the transaction will result in value, thus *anticipation* is effected.

Cognitive

Individuals use metaphors and narratives to reference and transcend contexts. This enables the establishment of shared meanings and interpretations of events. McKenzie (2002, p. 155) described a process where shared language, cultural norms, and shared mental models were used to funnel and condense complex knowledge into a form that allows the recipient to understand the knowledge and the context within which it has arisen. The recipient needs a sufficient context to interpret and reconstitute the received knowledge into a form that resembles the intended meaning (Shariq, 1998, p. 243). Thus, cognitive social capital provides the *capability* to effectively execute the exchange.

Individuals who develop common experience and communication history will acquire a knowledge base about the other individual (i.e., their context). This enables them to tailor their messages to that person and leverage their shared understanding. Therefore they are more likely to use "cues relevant to him or her, and containing information having a richer meaning for that communication partner (for instance, referring to shared experiences or using shared jargon)" (Carlson and Zmud, 1999, p. 156).

All three dimensions of social capital develop only after repeated and active social interactions over time (Vinding, 2002). This is especially true for cognitive social capital. Mutual interactions (outcome of structural social capital) and a desire to continue interactions (outcome of relational social capital) enable the development of shared knowledge. Continued interactions over time (only possible when structural and relational social capital are present) cause shared knowledge to evolve and become more refined and specific (Fiol, 2002).

Costs of Social Capital

Despite the benefits attributed to social capital there are some costs incurred when creating social capital. Resources and effort are required to maintain strong ties and locate weak ties. Maintaining ties can be time consuming and the reciprocal benefits are not always apparent. The time spent maintaining ties can be at the expense of more fruitful ventures.

Although both strong and weak ties can deliver significant benefits, the extent is reliant upon the arrangement of the other dimensions of social capital (Adler and Kwon, 2002, p. 13). Moreover, some expressions of social capital have the potential to restrict access to resources. A report by the Productivity Commission (2003) introduces the notion that social capital cannot be thought of as an unqualified good. Rather some costs and problems of social capital are mentioned. Some cohesive groups, often characterised by strong ties, can lead to conformity, by excluding influences and ideas from outside the group. Furthermore a group bound by strong ties without the necessary benevolence can be detrimental to the wider network.

Trust can make knowledge transfer less costly. Improved outcomes can be achieved when strong trusting ties are present. When only weak ties exist trust alone can substitute for attributes of strong ties to ensure the receipt of useful knowledge (Levin, Cross and Abrans, 2002, p.6). Bradach and Eccles claimed that "trust is a type of expectation that alleviates the fear that one's exchange partner will act opportunistically" (1989, p. 104) and take advantage of knowing about the vulnerability of the partner.

Through the development of trusting relationships within networks, particular actors build reputations as being trustworthy. These reputations serve as important information for other actors seeking sources of knowledge. However where the trustworthiness of a central actor is based primarily on benevolence trust with little regard for competence, the quality of the information and knowledge accessed could be significantly less than the optimal available form alternative actors. Therefore the reliance on trustworthiness can be detrimental to achieving high value outcomes.

Furthermore the rouge behaviour of an individual attributed with benevolence based trust, is unexpected where social capital is present. This is due to the co-establishment of the norm of generalized reciprocity, which acts to resolve collective action and binds communities (Adler and Kwon, 2002). However, due to the establishment of social norms and trust among actors, the network is vulnerable to opportunistic behaviour. Trust is established as an antecedent condition for cooperation. It is a type of expectation that is manifested to alleviate fear of opportunistic behaviour. Such a reliance on trust can be naïve given that networks are comprised of individuals with their own agendas and aspirations.

Within the context of social capital an important paradox exists that must be considered. It has been established that weak ties are the conduit to new information and knowledge. However the willingness of the network to incorporate and act upon this new knowledge is dependent on their capacity for change. Where the shared context in a network is strong and insufficient structural holes to peripheral networks are present the capability for sense making of new ideas is diminished.

So far, in this chapter we have defined social capital and how it is important to knowledge exchange and combination. From this discussion we have found some undeniable strengths of social capital for knowledge sharing. Like other forms of capital its acquisition and maintenance requires the investment of further resources, which can be costly, particularly if an over emphasis is placed on a single dimension without considering the purpose for having social capital and the intent of leveraging it. Therefore the appropriate form of social capital required in a network is dependent on the right mix of structural, relational and cognitive dimensions cognisant of the organisational context. Careful consideration must be given to how and why social capital is created and leveraged. The next section aims to discuss how social capital can be leveraged in the context of clinical healthcare settings.

Social Capital and Clinical Knowledge Management

Clinical care aims to achieve best practices. Such best practices are founded in having access to the best and most relevant information that includes the newest medical and scientific knowledge and which accounts for patient variability. Such knowledge is created by clinicians sharing their knowledge with others both within and across organisations. To facilitate the communication and sharing of such knowledge, the clinicians must record their discoveries and findings in the repositories accessible to others who would value such information.

For this outcome to be achievable clinicians are required to cooperate (Brakensiek, 2002) and communicate with others. This includes organisational management, clinicians, and all other staff that provide services to the patient. Through cooperation and communication the most comprehensive care for the patients will become more achievable (Van Beveren, 2002). The current culture of autonomy and "silos of knowledge" will need to be replaced. The new culture will emphasize cooperation, communication, education, and team building.

Healthcare sector organisations can be presented as simple environments for knowledge management that are so conducive to knowledge management that their existence has and continues to depend on it. The professions that combine to offer healthcare have their own set of values and directions often stated in a code of ethics. Government policy and legislation also offers direction and guidelines for those practicing professionals. The combination of all these sources offers a set of goals and strategic direction clearly understood and conveyed to all who provide healthcare delivery.

The techniques often stated in knowledge management as good human resource management techniques, which include providing a compelling vision and architecture with a coherent framework for guiding management decisions, together with appropriate bottom line measures, are applicable to both private and public sector organisations. The source of these techniques might need to be created within private sector organisations, while for public organisations they are provided from external sources, such as the community, government policy and stakeholders, or internally through the shared vision of the professionals that bring their professional codes of conduct to the organisation (Van Beveren, 2002).

Schneider (1993) analysed the effects of ideology on organisational structure and motivational consequences using the medical and community models of medical health delivery. For the medical model, decision-making is centralized and duties are specialized, whereas the community model practices decentralized decision-making and generalized functions. She found that the medical model of organisational design could slow down decision-making, cause passivity in patients and staff, and a fragmented treatment approach. Although the community model encourages teamwork, it could also lead to confusing responsibilities, authority, and job requirements. Therefore, it is unclear which model would be suitable for healthcare organisations as both are problematic. Schneider concluded that confusion resultant of either model results in an organisational structure being dictated by ideology, thus affecting the personal and professional identity, competence, responsibility, accountability, satisfaction, and motivation of patients and the medical staff. This finding has important implications for the development of organisational designs conducive to knowledge transfer and creation within healthcare organisations.

One might argue that the sharing and creation of knowledge in healthcare organisations need not alter the hierarchical structure, but creating informal network structures that overlay the existing structure might suffice. This might be achieved by establishing cross-disciplinary teams to work through apparent problems within the organisation. However this might also become an artificial situation that is overshadowed by the existing cultures of the organisation. Whilst many have identified that "socialization" activities support knowledge flow and the generation of new ideas and knowledge within private-sector organisations (Skyrme, 1998), it might be less simple in public sector organisations where power structures and hierarchical levels are culturally embedded.

Given that each healthcare organisation has its own unique culture and structure, the need to create an appropriate mix of social dimensions is paramount to facilitate effective knowledge sharing among its members. Therefore the development of strategies to further foster knowledge sharing and the management of knowledge an organisation has available, must be done cognisant of many contextual factors and usually one approach is not viable. Rather combinations of integrated strategies are necessary to provide an environment conducive to knowledge exchange between members of the organisation and with others external to the organisation, whish is necessary to renew and revise the current knowledge to achieve the outcomes of "best practice".

Social Capital and its Importance for the Sustainability of Meditute

According to Hansen, Nohria and Tierny (1999) the strategy of an organisation to manage knowledge leads to two approaches: a codification approach—capturing knowledge for many individuals for re-use by many others—suitable if your product or service is standardised; and a personalisation approach—relying on individuals sharing their intuition and know-how to create innovative or customised products and solutions. In this section we make the argument that the codification approach in isolation is useless, and that when a codification approach is adopted additional effort must be expended on the personalisation approach. In particular we will discuss the need for both approaches in developing and ensuring the sustainability of Meditute.

We aim to build on the foundation established in the previous sections, where the argument for the social capital perspective in clinical knowledge management has been established. The social capital perspective permits us to adopt a holistic approach to investigate the necessary conditions for effective knowledge sharing. An important contribution of the social capital perspective is the network analysis features that enable us to consider the input of individuals and the communication between them irrespective of whether such communication is in person or via a communication medium, such as online or via the telephone.

It is important to acknowledge at this point the fact that there are two extreme groups of thinking regarding the role of technology in relation to knowledge sharing. There are those who assert that no "real" social relationships can possibly be mediated by computer communication, and there are those who advocate the use of networks to create ideal democratic and transparent communication (Paccagnella, 2001). Adoption of the network approach via the social capital perspective permits us to avoid alignment with one of the extreme views or the other and adopt a one-dimensional and simplistic perspective.

Knowledge sharing within clinical healthcare organisations is complex, and calls for a complex approach to understand how the introduction of a computer mediated environment can help facilitate knowledge exchange. Previous studies (Garton, Haythornthwaite, & Wellman. 1998; Pickering and King, 1995; Wellman and Gulia 1999) have applied the theory of the strength of weak ties to the development of relationships over computer-mediated environments.

To foster social capital, some firms have adopted collaborative technologies such as shared knowledge repositories, for example Meditute. Adopting technology such as Meditute provides opportunity for knowledge exchange, however other factors such as motivation, ability and anticipation of value need to be considered. Furthermore efforts must be made to nurture environments conducive of providing impetus for participation with the technology. Lesser (2000) suggests that technologies assist in the establishment of more social ties, but building social capital requires also nurturing motivation and providing resources.

Environments hosted on the Internet provide opportunities for individuals to access weak ties (Lea and Spears, 1995). However the derivation of value from access to those ties relies on the establishment of trust and shared understanding of contexts. Often interpersonal relationships that are necessary for establishing trust and shared contexts are integrated through contacts that occur through several different media. Parks and Floyd (1996) and Hamman (2001) demonstrated people who meet via computer-mediated environments and form relationships often attempt to further develop stronger ties by meeting face to face.

Therefore building social capital to enhance participation in computer-mediated environments requires the establishment of opportunities to engage in multiple forms of contact. The creation of both strong and weak ties is essential for ensuring that participation is sustained through continual and renewed input. The information contained within the repository can only attract participation from clinicians if it is current and will provide value.

In this section, we have argued for the social capital perspective for investigating the adoption of technology to facilitate information exchange for knowledge sharing. Specifically, we identified that the computer-mediated environments are not substitutes for traditional forms of communication if social capital is to be established. Rather, computer-mediated environments provide opportunities for meeting potential contacts to establish relationships via tradition media such as telephone or mail and eventually in person. Furthermore, we established the need to consider further the formation of social capital cognisant of the goals and objectives of establishing Meditute in the first place. Next, we outline some important issues that must be considered when adopting technology for the sharing of information for knowledge exchange. Although these issues were formed from considering Meditute in the context of an Australian hospital, the issues could be equally applicable in other organisations or for other computer mediated environments for that matter.

Issues to Consider in Adopting Computer-Mediated Communication for Knowledge Sharing

This section outlines some key actionable strategies for implementing the Meditute system and improving the likelihood that it will be sustainable through continued participation and new input. The following points have been separated into those relevant to encouraging participation through providing incentives and those relevant to building social capital necessary for establishing value within the system from new input and refinement of content.

Motivation for Participation

- Demonstrate senior management "buy-in" for the system. The demonstration must come initially from recognising the value of the knowledge sharing activity. By making the knowledge sharing activity an explicit responsibility of staff ensures that it is incorporated into the values and culture of the organisation.

- Obviously physical access to the technology that connects participants to the Meditute program is necessary for participation to occur. Less obvious but nevertheless important is that the location for such access provides for convenience and timeliness to the participant needing to retrieve the information or make contributions.

- Rewards and incentives should be provided to those who participate and contribute to the success of the system. Such contributions can be the sharing of one's own knowledge or the establishment of weak ties which results in knowledge sharing.

- Awareness of the system and its value is important to achieve participation. Therefore, time must be set aside for those interested in participating with the system. Specific time should be allocated to learn how to use the system and make contributions.

- Mentoring and guidance is important for those wishing to make contributions. Provisions should be made for those wishing to contribute to the system to have access to a mentor. This person will share their understanding of the system and explain the need for common language and how information is classified to enable effective communication via the system.

Social Capital Building

- In existing social networks, key central actors should be identified and targeted to become champions of the system. This may require initial involvement by such people in developing the system and its implementation, as they must believe in the goals and objectives. The role of these champions is to advocate participation with the technology.

- The key users of the system and champions should be encouraged to attend conferences and other semi-social events to promote the systems and recruit and establish weak ties. These champions need to be armed with the success stories that have resulted in positive knowledge sharing and outcomes.

- As new knowledge is added to the system, participants need to be made aware of the additions. Such awareness increases the perception of value for the system and reminds them of the system as a resource. The notification that new knowledge has been added is achieved by email correspondence and alternative communications.

- Provide clear rules on the operation of the community. This involves both how the knowledge contributed to the system is evaluated and how the knowledge in the system is to be used. Open and transparent processes for the review, evaluation and revision of contributions are important for maintaining trust. Due acknowledgement of sources and the contributions made ensures that trust in the gatekeepers for the knowledge repository is maintained, but also encourages participation as members of the community establish healthy competitive rivalry.

To locate champions for the system, the existing social networks need to be understood and explored. Such exploration might involve mapping the network and perhaps auditing the existing social capital prior to development or implementation of the system. Without "taking stock" of the existing social capital available, the goals and objectives for the implementation of the system are impossible to articulate as such a system needs to fill a need for it to be successful.

For Meditute, the need arose from the strains on human capital due to the absence of staff during key moments after hours when situations occurred where the knowledge held by such staff was not available. Although the Meditute system is intended to provide access to such information and knowledge, its success in this matter is dependent on contributions of those with the knowledge and access to the knowledge by those who need it. As mentioned in this chapter such actions will only result when there is clearly opportunity to access the system, motivation to contribute, ability to contribute and interpret the knowledge, and anticipation that the knowledge will be valuable to meeting the need.

Conclusions

The social capital perspective is useful for investigating the necessary conditions for knowledge sharing. It provides a holistic framework from which to evaluate the important aspects of social network structures and associated relationship qualities and cultural values. Together, these dimensions form the concept of social capital.

In this chapter, we explained in detail the concept of social capital. We then demonstrated the relevance of social capital both clinical knowledge management. Furthermore, we illustrated how careful consideration in developing the dimensions of social capital is necessary when adopting a computer-mediated environment to provide opportunities for knowledge exchange. The importance of social capital as a direct effect on knowledge sharing was established. In addition, we established the importance of social capital to ensuring the sustainability of the computer-mediated environment of Meditute, which is reliant on the continual establishment of social capital as a source of information and participation.

References

Abou-Zeid, E. (2002). A knowledge management reference model. *Journal of Knowledge Management, 6*(5), 486-499.

Adler, P.S. & Kwon, S. (2002). Social capital: Prospects for a new concept. *Academy of Management Review, 27*(1), 17-46.

Allee, V. (2000). Reconfiguring the vale network. *Journal of Business Strategy, 21*(4), 36-39.

Andrews, S.A., Basler, C.R. & Coller, X. (1999). Organizational structures, cultures and identities: Overlaps and divergences. In S. Andrews & D. Knoke (Eds.), *Networks in and around organizations* (Vol. 16). Stamford: JAI Press.

Arenius, P.M. (2002). *Creation of firm level social capital: Its exploitation and the process of early internationalization.* Unpublished doctoral dissertation, Helsinki University of Technology, Helsinki.

Augier, M., Shariq, S.Z., & Thanning Vandelo, M. (2001). Understanding context: Its emergence, transformation, and role in tacit knowledge sharing. *Journal of Knowledge Management, 5*(2), 125-136.

Bourdieu, P. (1985). The forms of capital. In J. G. Richardson (Ed.), *Handbook of theory and research for the sociology of education* (pp. 241-258). New York: Greenwood.

Bradach, J.L. & Eccles, R.G. (1998). Markets versus hierarchies: From ideal types to plural forms. *Annual Review of Sociology, 15*, 97-118.

Brakensiek, J.C. (2002). Knowledge management for EHS professionals. *Occupational Health and Safety, 71*(1), 72-74.

Burt, R.S. (1992). *Structural holes: The social structure of competition.* Cambridge, MA: Harvard University Press.

Carlson, J.R. & Zmud, R.W. (1999). Channel expansion theory and the experiential nature of media richness perceptions. *Academy of management Journal, 42*(2), 153-171.

Coleman, J.S. (1998). Social capital in the creation of human capital. *American Journal of Sociology, 94*(Supplement), s95-s120.

Davenport, T.H. & Prusak, L. (1998). *Working knowledge: How organisations manage what they know.* Boston: Harvard Business School Press.

Fiol, C.M. (2002). Intraorganizational cognition and interpretation,. In J. A. C. Baum (Ed.), *The Blackwell companion to organizations.* Oxford: Blackwell.

Garton, L., Haythornthwaite, C., & Wellman, B. (1999). Studying online social networks. In S. Jones (Ed.), *Doing Internet research: Critical issues and methods for examining the net.* Thousand Oaks, CA: Sage.

Granovetter, M. (1973). The strength of weak ties. *American Journal of Sociology, 78*(May), 1360-1380.

Hamman, R. (2001). Computer networks linking network communities. In C. Werry & M. Mowbray (Eds.), *Online communities.* Upper Saddle River, NJ: Prentice-Hall.

Hansen, M., Nohira, N., & Tierney, T. (1999). Whats your strategy for knowledge management. *Harvard Business Review,* March/April, 104-116.

Hazleton, V. & Kennan, W. (2000). Social capital: Reconceptualizing the bottom line. *Corporate Communications: An International Journal, 5*(2), 81-86.

Heyman, R. (1994). *Why didn't you say that at the first place? How to be understood at work* (1st ed.). San Francisco: Jossey-Bass Publishers.

Johannessen, J.A., Olaisen, J., & Olsen, B. (2002). Aspects of a systemic philosophy of knowledge: From social facts to data, information and knowledge. *Kybernetes, 31*(7/8), 1099-1120.

Katzenbach, J. R., & Smith, D. K. (1993). *The wisdom of teams: Creating high performance organisations*. Boston: Harvard Business School Press.

Kogut, B. & Zander, U. (1992). Knowledge of the firm, combinative capabilities, and the replication of technology. *Organization Science, 3*(2), 383-397.

Lea, M. & Spears, R. (1995). Love at first byte? Building personal relationships over computer networks. In J. Wood & S. Duck (Eds.), *Under-studied relationships: Off the beaten tract*. Thousand Oaks, CA: Sage.

Lesser, E. & Cothrel, J. (2001). Fast friends: Virtuality and social capital. *Knowledge Directions*, Spring/Summer, 66-79.

Lesser, E. L. (2000). Leveraging social capital in organisations. In E. L. Lesser (Ed.), *Knowledge and social capital: Foundations and applications* (pp. 3-16). Boston: Butterworth-Heinemann.

Levin, D.Z., Cross, R., & Abrams, L. C. (2002). Why should I trust you? Predictors of interpersonal trust in a knowledge transfer context. Online www.kmadvantage.com. KM Advantage.

Maanen, J.V. & Barley, S. R. (1984). Occupational communities: Culture and control in organisations. *Research in Organizational Behaviour, 6*, 287-365.

Malhotra, Y. (2001). *Knowledge management for the new world of business* [Online Article]. BRINT Institute. Retrieved 21/03/03, from www.brint.com

McKenzie, K.M. (2002). *Exchanging payload knowledge: Interpersonal knowledge exchange within consulting communities of practice*. Unpublished doctoral dissertation, Swinburne University of Technology.

Nahapiet, J. & Ghoshal, S. (1997). Social capital, intellectual capital and the creation of value in firms. *Academy of Management Best Paper Proceedings* (pp. 35-39).

O'Dell, C. & Grayson, C.J. (1998). If only we knew what we know: Identification and transfer of internal best practices. *California Management Review, 40*(3), 154-174.

Paccagnella, L. (2001). Online Community Action: Perils and Possibilities. In C. Werry & M. Mowbray (Eds.), *Online communities*. Upper Saddle River, NJ: Prentice-Hall.

Parks, M.R. & Floyd, K. (1996). Making friends in cyberspace. *Journal of Computer-Mediated Communication, 1*(4).

Perry, C., Cavaye, A. & Coote, L. (2002). Technical and social bonds within business to business relationships. *Journal of Business and Industrial Marketing, 17*(1), 75-88.

Pickering, J.M. & King, J.L. (1995). Hardwiring weak ties: Interorganizational computer-mediated communication, occcupational communities, and organizational change. *Organization Science, 6*(4), 479-485.

Portes, A. (1998). Social capital: Its origins and applications in modern sociology. *Annual Reviews Sociology, 24*, 1-24.

Productivity Commission. (2003). *Social capital: Reviewing the concept and its policy implications* (Research Paper). Canberra, Australia: AusInfo.

Schneider, S.C. (1993). Conflicting ideologies: Structural and motivational consequences. *Human Relations, 46*(1), 45-64.

Shariq, S.Z. (1998). Sense making and artefacts: An exploration into the role of tools in knowledge management. *Journal of Knowledge Management, 2*(2), 10-19.

Skryme, D.J. (1998). Developing a knowledge strategy. *Strategy,* (January), 18-19.

Snowden, D. (2002). Complex acts of knowing: Paradox and descriptive self-awareness. *Journal of Knowledge Management, 6*(2), 100-111.

Tsai, W. & Ghoshal, S. (1998). Social capital and value creation: The role of intrafirm networks. *Academy of management Journal, 41*(4), 464-478.

Van Beveren, J. (2003). Does health care for knowledge management? *Journal of Knowledge Management, 7*(1), 90-95.

Vinding, A.L. (2002). *Interorganizational diffusion and transformation of knowledge in the process of product innovation.* Unpublished PhD thesis, Aalborg University, Aalborg.

Wellman, B. & Gulia, M. (1999). Net surfers don't ride alone: Virtual communities as communities. In P. Kollock & M. Smith (Eds.), *Communities in cyberspace.* New York: Routledge.

Woolcock, M. (1998). Social capital and economic development: Toward a theoretical synthesis and policy framework. *Theory and Society, 27*, 151-208.

Glossary

Accuracy of Information
This principle states that personal information should be as accurate, complete, and up-to-date as is necessary for the purposes for which it is to be used.

Artificial Intelligence
A broad term describing the field of developing computer programs to simulate human thought processes and behaviors.

Business Process Redesign
An examination of key business processes in order to identify radical changes to achieve improved results.

Cancer Information Service
The National Cancer Institute's link to the public, interpreting and explaining research findings in a clear and understandable manner, and providing personalized responses to specific questions about cancer.

Case Study
For the purposes of this text, a detailed analysis of a person or group from a social, psychological, or medical point of view.

Clinical Guidelines

Systematically developed statements to assist medical and patient decisions about appropriate healthcare.

Clinical Information System

An information system that can track individual patients as well as populations of patients. The care team can use the system to guide the course of treatment, anticipate problems, and track progress.

Clinical Protocol

A medical protocol intended to ensure and evaluate the efficacy of a new medical device and/or procedure.

Community of Practice

Networks of people who work on similar processes or in similar disciplines and who come together to develop and share their knowledge in that field for the benefit of both themselves and their organizations.

Computer-Based Training

Training carried out by interaction with a computer. The programs and data used are often referred to as "courseware".

Currency of Information

A term used to assist practitioners and healthcare professionals to determine how up to date an information source is.

Data Analysis

Typically used to sort through data in order to identify patterns and establish relationships.

Data Integrity

A state when data is unchanged from its source and has not been accidentally or maliciously modified, altered, or destroyed.

Data Model

The product of the database design process which aims to identify and organize the required data logically and physically.

Data Semantics
A connection from a database to the real world outside the database.

Data Sharing
Sharing and disseminating data with colleagues and collaborators, international entities, or making data available to the wider public.

Data Warehousing
A generic term for a system for storing, retrieving, and managing large amounts of any type of data. Data warehouse software often includes sophisticated compression and hashing techniques for fast searches, as well as advanced filtering.

Database
An organized body of related information.

Database Applications
One or more large structured sets of persistent data, usually associated with software to update and query the data.

Decision Support Systems
Information technology and software specifically designed to help people at all levels of the company make decisions.

Dicom
Digital Imaging and COmmunications in Medicine—a standard developed by the American College of Radiology—National Electrical Manufacturer's Association for communications between medical imaging devices. It conforms to the ISO reference model for network communications and incorporates object-oriented design concepts.

Distributed Healthcare Environment
A middleware which constitutes the basic functional infrastructure of the healthcare center, independent from the technological requirements of the organization.

Electronic Health Record
A secure, real-time, point-of-care, patient-centric information resource for clinicians. The EHR aids clinicians' decision-making by providing access to patient health record information where and when they need it and by incorporating evidence-based decision support.

Evidence-Based Medicine

Evidence-based medicine (EBM) involves integrating individual clinical experience with the best available external clinical evidence from systematic research when making decisions about patient care.

Expert System

A computer program developed to simulate human decisions in a specific field or fields. A branch of artificial intelligence.

Explicit Knowledge

Knowledge that can be shared by way of discussion or by writing it down and putting it into documents, manuals, or databases.

Health Informatics

The understanding, skills, and tools that enable the sharing and use of information to deliver healthcare and promote health.

Health Information Management

The planning, budgeting, control and exploitation of the information resources in a healthcare organization.

Health Information Network

Provides an integrative information resource, as well as training, service, and access to consumers and staff of important healthcare entities, such as treatment centers, service providers, consumer advocacy groups, and other affiliated organizations.

Healthcare

The prevention, treatment, and management of illness and the preservation of mental and physical well-being through the services offered by the medical and allied health professions.

Healthcare Information System

System consisting of the network of all communication channels used within a healthcare organization.

Health Care Infrastructure

Systematic provision of a society for the optimal well-being of its members.

Healthcare IS Research
Scientific investigation into systems and communication channels used within a healthcare organization.

Healthcare Standardization
Solving problems that are common across the healthcare communities.

Healthcare System
Organization by which an individual's health care is provided.

Hospital Information System
The aim of a HIS is to use a network of computers to collect, process and retrieve patient care and administrative information from various departments for all hospital activities to satisfy the functional requirement of the users.

Human Computer Interaction
The study of how humans interact with computers and how to design computer systems that are easy, quick, and productive for humans to use.

Information Processing
Sciences concerned with gathering, manipulating, storing, retrieving, and classifying recorded information.

Information Quality
Discerning which information sources are more useful and accurate than others.

Information Resource
Any entity, electronic or otherwise, capable of conveying or supporting intelligence or knowledge.

Information Retrieval
Actions, methods, and procedures for recovering stored data to provide information on a given subject.

Information Systems
The general term for computer systems in an organization that provide information about its business operations.

Interface

A connection (through a hardware device or through a software program) between different components of a computer system.

IT Capability

The capacity of IT to be used, treated, or developed for a specific purpose.

IT Management

The manner or practice of managing IT, specifically regarding handling, supervision, or control.

IT Strategy

An elaborate and systematic IT plan of action.

Knowledge Delivery

The act of conveying knowledge.

Knowledge Management

The creation and subsequent management of an environment which encourages knowledge to be created, shared, learned, enhanced, organized, and utilized for the benefit of the organization and its customers.

Knowledge Sharing

Mechanisms to communicate and disseminate knowledge throughout an organization.

Knowledge Worker

An employee whose role relies on his or her ability to find and use knowledge.

Knowledge-Based Systems

Computer programs designed to simulate the problem-solving behavior of human experts within very narrow domains or scientific disciplines—this discipline is a sub-set of Artificial Intelligence.

Learning Organization

An organization that views its success in the future as being based on continuous learning and adaptive behavior.

MIS

Information systems designed for structured flow of information and integration by business functions and generating reports from a database.

Neural Network

Computer architecture in which processors are connected in a manner suggestive of connections between neurons; can learn by trial and error.

Object-Oriented Approach

A design method in which a system is modelled as a collection of cooperating objects and individual objects are treated as instances of a class within a class hierarchy.

Organizational Learning

The ability of an organization to gain knowledge from experience through experimentation, observation, analysis, and a willingness to examine both successes and failures, and to then use that knowledge to do things differently.

Social Capital

Represents the degree of social cohesion which exists in communities. It refers to the processes between people which establish networks, norms, and social trust, and facilitate coordination and cooperation for mutual benefit.

Tacit Knowledge

The knowledge or know-how that people carry in their heads. Compared with explicit knowledge, tacit knowledge is more difficult to articulate or write down and so it tends to be shared between people through discussion, stories and personal interactions.

Virtual Community

Online meeting place for people. Designed to facilitate interaction and collaboration among people who share common interests and needs.

About the Authors

Rajeev K. Bali, BSc (Hons), MSc, PhD, PgC, SMIEEE, currently lectures and conducts research at Coventry University (UK). He is an invited contributor and reviewer for various international journals and conferences and is often called upon to deliver invited presentations at international conferences, workshops, and symposia. His involvement with the IEEE resulted in an appointment as publications chair for the Information Technology Applications in Biomedicine Conference in 2003. He is the invited guest editor of the special issue on Knowledge Management and IT in Healthcare for the *IEEE Transactions on Information Technology in Biomedicine*. He is the founder and head of the Knowledge Management for Healthcare (KMH) research subgroup and has a biographical entry in *Who's Who in the World*.

* * *

Juan Carlos Augusto graduated as Licenciado en Ciencias de la Computación from Universidad Nacional del Sur (Bahía Blanca - Argentina) in 1991 and then graduated as a PhD in computer science from the same university in 1998. In 2001, he left the Universidad Nacional del Sur and joined the Declarative Systems and Software Engineering Research Group at the University of Southampton as a research fellow. In 2003, he joined the School of Computing and Mathematics at the University of Ulster. He has produced about 50 research reports in the form of journal, conference, and workshop papers, edited volumes, or technical reports.

Alexander Berler was born in Lausanne, Switzerland in 1969. He received his degree in electrical engineering from the Aristotle University of Thessalonica, Greece (1995), and his MSc in biomedical engineering from the National Technical University of Athens,

Greece (1997). He is currently pursuing a PhD at NTUA in medical informatics focusing on the design and development of interoperable healthcare information systems towards the implementation of a citizen virtual medical record. He is working at Information Society SA as a project manager in large healthcare informatics government project co-funded by the European Union under the 3rd Community Support Framework.

Steve Clarke received a BSc in economics from The University of Kingston upon Hull (UK), an MBA from the Putteridge Bury Management Centre, The University of Luton (UK), and a PhD in human-centered approaches to information systems development from Brunel University (UK). He is professor of information systems and director of research for the University of Hull Business School. Steve has extensive experience in management systems and information systems consultancy and research, focusing primarily on the identification and satisfaction of user needs and issues connected with knowledge management. His research interests include: social theory and information systems practice; strategic planning; and the impact of user involvement in the development of management systems. Clarke's major current research is focused on approaches informed by critical social theory.

Ashish Dwivedi is currently a senior lecturer at Hull University Business School, UK. His primary research interest is in the application of information and communication technologies (ICT) and knowledge management (KM) paradigms on organizational decision-making, which was also his PhD research theme, received in 2004. One of the main areas of concern in KM is that there are no commonly accepted methodologies and a standard framework, despite the fact that the KM paradigm is recognized as an area of significant importance. He has additional interests in the use of ICT, data warehousing, decision support systems and intelligent data mining.

Huw Evans is a performance specialist and regional service lead for criminal justice with the Audit Commission, London. Prior to joining the Audit Commission, he served 30 years as a police officer with the Dyfed-Powys Police and the Hertfordshire Constabulary, being engaged in aspects of organizational development and operational policing. Huw's current role involves assessment of various aspects of public sector services. He has a specific interest in the use of group processes in decision-making and is working toward a PhD developing a framework for the transparent and critical mapping and evaluation of methodologies for the engagement of large numbers of people with Luton Business School, UK.

David Dagan Feng received his MSc in biocybernetics and his PhD in computer science from the University of California, Los Angeles (UCLA) (1985 and 1988, respectively). After briefly working as an assistant professor at the University of California, Riverside, he joined the University of Sydney as a lecturer, senior lecturer, reader, professor, and head of the Department of Computer Science. He is currently head of the School of Information Technologies and associate dean of the Faculty of Science at the University of Sydney. He is also honorary research consultant, Royal Prince Alfred Hospital, the largest hospital in Australia; and chair-professor of information technology, Hong Kong Polytechnic University. His research area is biomedical & multimedia information technology (BMIT). He is the founder and director of the BMIT Research Group. He has published more than 300 scholarly research papers, pioneered several new research directions, made a number of landmark contributions in his field with significant scientific impact and social benefit, and received the Crump Prize for Excellence in Medical Engineering from the United States. More importantly, however, is that many of his research results have been translated into solutions to real-life problems and have made tremendous improvements to the quality of life worldwide. He is a fellow of ACS, HKIE, IEE and IEEE, special area editor of *IEEE Transactions on Information Technology in Biomedicine*, and is the current chairman of IFAC-TC-BIOMED.

Aggelos Georgoulas was born in Arta, Greece in 1975. He received a degree in electrical & computer engineering from the National Technical University of Athens, Faculty of Electrical and Computer Engineering, Greece (1999). He is currently a PhD candidate in the same faculty. His expertise area is security in telemedicine. He has worked as a researcher for several EU-funded projects in the domain of heath informatics.

Spyretta Golemati received a degree in mechanical engineering in 1994 from the National Technical University of Athens, Greece, and an MSc and PhD in bioengineering from Imperial College, University of London (1995 and 2000, respectively). She then worked as a postdoctoral fellow in the National Technical University of Athens, Greece. She is currently a visiting associate professor at the Technical University of Crete. Her research interests include medical imaging, signal and image processing, biofluid mechanics, and surgical robotics. She has authored or co-authored more than 30 papers in refereed international journals, chapters in books and conference proceedings. She is a member of the Institute of Electrical and Electronic Engineers and the Technical Chamber of Greece.

Stuart Goose was born in 1968 and raised in the UK. He holds BSc (1993) and PhD (1997) degrees in computer science both from the University of Southampton, UK. He held a post doctoral position at the University of Southampton. At Siemens Corporate Research Inc. in Princeton (USA), he held a positions including a member of the technical staff, project manager and program manager in the Multimedia Technology Department. Currently, he is with Siemens Technology-To-Business Center in Berkeley, California. He has been an author on more than 40 technical publications and leads a research group exploring and applying various aspects of networking, mobility, multimedia, speech, and audio technologies. He serves as program committee member and reviewer for a number of ACM and IEEE conferences and journals.

J. David Johnson (PhD, Michigan State University, 1978) has been dean of the College of Communications and Information Studies at the University of Kentucky since August 1998. He has published three books: *Cancer-Related Information Seeking* (Hampton Press), *Information Seeking: An Organizational Dilemma* (Quorum Books), and

Organizational Communication Structure (Ablex). He has been recognized as among the 100 most prolific publishers of refereed journal articles in the history of the communication discipline, publishing articles in over 40 different refereed journals. His current research interests focus on information seeking, network analysis, innovation, and knowledge management.

Antonios Kordatzakis was born in Athens, Greece in 1976. He received his degree in electrical & computer engineering from the National Technical University of Athens, Faculty of Electrical and Computer Engineering, Greece (2003). He is currently a PhD candidate in the same school. His expertise area is DICOM and HL7 standards. He has worked as a researcher for several EU-funded projects in the domain of heath informatics and as a software analyst and developer in the domain of engineering.

Dimitris Koutsouris was born in Serres, Greece in 1955. He received his degree in electrical engineering in 1978 (Greece), DEA in biomechanics in 1979 (France), doctorate in genie biologie medicale (France), and doctorat d' etat in biomedical engineering in 1984 (France). Since 1986, he has performed research associated with USC (Los Angeles), Renè Dèscartes (Paris) and has served as associate professor at the School of Electrical & Computer Engineering at the National Technical University of Athens. He is currently a professor and head of the Biomedical Engineering Laboratory. He has published more than 100 research articles and book chapters and more than 150 conference communications. He is also president of the Greek Society of Biomedical Technology. Professor Koutsouris has been principal investigator in many European and National Research programs, especially in the field of telematics in healthcare.

Swamy Laxminarayan currently serves as the chief of biomedical information engineering at Idaho State University. Previous to this, he held several senior positions both in industry and academia. As an educator, researcher and technologist, Prof Laxminarayan has been involved in biomedical engineering and information technology applications in medicine and healthcare for over 25 years and has published more than 250 scientific and technical articles in international journals, books and conferences. Professor Laxminarayan is very actively involved in the IEEE activities for over 20 years. He is the founding editor-in-chief and editor *emeritus* of the *IEEE Transactions on Information Technology in Biomedicine*. He served as an elected member of the administrative and executive committees in the IEEE Engineering in Medicine and Biology Society and as the Society's vice president for two years. He currently serves as an elected member-at-large of the IEEE Publications Services and Products Board and on the administrative board of the International Federation for Medical and Biological Engineering. His contributions to the discipline have earned him numerous national and international awards. He is a fellow of the American Institute of Medical and Biological Engineering, a recipient of the IEEE 3rd Millennium Medal and a recipient of the Purkynje Award from the Czech Academy of Medical Societies, a recipient of the Career Achievement Award, numerous outstanding accomplishment awards and twice recipient of the IEEE EMBS distinguished service award.

About the Authors

Brian Lehaney is head of the Statistics and Operational Research Subject Group and professor of systems management at Coventry University, UK. His research is in the area of decision support for organizations in both the private and public sectors. It includes the theories and application of simulation modelling, intervention methodologies and knowledge management. Professor Lehaney has developed the mixed-mode modelling approach to decision support. He publishes widely in internationally renowned journals, including the *Journal of the Operational Research Society* and the *Journal of End User Computing*. His books include *Mixed-Mode Modelling: Mixing Methodologies for Organisational Intervention* (Kluwer). His latest book is titled *Beyond Knowledge Management* (2004, Idea Group). Professor Lehaney recently completed an EC-funded project on the development of tools and methodologies for knowledge sharing within organizations. Other related projects are in the process of development.

Efstathios (Stathis) Marinos was born in Patras, Greece in 1973. He received his degree in electrical & computer engineering from the National Technical University of Athens, Faculty of Electrical and Computer Engineering, Greece (1998). He is currently a PhD candidate in the same faculty. His expertise area is the electronic health record and medical informatics in general. He has also worked as a freelancer, developing software for several domains, including the health domain. He has wide experience in EU-funded projects in the domain of health informatics, where he has worked as a researcher. He has published seven papers in conference proceedings and one paper in journals.

George Marinos was born in Zakynthos, Greece in 1968. For the last 15 years he has been working as a system architect and developer. He is currently senior programmer in the Development Department of the National Bank of Greece. His expertise areas are the software engineering and system architectures. Since 2000 he has been the technical manager of Biomedical Engineering Laboratory of National Technical University of Athens (BEL). He has also worked for over 10 years as a freelancer; developing software for several domains, including the health domain. He has participated as technical consultant in a number of EU-funded research projects for the BEL.

Stavroula Gr. Mougiakakou was born in Hannover, Germany in 1974. She received her degree in electrical engineering and a PhD from the National Technical University of Athens, Greece (1997 and 2003, respectively). Her research interests include simulation of physiological systems, image processing, artificial neural networks, and genetic algorithms. She has co-authored more than 20 papers in international journals and conferences. Dr. Mougiakakou is a member of the Technical Chamber of Greece.

Raouf Naguib is head of the Biomedical Computing Research Group (BIOCORE) and professor of biomedical computing at Coventry University, UK. He has published more than 180 journal and conference papers and reports in many aspects of biomedical and digital signal processing, biomedical image processing and the applications of artificial intelligence and evolutionary computation in cancer research. He has also published a book on digital filtering, and co-edited a second book on the applications of artificial neural networks in cancer diagnosis, prognosis and patient management, which is his main area of research interest. Professor Naguib is a member of several national and international research committees and boards, and recently served on the administrative committee of the IEEE Engineering in Medicine and Biology Society (EMBS). He is actively taking part in a number of collaborative research projects with various partners and consortia in the UK (breast, colon, ovarian and urological cancers, and Hodgkin's disease), the EU (prostate and colorectal cancers), the U.S. (breast cancer and cancers of the oesophago-gastric junction) and Egypt (bladder cancer).

Konstantina S. Nikita (M'96-SM'00) received her degree in electrical engineering and her PhD degree from the National Technical University of Athens, Greece (1986 and 1990, respectively). She then received an MD from the University of Athens, Greece (1993). Since 1990, she has been working as a researcher at the Institute of Communication and Computer Systems, National Technical University of Athens. In 1996, she joined the Department of Electrical and Computer Engineering, National Technical University of Athens, where she is currently an associate professor. Her current research interests include medical imaging and image processing, biomedical informatics, health telematics, simulation of physiological systems, computational bioelectromagnetics, and applications of electromagnetic waves in medicine. Dr. Nikita has authored or co-authored 80 papers in refereed international journals and chapters in books, and more than 150 papers in international conference proceedings. She holds one patent. She has been the technical manager of several European and National Research and Development projects in the field of biomedical engineering. Dr. Nikita was the recipient of the 2003 Bodossakis Foundation Scientific Prize She is a member of the Institute of Electrical and Electronics Engineers, the Technical Chamber of Greece, the Athens Medical Association and the Hellenic Society of Biomedical Engineering.

Chris Nugent has a degree in electronic systems and a DPhil in biomedical engineering. Both were earned at the University of Ulster at Jordanstown. He is currently employed as a senior lecturer in the School of Computing of Mathematics at the University of Ulster. His research areas focus on the application of artificial intelligence to medical decision support systems, computerised electrocardiology, and the evolving usage of the Internet as a means of innovative healthcare delivery. He has published more than 80 articles in these research areas in journals, book chapters, and national and international conferences.

Jörg Ontrup studied physics and applied computer science in the natural sciences at the Faculty of Technology, Bielefeld University and at Trinity College, Dublin where he was supported by a European Erasmus scholarship. After graduating he became head of the information systems unit of Betron control systems and was involved in the development of industrial controllers based on artificial neural networks. Subsequently, he moved to the Neuroinformatics Group at the Faculty of Technology, Bielefeld University where he is currently working as a research assistant. His main interests are information systems which employ machine learning methods to create semantically meaningful structures on large unformatted data sets.

Martin Orr is a psychiatrist, a doctoral student at Southern Cross University, a senior lecturer at Auckland University and a clinical director of information services for the Waitemata District Health Board (Auckland, New Zealand). Dr. Orr is involved on a daily basis in the pragmatics of meeting the "opportunities and challenges" of developing health knowledge systems. His key research interests lie in working with the dedicated professionals that through the development, implementation and utilization of health knowledge systems, strive to "make a healthy difference" for their communities. Dr. Orr attempts to capture (in combination with related literature) key issues and knowledge gathered from working with these groups, in the form of visual and mnemonic models. His hope is that these evolving models will assist others in their development of health knowledge systems. Dr. Orr's expertise includes: health knowledge systems, privacy issues, conceptual and visual models.

Teemu Paavola has worked in teleoperator business development, in academic life, and as a member of technology strategy group of Sonera Plc (later TeliaSonera Plc). He is currently managing director and CEO of LifeIT Plc, a joint venture of a Finnish health district and private organizations like TietoEnator Plc. Paavola is a licentiate of technology from the Department of Industrial Engineering and Management of the Helsinki University of Technology. He has contributed to numerous articles on the development of telemedicine and information technology management. He has also served in several elected posts in the international science community.

Sotiris Pavlopoulos received his degree in electrical engineering from the University of Patras, Greece (1987) and his PhD in biomedical engineering (1992) jointly from Rutgers University and Robert Wood Johnson Medical School, New Jersey. He is a research associate professor at the Institute of Communication and Computer Systems, the National Technical University of Athens, Greece. He has published 5 book chapters, more than 30 journal articles and more than 45 refereed conference papers. His research interests include medical informatics, telemedicine, medical image and signal processing, and bioinformatics. Dr. Pavlopoulos has been active in a number of European and National R&D programs in the field of telematics applications in healthcare.

Maria Pragmatefteli was born in Athens, Greece in 1976. She received her degree in electrical & computer engineering from the National Technical University of Athens in 2001. She has worked as a computer engineer in the private health sector from 2001 to 2002. She is currently pursuing a PhD in biomedical engineering at Biomedical Engineering Laboratory. Her main areas of expertise are health informatics and image processing. She has participated in several European and national projects.

Sheila Price is a research associate based in the Department of Information Science at Loughborough University and secretary of its Health Informatics Research Group. She holds a BSc in information and library studies (1992). Her research interests focus on the use of informatics within the modern healthcare arena, including issues surrounding the deployment of the electronic care record in the UK, innovative uses for home based monitoring devices and child health informatics. She is a founder director of SP informatics Ltd, a spin out company from Loughborough University.

Srinivasa Raghavan works as a president of Krea Corporation, an IT and management consulting firm located in Illinois, USA. He also functions as an assistant professor at the School of Business, NorthCentral University. He has an MS in computer applications, an MBA in international business, an MS in project management and a PhD in organization and management. He currently chairs and is part of several research committees in the areas of decision support, quality management, and project management. His research interests include decision support systems, knowledge management, healthcare informatics, IT project estimation models, operation research, project management, and IT life cycle models.

Helge Ritter studied physics and mathematics at the Universities of Bayreuth, Heidelberg and Munich. After receiving a PhD in physics at the Technical University of Munich (1988), he visited the Laboratory of Computer Science at Helsinki University of Technology and the Beckman Institute for Advanced Science and Technology at the University of Illinois at Urbana-Champaign. Since 1990, he has served as head of the Neuroinformatics Group at the Faculty of Technology, Bielefeld University. His main interests are principles of neural computation, in particular for self-organization and learning, and their application to build intelligent systems. In 1999, Ritter was awarded the SEL Alcatel Research Prize and in 2001 the Leibniz Prize of the German Research Foundation DFG.

Dean F. Sittig received a PhD in medical informatics from the University of Utah in 1988. He has held numerous informatics' research and development positions at several of the leading academic medical centers in the U.S. (Yale, Vanderbilt, Harvard/Partners HealthCare System). He is currently the director of applied research in Medical Informatics at Kaiser Permanente, Northwest and a professor in the Department of Medical Informatics and Clinical Epidemiology at Oregon Health & Sciences University in Portland, Oregon. His research interests center on the design, development, implementation, and evaluation of all aspects of clinical decision support systems.

John Stoitsis was born in Poland in 1977. He received a degree in electrical and computer engineering from Aristotle University of Thessaloniki, Thessaloniki (2002) and an MSc in biomedical engineering from the Medical School, University of Patras (2004). He is currently working toward his PhD at the National Technical University of Athens. His research interests include image processing, medical imaging, computer graphics in medical applications, and neuroscience. He is a member of the Institute of Electrical and Electronics Engineers.

John Stoitsis was born in Poland in 1977. He received his degree in electrical and computer engineering from Aristotle University of Thessaloniki, Thessaloniki (2002) and an MSc in biomedical engineering from the Medical School, University of Patras (2004).

Copyright © 2005, Idea Group Inc. Copying or distributing in print or electronic forms without written permission of Idea Group Inc. is prohibited.

He is currently working toward a PhD at the National Technical University of Athens. His research interests include image processing, medical imaging, computer graphics in medical applications, and neuroscience. He is a member of the Institute of Electrical and Electronics Engineers.

Ron Summers is professor of information science at Loughborough University and director of its Health Informatics Research Group. He holds a BSc in biophysical science (1982), an MSc in information engineering (1985) and a PhD in artificial intelligence in medicine (1992). His research interests span the health informatics discipline from post-genomic medicine to health technology assessment. Informatics associated with capturing home monitored data and its place in a wider telehealth setting are particular research issues that are receiving his attention. He is a founder director of SP informatics Ltd, a spin-out company that exploits the business potential of research ideas.

Pekka Turunen holds the research director position of the Social and Health Information Technology Research Unit (Shiftec) at the University of Kuopio in Finland. He received his PhD from the Turku School of Economics and Business Administration in the evaluation of healthcare information systems, which has been his primary research interest in addition of management of ICT. He has been an honorary visiting research fellow in several research groups around Europe, including MIG in the University of Manchester and the Department of Medical Informatics of University Maastricht.

Ioannis Valavanis was born in Athens, Greece, in 1980. He received his degree in electrical & computer engineering from the National Technical University of Athens (NTUA), Greece, in 2003, and his is currently working toward his PhD at the NTUA. His research interests include medical image processing, computer aided diagnosis, artificial intelligence in medicine and biology as well as bioinformatics, including molecular surface modeling and biomolecular structure analysis. He has authored two papers in international conferences.

John Van Beveren is a lecturer in knowledge management and marketing at the University of Ballarat. Prior to his academic career he worked in information technology management and consulting. His research interests include consumer and customer behavior in online environments. He has published numerous journal articles, conference papers, and industry reports.

Subramanyam Vdaygiri was born in 1968 and raised in India. He holds a BS in chemical engineering from Osmania University, India (1989). He completed a dual MS (1995) in chemical engineering and system science from Louisiana State University, USA. He is currently at Siemens Corporate Research Inc., Princeton, New Jersey (USA) where he has held a few positions including member of technical staff and project manager in the Multimedia Technology Department. He leads a research group focusing on collaboration technologies and converged communications.

About the Authors

Jari Vuori is currently a professor in the Department of Health Policy Management at the University of Kuopio, Finland. He was been a visiting professor at Georgia Institute of Technology in the United States from 1998-1999 and at Nihon University in Japan from 2000-2001. His research interests include health management from organizational point of view, dialogical conflict management among doctors and nurses, and comparative performance measurement between public and private healthcare organizations and systems. During the next four years (2004-2008) Dr. Vuori will head an international research project at the Academy of Finland concerning public, private and hybrid forms of healthcare including measuring contracting-out.

Xiu Ying Wang received her BSc and MSc in computer science from Heilongjiang University, China (1994 and 1999, respectively). She was a visiting scholar in the School of Information Technologies, The University of Sydney from 2001 to 2002 and was a visiting researcher in the Department of Electronic and Information Engineering, Hong Kong Polytechnic University from 2002 to 2003. She is currently a PhD candidate at the University of Sydney and she is a student member of IEEE. Her research interests include image registration, multimedia, and computer graphics.

Jay Whittaker is a recent graduate of the University of Ballarat Honours program in the School of Business. He also has a bachelor of information technology. His research has focused on social capital and knowledge management. He has published several journal articles and conference papers.

Maurice Yolles is professor of systems management at Liverpool John Moores University, UK. His specialist area is management systems and managerial cybernetics. He has published more than 120 papers in refereed journals, conferences and book chapters. His main teaching area is in change and knowledge management. He is editor of the *International Journal of Organisational Transformation and Social Change* (OTASC), and director of the registered charity The Lentz Foundation for Peace and Conflict, which he established in 1975. He is also head of the Centre for Creating Coherent Change and Knowledge (C4K). He has undertaken a number of international research and development projects for the EU under various programs, and has been involved in TEMPUS projects with a variety of Central and Eastern European Countries. He is a visiting professor at Ostrava University in the Czech Republic, an honorary professor at Boatou University in China, and an honorary fellow of the Richardson Institute for conflict studies at Lancaster University. He is currently the vice president for research and publications in the International Society of Systems Science (ISSS), and honorary associate of the Centre for Leadership and Organisational Change at Teesside University, UK.

Index

A

affine transformation 165
algorithms 160
artificial intelligence 37, 252
autonomy 108
availability 140

B

biomedical image registration 159
biomedical registration transformations 165
biopsychosocial 74
business intelligence 120

C

client/server architecture 17
clinical applications 279
clinical governance 4
clinical guidelines 212
clinical image access service 16
clinical information system 54, 183, 214, 286
clinical knowledge management 160
Clinical Observations Access Service (COAS) 16
clinical practice 160
clinical protocols 199
clinical records management 183
collaboration 142
communication 142
community 300
community health information networks 5
complexity 100
cycle 106

D

data mining 120
data privacy 118
data warehouses 54, 120
data warehousing 119
decision algorithm 199
decision making 160
decision support systems 35, 197, 251, 286
DICOM 12
digital radiology 116
digital signal processing techniques 34
direct care 37
distributed applications 17
distributed healthcare environment 133
Downhill-Simplex algorithm 166

E

e-consent 74
e-learning 121
e-prescribing 36
electronic care communications 36, 40
electronic care records 184
electronic data capture 182
electronic health records 13, 182
electronic healthcare record 117
evidence based medicine (EBM) 3, 78
expert systems 197
explicit knowledge 6

F

feasibility 220

G

general practitioners' (GP) 3

H

health care (HC) 11
health information privacy 72
health knowledge management systems 73
health knowledge systems 72
health monitoring systems 53
healthcare common services (HCS) 13
healthcare informatics 118
healthcare information systems (HIS) 12, 130
healthcare managment 2, 96
HIPAA 118
home healthcare 152
hospital information systems 29

I

image registration methodology 163
information and communication technologies (ICT) 34, 117
information dissemination 183
information processing 243
information provision 35, 40
information retrieval 53
information seeking 236
information technology 2

instant messaging (IM) 143
integrated 10
integrated drug delivery system 182
integrated patient pathways 10
intelligent agent 120, 199
intelligent organization 96
intensity-based medical image registration 167
Internet 33
Internet based diagnosis 41
Internet based healthcare 35

K

knowledge alliance 81
knowledge exchange 299
knowledge management 1, 52, 119, 142, 160, 234, 279
knowledge management paradigm 96
knowledge maps 120
knowledge neurones 81
knowledge sharing 97, 196, 298
knowledge sharing standards 198
knowledge workers 122

L

learning 298
lexicon query service 15

M

magnetic resonance imaging 116
medical decision support system 196
medical image registration 163
medical knowledge 290
model of integrated patient pathways 3
multimedia 140

N

neural networks 53
Nonaka 10
nuclear medicine 116

O

OASIS 18
OpenScape 143

organization development 98

P

paradigm 101
patient compliance 186
patient privacy concerns 78
performance management 121
person identification service 15
positivism 110
Powell algorithm 166
privacy 74
privacy infringement 78

Q

quality 160
quality assurance 125

R

redundant information 108
requisite variety 108
resource access decision (RAD) 16

S

safety 162
security 17, 44, 74
semantics 56
sharing 220, 240
Sieloff 10
Siemens Corporate Research 139
social capital 299
software development 20

T

tacit knowledge 6, 273
telemedicine 141, 272
temporal reasoning (TR) 38
therapeutic alliance 82

U

UK National Health Service (NHS) 98
ultrasound (US) 160
unique health identification numbers 77
user presence 144

V

value analysis 75
viable systems 101

W

WAN (wide area network) 142
Web browsers 120
Web services 17
workflow applications 120
workflows 141

Instant access to the latest offerings of Idea Group, Inc. in the fields of
INFORMATION SCIENCE, TECHNOLOGY AND MANAGEMENT!

InfoSci-Online Database

- BOOK CHAPTERS
- JOURNAL ARTICLES
- CONFERENCE PROCEEDINGS
- CASE STUDIES

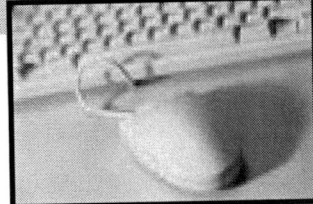

> **The Bottom Line:** With easy to use access to solid, current and in-demand information, InfoSci-Online, reasonably priced, is recommended for academic libraries.
>
> — Excerpted with permission from Library Journal, July 2003 Issue, Page 140

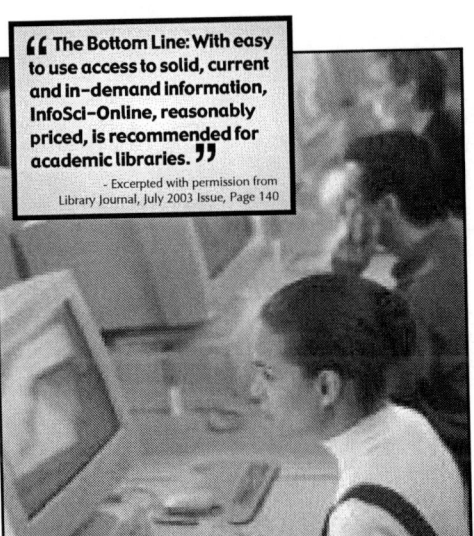

The InfoSci-Online database is the most comprehensive collection of full-text literature published by Idea Group, Inc. in:

- Distance Learning
- Knowledge Management
- Global Information Technology
- Data Mining & Warehousing
- E-Commerce & E-Government
- IT Engineering & Modeling
- Human Side of IT
- Multimedia Networking
- IT Virtual Organizations

BENEFITS
- Instant Access
- Full-Text
- Affordable
- Continuously Updated
- Advanced Searching Capabilities

Start exploring at www.infosci-online.com

Recommend to your Library Today!
Complimentary 30-Day Trial Access Available!

A product of:
 **Information Science Publishing***
Enhancing knowledge through information science

*A company of Idea Group, Inc.
www.idea-group.com

Books on Healthcare Information Systems

Effective Healthcare Information Systems

Adi Armoni, Ph.D., Tel Aviv University, Israel

Health and medical informatics encompass a very broad field that is rapidly developing in both its research and operational aspects. The discipline has many dimensions, including social, legal, ethical and economic. This book, **Effective Healthcare Information Systems** puts a special emphasis on issues dealing with the most recent innovations such as telemedicine, Web-based medical information and consulting systems, expert systems and artificial intelligence.

ISBN: 1-931777-01-2; eISBN: 1-931777-20-9; Copyright: 2002
Pages: 295 (s/c); Price: US $59.95

Strategies for Healthcare Information Systems

Robert Stegwee, Ph.D., University of Twente, The Netherlands
Ton Spil, University of Twente, The Netherlands

Information technologies of the past two decades have created significant fundamental changes in the delivery of healthcare services by healthcare provider organizations. **Strategies for Healthcare Information Systems** provides an overall coverage of different aspects of healthcare information systems strategies and challenges facing these organizations.

ISBN: 1-878289-89-6; Copyright: 2001; Pages: 232 (s/c); Price: US $74.95

Knowledge Media in Healthcare: Opportunities and Challenges

Edited by Rolf Grutter, Ph.D., University of St. Gallen, Switzerland

Because the field of healthcare reflects forms of both explicit and tacit knowledge such as evidence-based knowledge, clinical guidelines and the physician's experience, knowledge media have significant potential in this area. **Knowledge Media in Healthcare: Opportunities and Challenges** is an innovative new book which strives to show the positive impact that Knowledge Media and communication technology can have on human communication within the field of healthcare.

ISBN 1-930708-13-0; eISBN: 1-59140-006-6; Copyright: 2002; Pages: 296; Price: US$74.95

Excellent additions to your library–
Please recommend to your librarian.

Managing Healthcare Information Systems with Web-Enabled Technologies

Lauren Eder, Ph.D., Rider University, USA

Healthcare organizations are undergoing major adjustments to meet the increasing demands of improved healthcare access and quality. **Managing Healthcare Information Systems with Web-Enabled Technologies** presents studies from leading researchers focusing on the current challenges, directions and opportunities associated with healthcare organizations and their strategic use of Web-enabled technologies.

ISBN: 1-878289-65-9 ; eISBN:1-930708-67-X; Copyright: 2000
Pages: 288 (s/c); Price: US $69.95

Healthcare Information Systems: Challenges of the New Millennium

Adi Armoni, Ph.D., Tel Aviv University, Israel

Healthcare information sysstems are crucial to the effective and efficient delivery of healthcare. **Healthcare Information Systems: Challenges of the New Millennium** reports on the implementation of medical information systems.

ISBN: 1-878289-62-4; eISBN: 1-930708-55-6; Copyright: 2000
Pages: 256 (s/c); Price: US $69.95

It's Easy to Order! Order online at www.idea-group.com

or call 1-717-533-8845 ext.10!
Mon-Fri 8:30 am-5:00 pm (est) or fax 24 hours a day 717/533-8661

Idea Group Inc.
Hershey • London • Melbourne • Singapore • Beijing

New Releases from Idea Group Reference

Idea Group REFERENCE

The Premier Reference Source for Information Science and Technology Research

ENCYCLOPEDIA OF DATA WAREHOUSING AND MINING

Edited by: John Wang, Montclair State University, USA

Two-Volume Set • April 2005 • 1700 pp
ISBN: 1-59140-557-2; US $495.00 h/c
Pre-Publication Price: US $425.00*
*Pre-pub price is good through one month after the publication date

- Provides a comprehensive, critical and descriptive examination of concepts, issues, trends, and challenges in this rapidly expanding field of data warehousing and mining
- A single source of knowledge and latest discoveries in the field, consisting of more than 350 contributors from 32 countries
- Offers in-depth coverage of evolutions, theories, methodologies, functionalities, and applications of DWM in such interdisciplinary industries as healthcare informatics, artificial intelligence, financial modeling, and applied statistics
- Supplies over 1,300 terms and definitions, and more than 3,200 references

ENCYCLOPEDIA OF DISTANCE LEARNING

Four-Volume Set • April 2005 • 2500+ pp
ISBN: 1-59140-555-6; US $995.00 h/c
Pre-Pub Price: US $850.00*
*Pre-pub price is good through one month after the publication date

- More than 450 international contributors provide extensive coverage of topics such as workforce training, accessing education, digital divide, and the evolution of distance and online education into a multibillion dollar enterprise
- Offers over 3,000 terms and definitions and more than 6,000 references in the field of distance learning
- Excellent source of comprehensive knowledge and literature on the topic of distance learning programs
- Provides the most comprehensive coverage of the issues, concepts, trends, and technologies of distance learning

ENCYCLOPEDIA OF INFORMATION SCIENCE AND TECHNOLOGY
AVAILABLE NOW!

Five-Volume Set • January 2005 • 3807 pp
ISBN: 1-59140-553-X; US $1125.00 h/c

ENCYCLOPEDIA OF DATABASE TECHNOLOGIES AND APPLICATIONS

April 2005 • 650 pp
ISBN: 1-59140-560-2; US $275.00 h/c
Pre-Publication Price: US $235.00*
*Pre-publication price good through one month after publication date

ENCYCLOPEDIA OF MULTIMEDIA TECHNOLOGY AND NETWORKING

April 2005 • 650 pp
ISBN: 1-59140-561-0; US $275.00 h/c
Pre-Publication Price: US $235.00*
*Pre-pub price is good through one month after publication date

www.idea-group-ref.com

Idea Group Reference is pleased to offer complimentary access to the electronic version for the life of edition when your library purchases a print copy of an encyclopedia

For a complete catalog of our new & upcoming encyclopedias, please contact:
701 E. Chocolate Ave., Suite 200 • Hershey PA 17033, USA • 1-866-342-6657 (toll free) • cust@idea-group.com